Measurement of High-Speed Signals in Solid State Devices

SEMICONDUCTORS AND SEMIMETALS
Volume 28

Semiconductors and Semimetals

A Treatise

Edited by R.K. Willardson

ENICHEM AMERICAS INC.
PHOENIX, ARIZONA

Albert C. Beer

BATTELLE COLUMBUS LABORATORIES
COLUMBUS, OHIO

Measurement of High-Speed Signals in Solid State Devices

SEMICONDUCTORS AND SEMIMETALS

Volume 28

Volume Editor

ROBERT B. MARCUS

BELLCORE
RED BANK, NEW JERSEY

ACADEMIC PRESS, INC.

Harcourt Brace Jovanovich, Publishers

Boston   San Diego   New York
Berkeley   London   Sydney
Tokyo   Toronto

ACADEMIC PRESS, INC.
1250 Sixth Avenue, San Diego, CA 92101

United Kingdom Edition published by
ACADEMIC PRESS LIMITED
24-28 Oval Road, London NW1 7DX

09S46S

Library of Congress Cataloging-in-Publication Data

Semiconductors and semimetals.—Vol. 1-—New York: Academic
    Press, 1966-

        v.: ill.; 24 cm.

    Irregular.
    Each vol. has also a distinctive title.
    Edited by R.K. Willardson and Albert C. Beer.
    ISSN 0080-8784 = Semiconductors and semimetals

        1. Semiconductors—Collected works.   2. Semimetals—Collected works.
    I. Willardson, Robert K.   II. Beer, Albert C.
    QC610.9.S48   621.3815'2—dc19   85-642319
                                            AACR 2   MARC-S

    Library of Congress   [8709]
    ISBN 0-12-752128-3   (V.28)

Printed in the United States of America
90 91 92 93   9 8 7 6 5 4 3 2 1

Time goes, you say? Ah no!
Alas, Time stays, *we* go.

—Henry Austin Dobson, "The Paradox of Time," stanza I

Contents

Chapter 1   Materials and Devices for High-Speed Devices and Optoelectronic Applications

J. Frey and D.E. Ioannou

Chapter 2   Electronic Wafer Probing Techniques

H. Schumacher and E.W. Strid

Chapter 3   Picosecond Photoconductivity: High-Speed Measurements of Devices and Materials

D. H. Auston

Chapter 4   **Electro-Optic Measurement Techniques for Picosecond Materials, Devices, and Integrated Circuits**

J.A. Valdmanis

Chapter 5   **Direct Optical Probing of Integrated Circuits and High-Speed Devices**

J.M. Wiesenfeld and R.K. Jain

Chapter 6   **Electron-Beam Probing**

G. Plows

Chapter 7   **Photoemissive Probing**

A.M. Weiner and R.B. Marcus

List of Contributors

Numbers in parentheses indicate the pages on which the authors' contributions begin.

D.H. Auston, *Department of Electrical Engineering, Columbia University, 500 W. 120th Street, Seeley W. Mudd Building, New York, New York 10027* (85)

Jeffrey Frey, *Department of Electrical Engineering, University of Maryland, College Park, Maryland 20742* (1)

Dimitris E. Ioannou, *Department of Electrical Engineering, University of Maryland, College Park, Maryland 20742* (1)

Ravi K. Jain, *Amoco Research Center, P.O. Box 400, Naperville, Illinois 60566* (221)

R.B. Marcus, *Bellcore, 331 Newman Springs Road, Red Bank, New Jersey 07701* (383)

Graham Plows, *Cambridge Instruments, Ltd., Clifton Road, Cambridge CB1 3QH England* (335)

Hermann Schumacher, *Bellcore, 331 Newman Springs Road, Red Bank, New Jersey 07701* (41)

Eric W. Strid, *Cascade Microtech, P.O. Box 1589, Beaverton, Oregon, 97075* (41)

J.A. Valdmanis, *Ultrafast Science Laboratory, University of Michigan, EECS Building, 1301 Beal Avenue, Ann Arbor, Michigan 48109–2122* (135)

A.M. Weiner, *Bellcore, 331 Newman Springs Road, Red Bank, New Jersey 07701* (383)

J.M. Wiesenfeld, *Crawford Hill Laboratory, AT&T Bell Laboratories, Holmdel, New Jersey 07733* (221)

xi

Preface

A series of developments over the past few years has produced a number of new ways to interrogate high-speed electrical waveforms. These exciting new techniques have been in part stimulated by progress in the realization of electronic and optoelectronic devices operating with bandwidths greater than 100 GHz corresponding to speeds of a few picoseconds (ps) and with critical dimensions well below one μm. High operating speeds and small critical dimensions create severe restrictions on waveform-probing methods and dictate the need for "noninvasive" probes using new principles of waveform interrogation.

This book discusses the principles of operation and the current status of the various methods that have been developed to perform high-speed measurements. Most of the probe techniques discussed can be characterized as "contactless" in the sense that optical or electron beams interacting with the device under test (DUT) cause minimal loading and mechanical contact is avoided. (An exception comes in Chapter 2, which discusses conventional as well as new and exciting developments in high-speed contact probing.) Similarly, most make use of sampling techniques in which the probe-beam-pulse duration is very short compared to the duration of the waveform pulse under study, requiring synchronization and signal averaging in order to obtain a satisfactory signal–noise ratio. (Exceptions are discussed in Chapter 4, Section VI, and Chapter 5, Section V.C.1.)

Each chapter focuses on a different technique or subject and is authored by one or more experts in the field. Chapter 1 discusses basic principles of high-speed–electron and optoelectronic-device physics and thereby provides a tutorial background for the remaining chapters. The following six chapters present different probe methods; each chapter considers the principles of the method, methodology, advantages and limitations, necessary apparatus and instrumentation, voltage sensitivity, and temporal resolution, with a final section summarizing the technique with a discussion of future trends. We have accumulated considerable experience in "conventional" contact-probing methods, and in a sense the methodologies and principles derived by contact probing remain as the standard against which other techniques are compared. Chapter 2 discusses the principles and status of contact probing. The generation and measurement of signals by photoconductive methods are described in Chapter 3. Chapters 4 and 5 present the concept of electrooptic probing, with an emphasis in Chapter 4 on external (indirect) probing (where an electrooptically active sensing electrode is placed in close proximity to the DUT) and in Chapter 5 on internal (direct) probing (where the probe beam enters the electrooptically active substrate of the DUT). Electron-beam probing is a contactless method that has been commercially available for about 10 years, and the status of this technique (as of 1989) including recent developments is given in Chapter 6. Both electron-optic and photon techniques are combined in photoemissive probing, as described in Chapter 7.

This book is primarily intended as a basic reference to high-speed probing methods and instrumentation, and can be used either as a text or as a source of reference to details described elsewhere in the literature. Chapters 2–7 may be of interest to researchers, circuit designers, and process engineers concerned with device physics or signal propagation and measurement, while students at the advanced undergraduate or graduate level may find the entire book useful as a text. The need to generate short optical or electron pulses for waveform sampling has stimulated novel technological innovations in photon and electron optics that are also discussed in this book, and workers in laser and electron-beam optics should find the relevant chapters useful.

One of the yardsticks that we use to judge probe methods and that is, at the same time, a good measure of the advanced capabilities of these techniques is the temporal resolution. The best resolutions reported to date are:

for contact probing with a sampling scope, 4 ps (Marsland et al., 1988, Chapter 2),

photoconductive probing, 0.6 ps (Ketchen et al., 1986, Chapter 3),

external (indirect) electro-optic probing, 0.3 ps (Valdmanis, 1988, Chapter 4),

internal (direct) electro-optic probing, 3.5 ps (Madden et al., 1988, Chapter 5),

electron beam probing, 5 ps (May et al., 1987, Chapter 6),

and photoemissive probing, 5 ps (Weiner et al., 1987, Chapter 7).

No one of these techniques is suitable for all occasions, and there are circumstances under which each technique seems to have distinct advantages. Basic differences in the physical principles underlying the various probe methods lead to major differences in instrumentation and conditions of applicability as well as to different trade-offs with respect to voltage-sensitivity, spatial resolution, temporal resolution, ease of application, and degree of invasiveness. While the reader is left to judge the suitability of a

PREFACE xv

probe technique to a given problem, the basic information needed to make this judgment should be found in these chapters.

I wish to thank Albert C. Beer of Battelle Columbus Laboratories for encouraging me to undertake this project. I wish to also thank the following people for critical discussions and review of various parts of this book: Jerry Iafrate (U.S. Army Research Center at Fort Monmouth); Joe Abeles (Sarnoff Labs); John Hayes and Bob Leheny (Bellcore); Brian Kolner (Hewlett Packard); Jean-Mark Halbout (IBM Labs); and Phil Russell (North Carolina State University).

R.B. Marcus

—

CHAPTER 1

Materials and Devices for High-Speed and Optoelectronic Applications

Jeffrey Frey and Dimitris E. Ioannou

DEPARTMENT OF ELECTRICAL ENGINEERING
UNIVERSITY OF MARYLAND
COLLEGE PARK, MARYLAND

I. Introduction

High-speed devices—generally those that can switch on or off in picoseconds or less—are used to perform many analog and digital functions, including linear signal amplification, modulation or demodulation, or simple on–off switching in computers. As systems requirements increase the speeds at which these devices must operate, their sizes necessarily decrease, and effects formerly considered of second order (therefore, negligible) become important. For example, the large increase of electron

1

temperature over lattice temperature that can occur at the large electric fields existing in very small devices can lead to severe reliability problems. Therefore, it becomes necessary to understand the physics of the operation of these high-speed devices in ever-greater detail. In addition, new materials and/or entirely new types of devices are being explored in the search for new functionality and speed.

Optoelectronic devices are those that combine optical and electronic functions, such as the detection and subsequent demodulation of an information-carrying laser signal. They include photoelectronic devices (such as photodetectors and photodiodes), the operation of which depends on the interaction of light with electrons, and electro-optic devices (such as phase shifters), the operation of which depends on the effect of electric fields on the propagation of light. Used primarily for communications, such devices may have some computing applications. Demands for speed, sensitivity, and increased signal-to-noise ratio for these devices also increase continuously. Improved materials for their operation and improved device structures are continuously being sought.

In this chapter, we first discuss materials parameters which, in general, determine various aspects of the operation of high-speed and optoelectronic devices. Then, the devices of greatest interest today are described, and the ways in which these parameters affect them are explained. In the course of the discussion we shall try to illustrate various aspects of device performance or materials behavior that can be understood by applying high-speed probing techniques.

II. Important Materials Parameters

Depending on the acceptable level of approximation in the analysis of any device, various levels of exactness of definition of materials parameters are possible. Choice of the appropriate level will depend on the dimensions of the device under study, electric fields within, and time scales. For example, in describing carrier transport, the parameters at the various levels may be described as:

Level 1: Carrier mobility
Level 2: Carrier momentum or energy relaxation times; effective mass
Level 3: Coupling coefficients for intervalley scattering; band structure.

Level 1 (the least accurate approximate description of carrier transport) is appropriate for describing devices in which electric fields do not exceed the value at which mobility is no longer a constant. Level 3 is the most accurate of the three, and level 2 is an averaged version of the details of level 3.

Therefore, level 2 may be used for describing devices in which high-field effects are significant.

Because of the need to reduce parasitic capacitance (which depends on area) and carrier transit time (which depends on length), the higher the speed at which devices must work, the smaller they must be. Further, because signal-to-noise ratio must be maintained in the face of power-supply transients, or because established voltage-level standards are not easily changed, operating voltage levels cannot easily be reduced as device size is reduced. Therefore, peak electric fields inside devices invariably increase as speeds are increased, and proper understanding of high-speed devices requires that materials parameters be specified to at least level 2. Although many parameters can be deduced from purely electrical measurements (such as average conductivity), better techniques of measurement are available today using fast optical-switching techniques. Using these techniques, probes have been developed that operate on time scales similar to those of the "instantaneous" phenomena being measured. These probes are described in Chapters III–VII.

Both materials *transport parameters* and *band-structure parameters* affect the performance of high-speed and optoelectronic devices. The former may be considered direct determinants of switching times or maximum frequency of amplification or detection. The latter affect both carrier transport and the optical frequencies to which the device of interest responds or which it generates.

A. Transport Parameters

1. Drift Parameters

A *majority-carrier* device is one in which the switching performance depends on the transport of majority carriers. A field-effect transistor is a majority-carrier device; a bipolar transistor is not, because while they are moving through the modulating region (the base), the current-carrying particles are minority carriers. The speed with which a majority-carrier device operates is primarily determined by the transit time of carriers across some modulation length, roughly

$$t = L/\langle v \rangle \tag{1}$$

where t is a measure of switching speed or inverse of some maximum frequency of amplification; L is the active length of the device; and $\langle v \rangle$ is the average carrier drift velocity across the active length. The speed of bipolar transistors cannot, unfortunately, be described as simply as in Eq. (1), because their performance is determined by factors other than simple speed, such as parasitic base-collector capacitance, injection efficiency across the emitter-base junction, and base resistance. At low fields, the average velocity can be related to drift velocity by the basic level 1 transport parameter *mobility*, classically taken as a constant, that relates electron or hold velocity v to electric field $\mathscr{E}$ as follows:

$$v = \mu \mathscr{E} \tag{2}$$

However, the approximation that mobility is a constant—although a good one for most materials at fields under a thousand V/cm—is not good enough to describe the behavior of high-speed devices, in which electric fields may reach values in excess of 10^5 V/cm. To simplify device analysis, empirical formulas are often used to express the variation of mobility with temperature, impurity concentration, field, and field anisotropy. However, these curve-fitting formulas are not based on physics, which at the second level of approximation specifies that:

$$\mu = e\tau/m^* \tag{3}$$

where e = electronic charge, m^* is the carrier's effective mass, and τ is the average relaxation time. The effective mass, if properly described as a tensor function of electron energy, is a shorthand description of the band structure. If there is more than one scattering process, and all may be described by a relaxation time, the overall average relaxation time is obtained from the formula:

$$\frac{1}{\tau} = \sum_{i=1}^{n} \frac{1}{\tau_i}. \tag{4}$$

The meaning of Eq. (4) is:

Consider that a perturbing force, arising perhaps from the application of an electric field, is applied to a quantity of electrons with some initial distribution of momentum and position. The force perturbs this distribution, and the electrons accelerate and move. When the perturbation (field) is removed, the distribution relaxes to its original state under the influence of n processes of energy and momentum exchange. τ is a measure of the rate at which the distribution relaxes.

By making certain approximations, expressions for scattering times (and, hence, mobility) may be derived by analyzing the major scattering processes suffered by charge carriers. Sample derivations are given by Nag

(1972, Chap. 4), and result in the following expressions:

For *acoustic phonon scattering* in materials with spherical constant energy surfaces (e.g., GaAs central valley),

$$\frac{1}{\tau_a} = \frac{m^* E_1^2}{4\pi\varrho\hbar^2 2k^3} \frac{2kT}{\hbar c_l^2} \frac{(2k)^4}{4} = \frac{\sqrt{2}}{\pi} \frac{E_1^2 m^{*3/2} kT}{\hbar^4 \varrho c_l^2} E^{1/2} \tag{5}$$

For *acoustic phonon scattering* in materials with ellipsoidal constant energy surfaces (e.g., Si),

$$\frac{1}{\tau_{\parallel a}} = \frac{3\pi}{\chi_1} \frac{m_D^{3/2}}{2^{3/2}} \frac{kTE^{1/2}}{\pi^2\hbar^4} [\xi_\parallel \Xi_d^2 + \eta_\parallel \Xi_d \Xi_u + \zeta_\parallel \Xi_u^2] \quad \text{and}$$

$$\frac{1}{\tau_{\perp a}} = \frac{3\pi}{\chi_1} \frac{m_D^{3/2}}{2^{3/2}} \frac{kTE^{1/2}}{\pi^2\hbar^4} [\xi_\perp \Xi_d^2 + \eta_\perp \Xi_d \Xi_u + \zeta_\perp \Xi_u^2]. \tag{6}$$

For *piezoelectric scattering*,

$$\frac{1}{\tau_p} = \frac{e^2 m^{*1/2} kT}{2^{3/2}\pi\hbar^2\varepsilon^2 E^{1/2}} \left\langle \frac{\hbar^2}{\varrho c_1^2} \right\rangle \tag{7}$$

For *optical phonon scattering*,

$$\frac{1}{\tau_{op}} = \frac{(2m^*)^{3/2} D_0^2}{4\pi\hbar^3\varrho\omega_0} [(n_0 + 1)(E - \hbar\omega_0)^{1/2} + n_0(E + \hbar\omega_0)^{1/2}]. \tag{8}$$

For *polar optical phonon scattering*,

$$\frac{1}{\tau_{po}} = \frac{e^2\omega_1}{4\sqrt{2\pi\varepsilon_0}} \left(\frac{1}{K_\infty} - \frac{1}{K_1}\right) \frac{m^{*1/2}}{\hbar\sqrt{E}} (2n_0 + 1). \tag{9}$$

For *intervalley phonon scattering*,

$$\frac{1}{\tau_{jv}} = \sum_i^{i\neq j} \frac{(2m_D)^{3/2}}{4\pi\hbar^3\varrho} \frac{D_{ji}^2}{\omega_{ji}^2} [n_j(E + \hbar\omega_{ji})^{1/2} + (n_j + 1)(E - h\omega_{ji})^{1/2}]. \tag{10}$$

For *ionized impurity scattering*,

$$\frac{1}{\tau_{ii}} = \frac{Z^2 e^4 N_i C}{16\pi(2m^*)^{1/2}\varepsilon^2} E^{-32} \tag{11}$$

In the preceding expressions, the symbols are:

E_1 = energy eigenvalue for state 1

E = total electron energy

e = electronic charge

k = Boltzmann's constant

h = Planck's constant

$$\hbar = \frac{h}{2\pi}$$

ρ = density

c_l = acoustic velocity

ε = dielectric constant

Ξ_d, Ξ_u = deformation potentials for dilation and uniaxial stress, respectively

K_1, K_∞ = high-and low-frequency dielectric constant, respectively

χ_1 = average elastic constant for longitudinal waves

m^*, m_D = density-of-states effective mass

ξ, η, ζ = arc fitting constants

D_0 = optical deformation potential

n_0 = number of optical phonons

n_j = intervalley phonon density

ω_0 = optical phonon frequency

ω_{ij} = frequency of ij intervalley phonon

D_{ij} = intervalley scattering deformation potential

N_i = scattering centers/unit volume

C = a constant, $1.4 < C < 2$

Z = 1 for singly charged impurity, etc.

The relaxation of a distribution function after some perturbation can be described in terms of relaxation times only if the scattering processes responsible for the relaxation do not interact and are either randomizing or elastic. Polar optical phonon scattering—the most important scattering process in compound semiconductors at even moderately high fields—is neither, although it is sometimes represented approximately by a relaxation time.

2. Diffusion

Carrier transport by diffusion is related to relaxation of a distribution function, as is drift. For carrier concentrations that are small compared to the density of states of carriers, and for low fields, the connection is via the Einstein Relation (Smith, 1979, p. 172):

$$D = \mu k T / e \tag{12}$$

where

D = diffusion coefficient,

μ = carrier mobility, and

T = temperature.

B. BAND-STRUCTURE PARAMETERS

The band structure (Bube, 1974, Chaps. 4, 5) of a crystalline solid relates the energy (E) of charge carriers to their momentum ($\hbar k$) and is determined by setting up Schroedinger's equation for the solid in question and determining the conditions (i.e., relationship of energy to momentum) for which solutions exist. For all semiconductors, regions in the E-k plane are found in which solutions do not exist. In such a material, the difference in energy between the maximum energy of the highest allowed energy band filled with carriers at 0 K (the valence band) and the minimum energy of the next allowed band above it (the conduction band) is called the *energy gap*.

The energy gap is an important semiconductor parameter. In those semiconductors capable of it, the frequency of light emission by recombination of electrons in the conduction band with holes in the valence band is proportional to the energy released in the recombination process. In this process, a conduction-band electron recombines with a valence-band hole. Because under most conditions both carriers exist near the band edges (i.e., at the top of the valence band or the bottom of the conduction band), the energy lost is equal to the band gap.

The shape of the band determines the effective mass of carriers. Because the second derivative of the E versus k curve is in general a function of energy, the effective mass is also a function of energy. However, because most carriers under most conditions have energies near the band edges, the variation of this derivative with energy is usually ignored, and the effective mass is treated as a constant. As with the transport parameters, though, in small devices high fields give rise to electrons so energetic that the variation of their effective mass with energy may be important.

The shape of the band determines whether a semiconductor will be capable of light emission as a laser or light-emitting diode (LED). An emitted photon has much less momentum than the electron whose recombination will give it life; momentum must be conserved in the process, and thus the electron's momentum after recombination must be very close to its initial momentum. Because the recombining electron initially must have been near the bottom of the conduction band and would recombine with a hole near the top of the valence band, those two band extrema must occur at the same value of momentum in the E-k diagram. A material in which the band structure has this property is called a *direct-bandgap material*; one without this property is called an *indirect-bandgap material*. As shown in Fig. 1, GaAs is a direct-gap material, and Ge and Si are indirect-gap materials (Chelikowsky and Cohen, 1976).

C. Optical Parameters

The relative difficulty with which a solid may be polarized (i.e., with which its positive and negative charges can be separated) is measured by the dielectric constant. The dielectric constant thus appears in the expression for polar optical phonon scattering, Eq. (9), and affects carrier transport. In addition, the dielectric constant is related to the index of refraction, n, and the optical absorption coefficient, α, through (Bube, 1981, Chap. 4):

$$n^2 = \frac{1}{2}\left[\varepsilon_r \mu_r + \left(\frac{\varepsilon_r^2 \mu_r^2 + \sigma^2 \mu_r^2}{\omega^2 \varepsilon_0^2} \right)^{2,2} \right] \tag{13}$$

$$n^2 = \varepsilon_r \mu_r + \left(\frac{c^2 \alpha^2}{4\omega^2} \right). \tag{14}$$

In these expressions,

ε_r = relative dielectric constant;

μ_r = relative magnetic permeability, and

σ = electrical conductivity.

The index of refraction n is often deduced from measurements of another important optical parameter, the reflection coefficient R. For an interface between a vacuum and a material without absorption on either side (Bube, 1981, Chap. 8),

$$R = \frac{(n-1)^2}{(n+1)^2}. \tag{15}$$

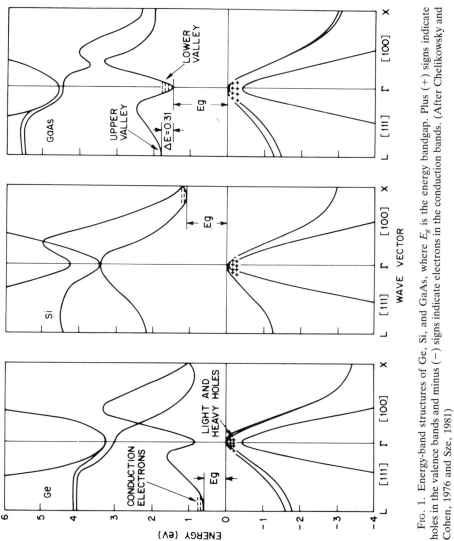

Fig. 1. Energy-band structures of Ge, Si, and GaAs, where E_g is the energy bandgap. Plus (+) signs indicate holes in the valence bands and minus (−) signs indicate electrons in the conduction bands. (After Chelikowsky and Cohen, 1976 and Sze, 1981)

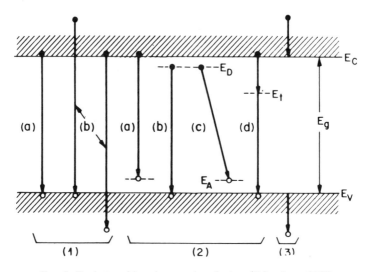

FIG. 2. Basic transitions in a semiconductor. (After Ivey, 1966)

In optoelectronic devices, the optical-absorption coefficient is important for obvious reasons. The dielectric constant is important because the propagation of light may be confined in a region with a dielectric constant higher than that of its surroundings; many optoelectronic devices make use of this valuable dielectric waveguide property.

D. Generation and Recombination Rates

Recombination and generation of carriers are important processes in certain devices. Recombination in the base determines the operation of bipolar transistors, and the light emitted by LEDs and lasers is an alternative form of the energy lost by recombining electrons. Figure 2 shows the most important transitions in semiconductors, where of course the reverse transitions may also take place under appropriate excitation. Each of these processes may be characterized by a lifetime, τ, which relates the rate of the process to the densities of the involved carriers. The basic description of recombination through the agency of a single trap level within the energy gap was given by Shockley and Read, 1952; Hall, 1957:

$$U = \frac{\sigma_p \sigma_n v_{th}(n_h - n_i^2)N_t}{\sigma_n\left[n_e + n_i \exp\left(\dfrac{E_t - E_i}{kT}\right)\right] + \sigma_p\left[n_h + n_i \exp\left(-\dfrac{E_t - E_i}{kT}\right)\right]}. \quad (16)$$

Here,

U = net recombination rate of electron–hole pairs,

σ_p = hole-capture cross section,

σ_n = electron-capture cross section,

n_h, n_e = hole and electron densities, respectively,

n_i = intrinsic carrier density,

N_t = trap density, and

E_t = trap energy

Generation may occur thermally, under the influence of high electric fields, or because of photoexcitation and various other forms of irradiation, (for example, electron irradiation as in a Scanning Electron Microscope (SEM)). The operation of the microwave Impact Avalanche Transit Time (IMPATT) diode necessarily depends on carriers generated in a back-biased p–n junction. However, generation by impact ionization is the bane of designers of short-channel field-effect transistors in which the resulting electrons may produce parasitic substrate or gate currents. Photogeneration makes photodetectors work, whereas the following formula applies when an electron beam (such as that of an SEM) is incident on a semiconductor sample (Newbury et al., 1986, Chap. 2):

$$\langle g \rangle = \frac{E_0(1 - \eta)I_b}{E_{eh} \cdot e}. \tag{17}$$

Here $\langle g \rangle$ represents the mean rate of carrier generation, I_b is the incident beam current in amperes, e is the electronic charge (1.6×10^{-19} coulomb), E_0 is the incident beam energy (in eV) E_{eh} is the energy (in eV) required to produce one electron–hole pair, and η is the bulk backscattering coefficient.

Recombination processes that give rise to radiation occur simultaneously with processes that don't. A measure of how efficiently the fraction of the total number of recombining carriers that recombine radiatively. The quantum efficiency of a photodetector, which determines its sensitivity, is given by the number of carriers generated per photon.

III. Device Parameters

It is often useful to know the dynamic behavior of various physical quantities internal to high-frequency devices so as to verify design concepts or sort out problems (Muller and Kamins, 1986, Sze, 1981, Parker, 1985). The following information can be used for a complete understanding of device design:

1. Charge distributions
2. Electric fields
3. Potentials
4. Energies of charge carriers; energy distributions

The dynamic behavior of these distributions determines the waveforms of the normally measured external signals (e.g., currents and voltages). Although optical probing, with its capacity for better-than-picosecond time resolution, can be very useful in determining these external waveforms, it may also be useful in studying the internal waveforms and thus in increasing our understanding of the basic principles of high-speed devices.

Theoretical profiles of the distributions of the quantities just listed can be obtained for virtually any device by means of two-dimensional or three-dimensional simulation. However, experimental measurements are obviously necessary for verification.

IV. Devices for Switching and Amplification

The high-frequency semiconductor devices that we shall consider here switch or amplify at microwave frequencies in excess of 100 GHz. Such devices include field-effect transistors, both metal-insulator-semiconductor (MOSFET, if the insulator is silicon dioxide) and metal-semiconductor or Schottky barrier (MESFET) types; bipolar transistors; and some new types of devices in which electron energies are filtered by strategically placed energy barriers so as to cull slow-moving carriers from the active region. Included in this category of new devices are the heterojunction MESFET (or high electron mobility transistor (HEMT)) and the heterojunction bipolar transistor (HBT).

To describe qualitatively how the operation of these devices depends on the materials parameters set forth in Section II, we describe them at level 1 of approximation. Level-2 and level-3 details will be added where they are important, but could be overlooked (e.g., in studying the effects of hot-electron transport on charge accumulation, hence, parasitic capacitance of MESFETs, and of electron heating on carrier retention in both MOSFET and HEMT channels). The importance of level-2 and level-3 parameters on other aspects of device performance should be obvious from the device description.

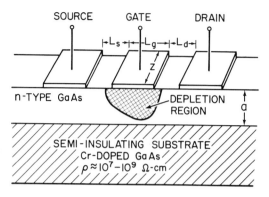

FIG. 3. Schematic diagram of a MESFET.

A. FIELD-EFFECT TRANSISTORS

Field-effect transistors are three-terminal switching or amplification devices in which the flow of current from one electrode (the source) to another (the drain) either is switched on and off by a voltage applied to a third control electrode (the gate) or is modulated in a less drastic fashion by a voltage on the gate smaller than that required to fully shut off the current. Because the change in current flowing to the drain (ΔI_D) depends, ideally, only on the change in voltage on the gate (ΔV_G) and is independent of the drain voltage, voltage gain is possible. The figure of merit of this device is then the mutual transconductance,

$$g_m = \partial I_D / \partial V_G. \tag{18}$$

1. MESFET

Figure 3 shows a MESFET (Mead, 1966, Hooper and Lehrer, 1967; Pucel et al., 1975; Liechti, 1976) in which the gate electrode is a rectifying metal/n-type semiconductor contact. Figure 3 is a very simplified representation in which a negative gate voltage reverse-biases the contact, creating a clearly defined region below the gate in which no mobile carriers exist and therefore cutting off the device. The classical (level-1) analysis of this structure uses the following simplifications:

1. Mobility is constant.
2. Diffusion is neglected, so the edge of the depleted region is abrupt.
3. Electric field in the longitudinal direction (along the channel) is neglected.

The result of this analysis is the following expression for drain current in terms of drain and gate voltages and device and materials parameters. (Muller and Kamins, 1986, p. 207):

$$I_D = G_0 \left\{ V_D - \frac{2}{3} \left(\frac{2\varepsilon_s}{N_d t^2} \right)^{1/2} [(\phi_i - V_G + V_D)^{3/2} - (\phi_i - V_G)^{3/2}] \right\}. \quad (19)$$

In this equation, ϕ_i is the built-in-barrier voltage at the gate, and N_d is the doping density in the fully ionized n-type active layer.

The speed of a field-effect transistor depends on how fast it can charge or discharge the capacitance loading it (i.e., connected to the drain) and the minimum time in which an interruption to current in the channel can make its presence felt at the drain. The mutual transconductance (Eq. 18) is a measure of the first parameter, while the transit time (Eq. 1), with the critical length usually taken as channel length L, is a measure of the second. Both these equations indicate that to first order electron velocity in the channel is the most important consideration in determining device speed. As shown in Fig. 4, however, parasitic resistances and capacitances may influence the device speed and must not be overlooked when designing and fabircating the device (Chand and Morkoc, 1985).

As compound semiconductor MESFETs are reduced in size to below about one micron in channel-length, level-2 analysis must be performed to properly describe the device. The differences between level-1 analysis and the more physically accurate level-2 result are shown in Fig. 5. Figure 5 shows the theoretical charge and electron-velocity distributions in a 0.25-micron channel-length GaAs MESFET, as determined by both types of analyses (Buot and Frey, 1983). The velocity distribution is much smoother in the more accurate analysis, and the peak velocity is about three times higher. Further, the level-2 analysis shows not only that the abrupt depletion region indicated in Fig. 3 does not exist, but also that dipole layers form between the gate and the drain, greatly affecting the device's parasitic capacitances. These internal waveforms certainly affect externally measured performance—particularly frequency response—greatly, in ways that could be clarified by optical probing. In addition, properly designed devices might be useful to optical probes for internal carrier generation to clarify the internal processes illustrated in Fig. 5.

2. MOSFET

The MOSFET (Kahng, 1976; Kahng and Atalla, 1960; Brews, 1981; Tsividis, 1987) is popular among both designers and manufacturers of VLSI circuits. Designers like this device because it consumes little power

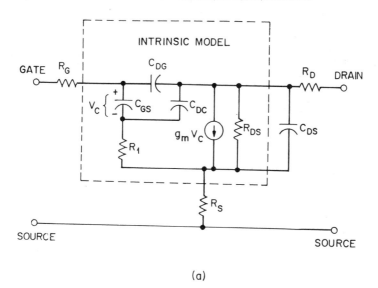

(a)

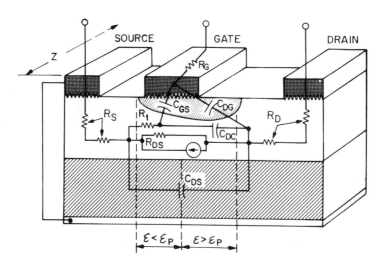

(b)

FIG. 4. (a) Equivalent circuit of a MESFET. (b) Physical origin of the circuit elements. (After Liechti, 1976; and Sze, 1981)

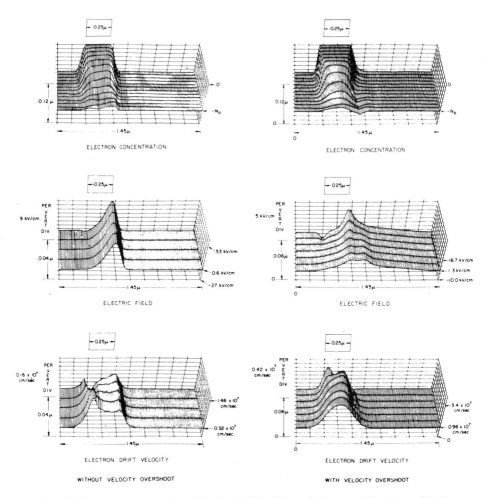

FIG. 5. Electron concentration (n), electric field ($\mathscr{E}$), and electron drift velocity (v) distributions for a 0.25μ GaAs MESFET at $V_{DS} = 1.0$V and $V_{GS} = +0.1$V (with velocity overshoot), $V_{GS} = +0.2$V (without velocity overshoot). The $\mathscr{E}$ and v distributions are for part of the epi thickness from the bottom of the epilayer. (After Buot and Frey, 1983)

and is very small. Figure 6 depicts an n-channel MOSFET in which electrons carry current between the source and the drain. In this device, the source-substrate junction is forward-biased, and the drain-substrate junction is biased. With no voltage on the gate, current cannot flow between source and drain. Making the gate more positive than a threshold voltage required to overcome charges trapped at the silicon–oxide interface and

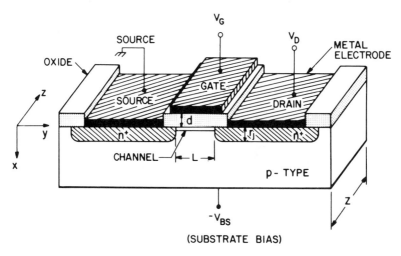

FIG. 6. Schematic diagram of MOSFET. (After Kahng and Atalla, 1960; and Sze, 1981)

within the oxide itself, as well as work function differences between the gate material and the silicon, causes an appreciable number of electrons to be drawn from the bulk material to the surface, forming a conducting channel from source to drain. Thus, the amount of current that flows from source to drain depends on both the drain voltage, which determines the longitudinal electric field within the channel that accelerates electrons, and the gate voltage, which determines the number of electrons available to conduct. The main reason for the small size of MOS circuits lies in the back-biased source and drain junctions; because of them, conduction is impossible between adjacent devices without special isolation precautions necessary with other types of devices.

MOSFETs may be made using p-type instead of n-type semiconductor active layers. In these p-MOSFETs, conduction is by holes, so that the required gate turn-on voltage is negative. If both p-type and n-type devices are available, complementary circuits are possible. In these circuits, a p-type transistor is arranged to be turned off whenever an n-type transistor is turned on, and vice versa, so that voltage connections can be switched without the flow of current. Complementary circuits can be made to consume very little power.

The various MOSFETs share the generic expressions for mutual transconductance and switching speed with the MESFET, and depend, to first order, on the same parameter of transit time as a factor determining switching speed or maximum amplification frequency. Figure 7 shows a

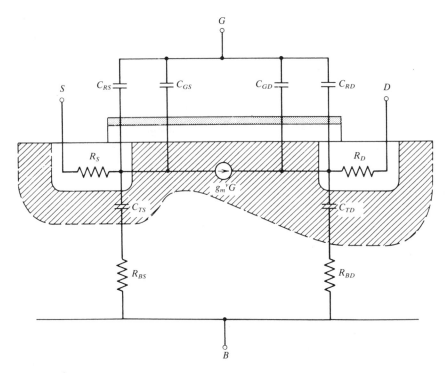

FIG. 7. Small-signal equivalent circuit for the MOSFET showing parasitic elements. (After Muller and Kamins, 1986)

simplified circuit diagram of the MOSFET, depicting various parasitic elements that may also influence the speed of the device and should be considered during device design and fabrication. Because MOSFETs have different geometry and are made using different technologies than MES-FETs, the means by which second-order effects impact high-frequency performance may be quite different. In addition, internal dynamic processes in silicon occur at only about one-tenth the speed as in GaAs, so that the necessity for using a level-2 analysis, even in MOSFETs with submicron channel lengths, is much smaller than for GaAs MESFETs.

Optical probing of the type discussed in section III of Chapter V has been useful in measuring at least one very important MOSFET parameter: the surface carrier mobility. Because of the peculiar behavior of electrons at crystal surfaces, the mobility of carriers there is less than in bulk material, and this mobility must be considered in designing devices. The measurement technique shown in Fig. 8 has been used to determine surface mobility (Cooper and Nelson, 1981). A resistive (rather than a metal-

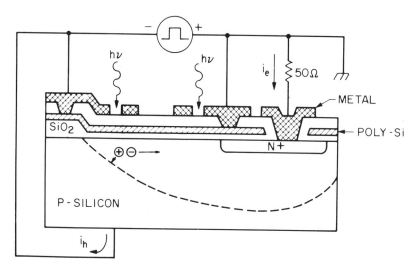

FIG. 8. Use of pulsed laser light to determine the velocity of electrons in a silicon-inversion layer. Biasing pulses are applied across the resistive gate, establishing a uniform field along the silicon surface. A pulsed laser, shining through two openings in the overlying metal, creates electrons and holes under both apertures. While the holes flow into the substrate, the electrons drift along the surface under the influence of the uniform field, giving rise to two pulses in the sampling resistor when they arrive at the drain junction. The difference in time between the arrival of the electron packets is inversely proportional to the velocity of the electrons. (After Cooper and Nelson, 1981)

lic) gate was used to provide a uniform longitudinal field and pulsed laser beams were used to generate packets of electrons at approximately the two ends of the gate. The difference in time of arrival of these packets at the drain, which could be measured electrically, could be directly translated into electron or hole velocity.

3. Heterostructure Field-Effect Devices

Carrier transit time may be reduced by increasing average carrier velocity or shortening average length. To increase velocity, engineers have attempted to use materials in which electron mobility is higher than in Si or GaAs. The level-1 basic equation (Eq. 17) indicates that transconductance is proportional to mobility. Thus, GaAs FETs are used as high-frequency amplifiers in preference to Si devices, and there are continuing attempts to find applications for GaAs fast-switching devices. However, at least in the switching area, improvements in speed using GaAs instead of Si have not reached anywhere near the ratio of the low-field mobilities in the two

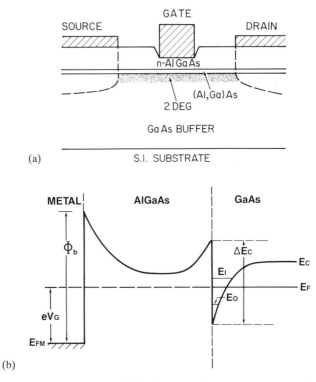

FIG. 9. (a) Schematic of a typical HEMT, incorporating a "spacer" layer. (b) The corresponding energy-band diagram.

materials, for reasons to be explained shortly when level-2 analysis is described.

The high electron mobility transistor (HEMT), also known as modulation-doped field-effect transistor (MODFET), two-dimensional electron gas field-effect transistor (TEGFET), and selectively doped heterostructure transistor (SDHT) (Morkoc, 1985; Morkoc and Solomon, 1984; Delagebeaudeuf and Linh, 1982) is another attempt to improve switching speed by increasing carrier mobility, and its history illustrates the dangers in assuming that FET speed is determined solely by this low-field parameter. The HEMT, shown in Fig. 9 along with the corresponding energy-band diagram, consists in its most elementary form of a layer of doped AlGaAs epitaxially grown over a layer of undoped GaAs. To form the gate, a Schottky barrier is formed on the top surface of the AlGaAs. The heterojunction between the two dissimilar materials supports a built-in energy difference of about 0.32 eV (GaAs on the lower side) because of differences in band gap between the two materials. Conduction electrons from the n-type dopant in AlGaAs are swept into the GaAs by this energy difference. There, these electrons neither drift/diffuse away from the interface into the GaAs bulk (because of the attractive Coulomb force of the depleted donors in the AlGaAs) nor do they go back into the AlGaAs because of the presence of the heterojunction built-in energy barrier. They thus accumulate near the interface, where they are confined in a nearly triangular potential well as a two-dimensional electron gas (2 DEG), and form the transistor channel. The current between the source and the drain flows parallel to the interface and, depends on, among other things, the electron density, electron velocity, and electron mobility. This current can be controlled by the gate bias, as this bias directly controls the 2 DEG density (Delagebeaudeuf and Linh, 1982). Application of a negative gate bias leads to a reduction of the current, and eventually, as the negative bias is increased, channel pinch-off occurs. The currrent flow can be restored, by appropriately reducing this negative gate bias. In the GaAs, these electrons presumably are free to travel unhindered at least by ionized-impurity scattering, because all the ionized impurities are in the GaAlAs. In fact, it is just for this reason that an undoped thin "spacer" layer of AlGaAs (also shown in Fig. 9) is routinely placed between the doped AlGaAs and the undoped GaAs layers, so that the 2 DEG is further removed from the potential donor scattering centers in the AlGaAs. Without such scattering, the low-field mobility should be higher than it would be in doped GaAs.

However, this argument contains two flaws:

1. In a device operating with practical voltages, the longitudinal electric field in most of the channel is considerably higher than that required for velocity saturation, so that low-field mobility is meaningless except insofar as it determines source resistance.
2. A 0.32eV energy difference is not sufficient to keep the conduction electrons in the GaAs as they become heated by the channel electric field, so that conduction properties near the drain will be determined by the AlGaAs properties in any case.

As discussed previously, HEMTs based on $Al_xGa_{1-x}As/GaAs$ heterostructures have many attributes required by high-speed systems. HEMTs, however, are subject to two serious limitations (Morkoc, 1985):

1. When the device is cooled to 77 K without exposure to light, the current in the channel drops to near zero, for drain-to-source voltages less than about 0.5 V, because $Al_xGa_{1-x}As$ with large Al mole fractions x contains defects induced by donors, referred to as DX centers.

2. The sheet carrier concentration of the 2 DEG is limited to below 10^{12} cm^{-2} per interface, which sets up an upper limit for the current. To increase the sheet carrier concentration, larger mole fractions are required, which in turn result in increased defect concentrations and donor binding energy.

Efforts are being made in various laboratories to solve these problems by trying to reduce the DX trap content through improved crystal-growth conditions, and by seeking new materials and compounds for the fabrication of more efficient heterojunctions. New devices with different materials or lower Al concentration can get around this problem (Narayanamurti, 1987; Rosenberg et al, 1985).

B. BIPOLAR DEVICES

A bipolar junction transistor (BJT) (Muller and Kamins, 1986; Shockley, 1949, 1984) is a current-controlled device that consists of two p–n junctions close to each other, as in Fig. 10. Current of electrons or holes

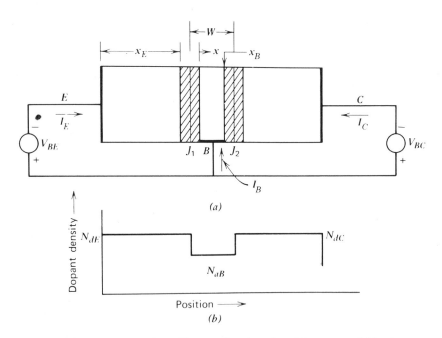

FIG. 10. (a) Prototype transistor. Two p–n junctions J_1 and J_2 are spaced W units apart. (b) Each region has a constant doping density. The quasi-neutral portion of the middle p-region is bounded by the edges of space-charge regions at $x = 0$ and $x = x_B$, respectively. (After Muller and Kamins, 1986)

with the doping and bias polarities chosen appropriately is emitted from the forward-biased junction formed by the emitter and base, into the base. The charge carriers composing this current are minority carriers in the base and may recombine there. For this reason, care is taken during the fabrication of the transistor to obtain high carrier lifetime values in the base; the higher the lifetime, the larger the fraction of emitted current that reaches the next junction, between the collector and base, and can be collected there. Because the current that flows into the collector is independent of load impedance, voltage amplification of a small base-collector input signal is possible. In addition, because sufficient majority carriers to completely cut off the collected current may easily be injected into the base, simply binary switching is also possible.

1. BJT Figures of Merit

The basic equation that describes the physics of homojunction bipolar transistor action is (Muller and Kamins, 1986):

$$J = J_S \left[\exp\left(\frac{eV_{BC}}{kT}\right) - \exp\left(\frac{eV_{BE}}{kT}\right) \right] \tag{20}$$

$$J_S = \frac{e^2 n_i^2 \tilde{D}_n}{Q_S}.$$

In these equations, the device and materials parameters in the base are Q_S, the total majority-carrier charge density per unit area; n_i^2, the square of the intrinsic carrier density; and $\tilde{D}_n$, the average diffusion constant for minority carriers (electrons, here). In a BJT, increased switching or amplification speed depends to first order on being able to decrease the minority carrier transit time τ_B ($= x_B^2/2\tilde{D}_n$) across the quasi-neutral base, and being able to increase the current available to drive a load. These may be achieved by decreasing doping in the base (Q_B), decreasing base width (x_B), and increasing the average diffusion constant. Parasitic elements (shown in Fig. 11) may also play an important role in determining speed and must not be ignored.

The magnitude of the ratio of the collector current I_C to the emitter current I_E under active bias is given the symbol a_F, [16] and is the product of the base transport factor α_T, which describes how much of the emitted current is lost to recombination in the base:

$$\alpha_T = 1 - \frac{x_B^2}{2D_n \tau_n} = 1 - \frac{x_B^2}{2L_n^2} \tag{21}$$

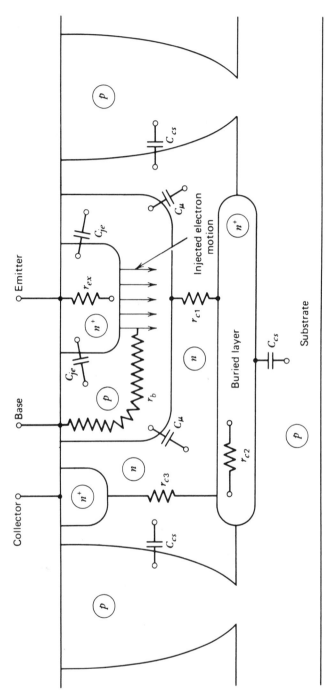

FIG. 11. Integrated-circuit n–p–n bipolar transistor structure showing parasitic elements. (After Gray and Meyer, 1984)

and the emitter efficiency γ, a measure of how much of the current emitted is useful in producing gain (i.e., traceable to the emitter terminal):

$$\gamma = \frac{I_{n\mathrm{E}}}{I_{n\mathrm{E}} + I_{p\mathrm{E}}} = \frac{1}{1 + \dfrac{x_B N_{aB} D_{p\mathrm{E}}}{x_E N_{d\mathrm{E}} D_{n\mathrm{B}}}}. \tag{22}$$

Equations (21) and (22) are written for an npn device. L_n is the average diffusion length of electrons in the base; τ_n is the minority carrier (electron) lifetime; $D_{n\mathrm{B}}$ and $D_{p\mathrm{E}}$ the electron and hole diffusion coefficients in the base and in the emitter, respectively; $I_{n\mathrm{E}}$ and $I_{p\mathrm{E}}$ are the currents flowing from emitter to base because of electron and hole flow, respectively (only the former is useful); $N_{a\mathrm{B}}$ and $N_{d\mathrm{E}}$ are acceptor doping in the base and donor doping in the emitter, respectively; and the x's are base and emitter widths. The current gain β_{F} for the case in which the input current flows between the base and emitter and the output current flows into the collector is related to a_F by:

$$\beta_{\mathrm{F}} = \frac{a_{\mathrm{F}}}{1 - a_{\mathrm{F}}}. \tag{23}$$

For our purposes, it is important to see, from Eqs. (21), (22), and (23) that any BJT transistor's a_{F} and β_{B} will be increased if recombination of minority carriers in the base and the flow of holes into the emitter from the base, are minimized.

2. Improving BJT Performance

Reduction in base width is a technological problem that may be solved, to a point, by developing new shallow-doping or epitaxial growth methods. The flow of majority carriers from the base into the emitter, however, while a function only of the ratio $N_{a\mathrm{B}}$ to $N_{d\mathrm{E}}$ in the homojunction transistor, can be affected by adjusting the built-in potential difference across the emitter-base junction. This difference can be adjusted by making the emitter and the rest of the transistor from different semiconductors (i.e., by building a heterojunction device). If the emitter of such a device were n-type AlGaAs and the remainder GaAs, a barrier to hole flow from base to emitter equal to the band-gap difference between the GaAs and GaAlAs, 0.32 eV, would exist. The consequent reduction of useless hole flow, and improvement of emitter injection efficiency, would be so great that base doping could be much larger than in a homojunction BJT. The heterojunction bipolar transistor should be better than the BJT purely because of the band-gap energetics.

Technological problems make fabrication of a high-performance HBT difficult (Chand and Morkoc, 1985; Asbeck et al., 1984). In particular, heteroepitaxial methods for growing the GaAlAs on the GaAs substrate are difficult to control, particularly if localized doping or growth must be performed to produce a planar device. Finally, a more detailed analysis of the HBT high frequency characteristics and behaviour is needed because, as in the GaAs MESFET, the time constants of the performance-determining parameters may be comparable to those of the signals to be amplified or switched. Therefore, velocity overshoot may play an important role in determining HBT performance.

V. Microwave Two-Terminal Devices

Before it became technologically feasible to make three terminal devices (such as MESFETs) small enough to work satisfactorily at microwave frequencies, several two-terminal devices were developed for this purpose. In these devices, the maximum operating frequencies were not determined primarily by photolithographic techniques, as in the FET, but, rather, by some basic materials time constants as well as the time of drift across an active region. The active region was determined by the thickness of an epitaxial layer or a relatively long length of bulk material. These dimensions were far easier to control, before about 1980, than the one-micron channel length required for a microwave FET.

A. IMPACT AVALANCHE TRANSIT TIME DIODES

The Impact Avalanche Transit Time diode (De Loach, 1976; Read, 1958; "Special Issue on Solid State Microwave Millimeter-Ware Power Generations, Amplification, and Control," 1979) exhibits a negative resistance at microwave frequencies. This resistance arises from a time delay between the AC voltage across the device terminals and the current flowing through these terminals. The simplest IMPATT diode is a p–n junction with a DC reverse bias just below the breakdown voltage. If an AC voltage is superimposed on the DC bias, whenever the AC amplitude brings the total bias above the value required for breakdown, impact ionization will occur and will continue until the AC voltage swing brings the overall bias below the breakdown level. During the high-bias period, a quantity of electrons and holes will be created by impact ionization; in a p^+n IMPATT the electrons will drift through the n-region to its contact, and the holes will drift through the p^+ region. The total delay giving rise to the negative resistance is roughly the difference between the time at which the applied voltage and the current collected at the n contact were at their peaks. The components of this delay are the avalanche delay, which is the time required to build up the electron density, and the transit-time delay, which is the time required for the generated electrons to traverse the n region. The AC resistance of the diode is negative at the frequency at which the total delay is equal to one-half of an AC cycle time.

The IMPATT diode is a simpler device to manufacture than the FET or BJT, and it can perform credibly as a generator of moderate amounts of microwave energy well into the millimeter wavelength range. Because of noise in the avalanche process, however, IMPATT diodes find less favor as amplifiers. Furthermore, because of the high internal fields that must exist to produce impact ionization, assurance of IMPATT diode reliability may be difficult.

B. TRANSFERRED-ELECTRON DIODES

Transferred-electron diodes (TEDs)(Ridley, 1977; Ridley and Watkins, 1961; Gunn, 1963; Hilsum, 1962, 1978; Thim, 1980) are extremely simple devices: just chunks of bulk n-type compound semiconductor with two ohmic contacts placed a suitable distance apart. The operation of these devices is based on the fact that the conduction band of GaAs (and InP) consists (Fig. 12) of a central low-energy valley in which electron effective mass is very low; and higher-energy valleys, displaced from the center of the Brillouin zone and lying upwards of 0.35 eV above the central valley in energy, in which electron effective mass is very high (Aspnes, 1976; Rees and Gray, 1976). At low applied fields, all the conduction-band electrons will be in the central (lowest-energy) valley, but when the field reaches about 3,000 V/cm, some electrons acquire energies sufficient to transfer to the high-mass upper valleys. As the field increases further, more and more electrons are excited to the upper valleys, where the acceleration is considerably slower than in the central valley. On average, the electron velocity in the material decreases. Because this decrease in velocity is not accompanied by an increase in charge density, the terminal current must decrease as the electrons are slowed down a negative resistance at the terminals.

Various types of negative differential resistivity (NDR) may arise from the valley-transfer phenomenon. However, the basic parameters in all of these are:

- Energy separation between the valleys,
- Effective masses in the valleys,
- Velocity-field characteristics of the material, which implicitly incorporate the preceding parameters as well as the influence of electric field on electron occupancy in the upper and lower valleys.

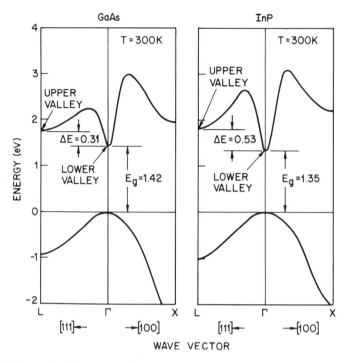

FIG. 12. Energy-band structures of GaAs and InP. The lower conduction valley is at $k = 0$ (Γ); the high valley is along the $\langle 111 \rangle$ axis (L). (After Aspnes, 1976; Rees and Gray, 1976; and Sze, 1981)

VI. Integrated Circuits

Monolithic integrated circuits (Grinich and Jackson, 1975; Hamilton and Howard, 1975; Glaser and Subak-Sharpe, 1977; Gray and Meyer, 1984; Hodges and Jackson, 1988; Mukherjee, 1986) are formed when numbers of devices, such as those previously discussed, are produced simultaneously in a single crystal of semiconductor and interconnected using deposited conductors to perform specific functions. Circuits may be integrated in Si, GaAs, or other materials, and they may perform digital, linear (low-frequency), microwave, or optoelectronic functions as well as various combinations of the above.

One of the most important possible uses of optical and electron probing is in the nondestructive testing of integrated circuits. Because most mechanical probe tips are considerably larger than the width of metallic lines within an integrated circuit, it is impossible to make metallic contact to the interior nodes of an integrated circuit. ICs are thus tested functionally, and if they do not perform their designed functions correctly, engineers must work back from functional results to determine what physical parameter has gone wrong. Contactless probing offers a possibility for internally probing, on a microscopic scale, to locate fabrication or design flaws.

Integrated circuits may be made using either MOS or bipolar technology, although some circuits are appearing with both technologies on the same chip. MOS and CMOS technologies are now used for most integrated circuits where density, cost per function (which depends on density and ease of fabrication), and low power consumption are most important (Hodges and Jackson, 1988; Mukherjee, 1986). Bipolar technology is used where speed is of primary importance (Gray and Meyer, 1984).

In ICs of the late 1980s, millions of logic gates or memory cells may be interconnected. Therefore, IC design is an expensive and complicated undertaking, which proceeds with the help of computer-aided design programs and verification of simulation. Despite these aids, the design process is not infallible and optical and electron probing can be very helpful in determining, on a single-transistor or single-gate level, where on a chip errors may have arisen.

A. MEMORIES

Digital integrated circuits process and manipulate data using transistors as switches. Memory chips (Glaser and Subak-Sharpe, 1977; Hodges and Jackson, 1988) store this data, which may represent both information being processed and instructions for operating the chip itself. *Random-access memories* (RAMs) store data or instructions only temporarily. *Dynamic RAMs* (DRAMs) are RAMs in which data is stored in the form of the presence of charge on a capacitor plate. Requiring only a capacitor and an access transistor per bit, these memories are the densest available. However, the DRAM must periodically be taken out of operation so that the stored charge, which normally leaks off the plates, can be refreshed. The periodic read-and-refreshing cycle is what makes these devices dynamic. This cycle also causes these devices to be slow, because time must be devoted to refreshing.

Static RAMs (SRAMs) store information in the state of a bistable circuit. While speed is increased because no refreshing is necessary, SRAMs require up to six transistors to store one bit and are considerably less dense than DRAMs.

Read-only memories (ROMs) store information more or less permanently. Standard ROMs store information in the form of fixed conduction paths and are programmed during manufacture using a specific interconnection pattern. ROMs programmed by thermal or optical destruction of specific interconnections in a matrix array are called *programmable ROMs* (PROMs). Programming is not destructive in erasable programmable ROMs (EPROMs), which may be erased (e.g., by flooding with ultraviolet light) and then reprogrammed.

B. LOGIC CIRCUITS

Logic chips (Hodges and Jackson, 1988) process instructions and data, and are the key components of all of today's modern electronic devices. Logic may be packaged as separate microprocessor and peripheral chips (e.g., a CPU chip that performs basic operations and support chips that perform clocking functions, control input–output devices, etc.) or these functions may be integrated on a single chip to reduce cost. Many standard logic chips, which contain relatively few generic logic gates, are available at low cost to support the high-cost high technology microprocessor chips.

As technology develops, system designers are beginning to replace the circuit-board–level integration of microprocessor and peripheral chips with chip-level integration, to perform specific functions. Thus, application-specific (or custom) ICs are being developed for use in proprietary products.

C. LINEAR INTEGRATED CIRCUITS

Linear integrated circuits (Gray and Meyer, 1984) perform analog or continuous functions, in some cases to make the output of a digital circuit usable by an analog device, such as a video screen or a loudspeaker. Low-frequency linear circuits as of 1989 usually employ silicon bipolar device technology. Microwave linear circuits employ GaAs MESFET technology.

VII. Optoelectronic Devices

Optoelectronic devices of interest (Bergh and Dean, 1976; Thomson, 1980; Wilson and Hawkes, 1983; Yariv, 1985; Yariv and Yeh, 1984) include LEDs and lasers (which emit radiation) and photodetectors (which convert radiation into electrical voltages or currents). Solar cells are photodetectors on a relatively massive scale. Also important are monolithic (but perhaps heterostructural) optoelectronic integrated circuits, which combine emitters, detectors, and signal-processing circuits.

Although both LEDs and laser emit radiation, the spectral width of their emissions differs. LEDs can be as simple as p–n junction diodes made in direct-transition material, and emit radiation within a band of perhaps

500 Å. Lasers are junction devices that are precisely made to ensure that an internal resonance occurs at a particular output frequency. Lasers emit radiation that may have a bandwidth of considerably less than 1 Å. As growth technology has advanced, new and often rather complicated hetero-structure versions developed for both LEDs and lasers have led to great increases in the output power and efficiency of these devices.

A. EMITTING DEVICES

In semiconductor lasers and LEDs, "optical" radiation that may lie far outside the visible-light range is emitted when carriers are injected across a forward-biased junction. In an appropriate material, the subsequent re-combination of these carriers will lead to the desired emission. To narrow the spectrum of the emitted radiation and make a laser, the junction is formed into a Fabry-Perot cavity (i.e., a specific length of active junction is confined between two parallel planes, defined by cleaving, etching, or polishing). Because of cavity resonance, emission at the optical frequency, which is the resonance frequency of the cavity, grows at the expense of radiation at other frequencies.

1. Radiative Transitions

The light-emitting process results from a transition of an electron from a high-energy state to a lower-energy state, in which the energy difference is radiated as a photon (Fig. 13). This process is called *radiative transition*. As

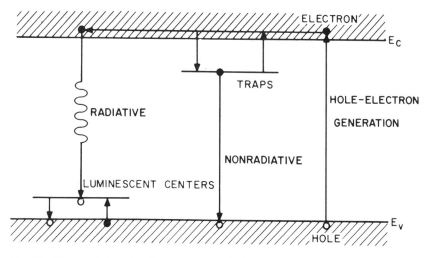

FIG. 13. Representation of radiative and nonradiative recombinations. (After Ivey, 1966)

seen in Fig. 13, recombination across the band gap of conduction-band electrons and valence-band holes is not the only transition that can lead to radiation; radiative transitions may also occur between donor or acceptor levels and the valence or conduction bands, but not all such transitions may be radiative. Indeed, the major frequency of emission, spectral width, and luminescent efficiency of a laser or LED depend on the semiconductor used and its intentional and unintentional doping.

2. Light-Emitting Diodes

The range of semiconductors that may be used for visibly emitting devices is shown in Fig. 14. The bar-like continuous representation for the

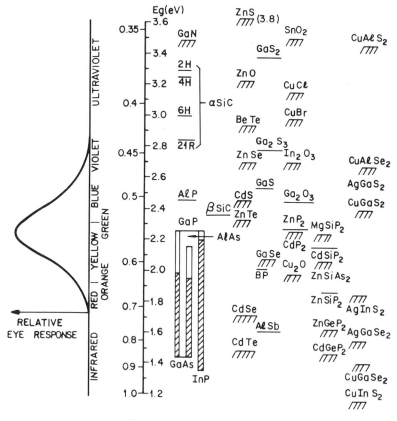

FIG. 14. Semiconductors of interest as visible LEDs, including the relative luminosity function of the human eye. (After Bergh and Dean, 1976)

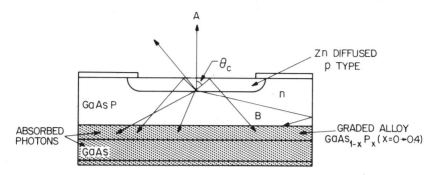

FIG. 15. Schematic of a typical light-emitting diode. (After Gage et al. 1977)

GaAs/InP/GaP/AlAs system represents the possibility to grow ternary and quaternary alloys of these materials (e.g., $GaAs_{1-x}P_x$) to produce a device with a specific output wavelength. However, as is also indicated in Fig. 14, the energy gap is not direct and thus emissive behavior is not obtained for all possible combinations of materials. In such cases (e.g., for the preceding alloy with $x > 0.45$), impurities are intentionally incorporated during growth to act as centers for radiative recombination.

A typical LED is shown in Fig. 15. This particular device uses an active layer of $GaAs_{1-x}P_x$, with x chosen to achieve the desired emission frequency. The active layer is grown epitaxially on a GaAs substrate because bulk GaAs is relatively easily obtainable. To minimize crystalline strain in the active layer, caused by mismatch between the GaAs and GaAsP lattices, a graded alloy of GaAsP is usually grown between the substrate and active layer. The device is completed by diffusion of Zn through a mask to create a planar p–n junction.

Photons are emitted from the LED of Fig. 15 in all directions, so that not all are visible. Therefore, the visual output of this device may be increased by using a substrate material that does not absorb the photons (as does GaAs), backed by a reflector; and by changing the geometry of the device such as by making the n-region hemispherical and using a nonabsorbing substrate (Bergh and Dean, 1976; Thompson, 1980).

3. Junction Lasers

Figure 16 depicts the basic structure of a homojunction laser. In operation, forward bias is applied to the junction and current flows, accompanied by LED-type anisotropic radiative emission. At a certain value of current—the threshold current—the positive feedback at the desired emission

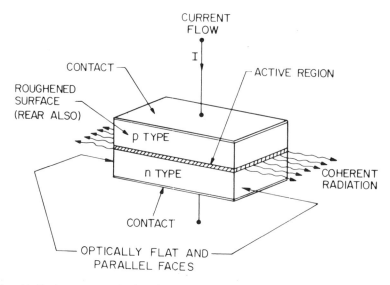

Fig. 16. Basic structure of a junction laser in the form of a Fabry-Perot cavity. (After Sze, 1981)

wavelength arising from multiple and reinforcing reflections within the cavity stimulates this emission, and action begins.

The preferential stimulation of emission requires several radiative recombinations (i.e., a large number of carriers injected across the junction). Threshold current density for a GaAs homojunction laser is sufficiently large to be impossible to sustain in DC operation. The carrier-confining properties of heterojunctions may be used to overcome this problem. Figure 17 compares some characteristics of homostructure, single-heterostructure, and double-heterostructure (DH) lasers (Panish et al., 1970). It is seen that in addition to electron confinement between the heterojunctions in the DH device, caused by the band-gap differences, the index of refraction of the emitting GaAs is higher than that of the confining GaAlAs, serving to further confine the optical output in that region. Room-temperature threshold currents in DH lasers can be more than an order of magnitude lower than those in homojunction lasers.

B. Photodetectors

In *photodetectors*, incident illumination of the proper energy (wavelength) creates electron–hole pairs that cause current to flow in an external circuit (Anderson et al., 1970; Lee and Li, 1979; Sze, 1981). The process is thus

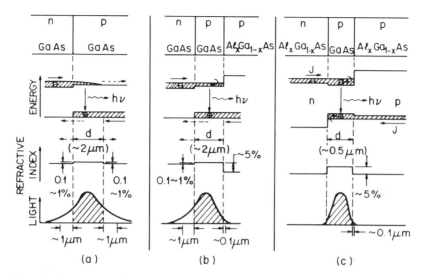

FIG. 17. Comparison of (a) homostructure and (b) double-heterostructure lasers. Top row shows band diagrams with forward bias. Bottom row shows area over which light is emitted, hence greater confinement in heterostructure. (After Panish et al., (1970) and Sze, 1981)

the direct inverse of that occuring in *photoemitters*, where current flow in an external circuit leads to recombination of electron–hole pairs with consequent photoemission. The generation of carrier pairs by photons is a much more probable occurrence than the generation of photons by recombination of carrier pairs, so that photodetectors do not need direct-gap semiconductors. To achieve the desired photodetector characteristics of high sensitivity and speed, photodetectors may be made in the form of heterostructure devices, including phototransistors. To lower systems costs and improve performance, detectors and emitters may be fabricated together as optoelectronic integrated circuits (OEICs).

The quantum efficiency of a photodetector, Y, is the number of carriers generated per incident photon. This parameter can be used to calculate the primary photocurrent, which is the product of the electron charge e and the total number of carriers generated per second in the photoconductor. The gain of the photoconductor is defined as the ratio of the photocurrent flowing in the external circuit to the primary photocurrent.

For a material with an electron mobility μ_n, carrier lifetime τ, applied electric field $\mathscr{E}$, and length L between contacts, and with the illumination falling evenly over the entire length, the gain is given by (Sze, 1981, Chap. 13).

$$Gain = \mu_n \tau \mathscr{E} / L \qquad (24)$$

Thus, improvements in gain follow from reductions in carrier transit time across any region of possible recombination $\left(tr = \dfrac{L}{\mu_n \mathscr{E}} \right)$ and increases in carrier lifetime.

Another detector figure of merit is the detectivity, D^*, defined as

$$D^* = \frac{A^{1/2}B^{1/2}}{NEP} \tag{25}$$

where A is illuminated area, B is detector bandwidth, and noise-equivalent power (NEP) is the incident rms optical power required to produce a signal-to-noise ratio of one in a one-Hz bandwidth,

$$NEP = 4\sqrt{2} \frac{h\nu}{mY} \left[1 + \frac{kT}{e} \frac{t_r}{\tau} (1 + \omega^2 \tau^2) \frac{G}{I_0} \right], \tag{26}$$

where G is the device's dark conductance, I_0 is the steady-state output current when illuminated and m and ω are the modulation index and modulation frequency of the intensity-modulated optical signal. (Sze, 1981, Chap. 13).

The speed of a photodetector is determined by carrier transit time and size (i.e., magnitude of parasitic capacitance). Therefore, in simplest photodetectors (i.e., reverse-biased p–n junctions in which light is absorbed), a large quantum efficiency that requires a wide depletion region to ensure maximum photo-induced carrier collection is inimical to high speed, which requires a short transit time of carriers across this depletion region. To facilitate balancing the trade-offs, it is useful to add some freedom to the device design. This may be accomplished by fabricating a $p^+–i–n^+$ diode instead of a simple p–n diode. The length of the intrinsic region may be adjusted to optimize both quantum efficiency and speed in a balanced way. In such diodes, illumination may be incident through the p^+ region or directly into the i region; further, the p^+ region may be replaced by a Schottky-barrier contact. Further, heterojunction photodiodes can have a greater quantum efficiency than homojunction types; a nonabsorbing material can be used for the region through which illumination is incident.

By increasing the reverse bias to just below the point of avalanching, impact ionization will result in an effective gain in the detection process of a photodiode. This increase of gain means that a detection region of reduced length will still give rise to sufficient output signal; therefore, avalanche photodetectors (APDs) can have very good high-frequency response, at the cost of some increased output noise. Heterostructure APDs theoretically can have further increased performance, arising from reduced absorption and drift times, but difficulties exist in producing heterojunctions with defect densities low enough to eliminate microplasma formation over the junction area.

The gain of transistor structures can be used to advantage by arranging a transistor so that illumination can fall either on the collector-base junction of a bipolar device or on the channel of a MOSFET or MESFET. In the bipolar transistor, if the phonon-induced current in the collector is I_{ph} with the base unconnected (i.e., floating), the total collector current I' will be

$$I' = (1 + h_{FE})I_{ph} \qquad (27)$$

where h_{FE} is the DC common-emitter current gain, which can be much larger than unity, and the effective quantum efficiency is $(1 + h_{FE})$ times larger than that of the base-collector photodiode (Sze, 1981, Chap. 13). However, overall noise performance will degenerate because the AC transistor gain amplifies the noise signal.

In an FET, photodetection with gain may be obtained by illuminating the channel, thus modulating its conductance. Sensitivity is greater when modulating the channel of a depletion-mode device.

VIII. Conclusions

We have attempted to present the physics of both the materials and the devices for which greater understanding may be achieved through optical and electron-probing techniques. The usefulness of optical probing for device or materials analysis derives (1) from the ability to *localize* an optical probe and *adjust its wavelength*, allowing selection of the internal material or device energy with which interaction is desired and (2) from the *speed* with which an optical probe can be switched. Electron-probing techniques (discussed in Chapter VI), although slower than optical ones, take advantage of highly advanced scanning electron microscopes, which are now readily available in many laboratories.

Localization with submicron precision is desirable because internal device or materials phenomena occur within distances less than a micron. In integrated circuits of any form, perceptible electrical signals change over distances comparable to device sizes, which are now approaching the submicron scale. Wavelength *turnability* is useful particularly in materials analysis, in which the energetics of internal processes are well defined. Unwanted internal phenomena may be excluded from measurements, therefore, by tuning the interacting probe's frequency to resonate with that of the internal phenomenon to be studied. *Speed* is important because the phenomena now being studied in both materials and devices are occurring on a subpicosecond time scale.

For these reasons, our understanding of the materials and devices described here should be considerably enhanced by applying high-speed probing techniques.

References

Anderson, L.K., DiDomenico, M. Jr., and Fisher, M.B. (1970). "High-Speed Photodetectors for Microwave Demodulation of Light," in L. Young, ed., *Advances in Microwaves*, **5**, Academic Press, New York, 1–122.

Asbeck, P.M., Gupta, A.K., Ryan, F.J., Miller, D.L., Anderson, R.J., Liechti, C.A., and Eisen, F.H. (1984). "Microwave Performance of GaAs/AlGaAs Heterojunction Bipolar Transistors," *1984 IEEE Int. Electron Device Meeting Proceedings*, 864.

Aspnes, D.E. (1976). "GaAs Lower Conduction Band Minimum: Ordering and Properties," *Phys. Rev.* **14**, 5331.

Bergh, A.A., and Dean, P.J. (1976). *Light-Emitting Diodes*. Clarendon, Oxford, England.

Brews, J.R. (1981). "Physics of the MOS Transistor," in D. Kahng, ed., *Applied Solid State Science*, **Suppl. 2A**. Academic Press, New York.

Bube, R.H. (1974). *Electronic Properties of Crystalline Solids: An Introduction to Fundamentals*. Academic Press, New York.

Bube, R.H. (1981). *Electrons in Solids: An Introductory Survey*, Academic Press, New York.

Buot, F., and Frey, J. (1983). "Effects of Velocity Overshoot on Performance of GaAs Devices, with Design Information," *Solid State Eletron.* **26**, 617.

Chand N., and Morkoc, H. (1985). "Prospects of High-Speed Semiconductor Devices," in G.A. Mourou, D.M. Bloom, and C.H. Lee, eds., *Picosecond Electronics and Optoelectronics*, Springer Series in Electrophysics **21**. Springer-Verlag, Berlin, 9.

Chelikowsky, J.R., and Cohen, M.L. (1976). "Nonlocal Pseudopotential Calculations for the Electronic Structure of Eleven Diamond and Zinc-Blende Semiconductors," *Phys. Rev.* **B14**, 556.

Cooper, J.F., and Nelson, D.F. (1981). "Measurement of the High-Field Drift Velocity of Electronics in Inversion Layers on Silicon," *IEEE Electron Device Lett.* **EDL-2**, 171.

Delagebeaudeuf, D., and Linh, N.T. Metal-(n) AlGaAs/GaAs Two Dimensional Electron Gas FET," *IEEE Trans, Electron Devices* **ED-29**, 955.

DeLoach, B.C. Jr., (1976). "The IMPATT Story," *IEEE Trans. Electron Devices* **ED-23**, 57.

Gage, S., Evans, D., Hodapp, M., and Sorenson, H. (1977). *Optoelectronics Applications Manual*. McGraw-Hill, New York.

Glaser, A.B., and Subak-Sharpe, G.E. (1977). *Integrated Circuit Engineering*, Addison-Wesley, Reading, Mass.

Gray, P.R., and Meyer, R.G. (1984). *Analysis and Design of Analog Integrated Circuits*. 2nd ed. Wiley, New York.

Grinich, V.H., and Jackson, H.G. (1975). *Introduction to Integrated Circuits*. McGraw-Hill, New York.

Gunn, J.B. (1963). "Microwave Oscillation of Current in III-V Semiconductors," *Solid State Commun.* **1**, 88.

Hall, R.N. (1985). "Electron-Hole Recombination in Germanium," *Phys. Rev.* **87**, 387.

Hamilton, D.J., and Howard, W.G. (1975). *Basic Integrated Circuit Engineering*. McGraw-Hill, New York.

Hilsum, C. (1962). "Transferred Electron Amplifiers and Oscillators," *Proc. IRE* **50**, 185.

Hilsum, C. (1978). "Historical Background of Hot Electron Physics," *Solid State Electron.* **21**, 5.

Hodges, D.A., and Jackson, H.G. (1987). *Analysis and Design of Digital Integrated Circuits*, 2nd ed. McGraw-Hill, New York.

Hooper, W.W., and Lehrer, W.I. (1967). "An Epitaxial GaAs Field-Effect Transistor," *Proc. IEEE* **55**, 1237.

Ivey, H.F. (1966). "Electroluminescence and Semiconductor Lasers," *IEEE J. Quantum Electron.* **QE-2**, 713.

Kahng, D., (1960). "A Historical Perspective on the Development of MOS Transistors and Related Devices," **ED-23**, 655.

Kahng, D. and Atalla, M.M. (1960). "Silicon-Silicon Dioxide Field Induced Surface Devices," IRE Solid-State Device Res. Conf., Carnegie Institute of Technology, Pittsburgh, Pa.

Lee, T.P., and Li, T.Y. (1979). "Photodetectors," in S.E. Miller and A.G. Chynoweth, eds., *Optical Fiber Communications*. Academic Press, New York, Chapter 18.

Liechti, C.A. (1976). Microwave Field-Effect Transistors-1976, IEEE Trans. Microwave Theory Tech. **MTT-24**, 279.

McKelvey, J.P. (1966). *Solid State and Semiconductor Physics*. Original publishers: Harper & Row, New York. Reprinted by Krieger Publishing Company, Malabar, Florida, 1986.

Mead, C.A. (1966). "Schottky Barrier Gate Field-Effect Transistor," *Proc. IEEE* **54**, 307.

Morkoc, H. (1985). "Modulation Doped AlGaAs/GaAs Heterostructures," in E.H.C. Parker, ed., *The Technology and Physics of Molecular Beam Epitaxy*. Plenum, New York, 185.

Morkoc, H., and Solomon, P.M. (1984). "The HEMT: A Superfast Transistor," *IEEE Spectrum* (February), 34.

Mukherjee, A. (1986). *Introduction to n-MOS and CMOS VLSI Systems Design*. Prentice-Hall, Englewood Cliffs, N.J.

Muller, R.S., and Kamins, T.I. (1986). *Device Electronics for Integrated Circuits*, 2nd ed., Wiley, New York.

Nag, B.R. (1972). *Theory of Electrical Transport in Semiconductors*. Pergamon Press, Oxford, England.

Narayanamurti, V. (1987). "Novel Heterojunction Devices," *IEEE 1987 IEDM*, 60.

Newbury, D.E., Joy, D.C., Echlin, P., Fiori, C.E., and Goldstein, J.J. (1986). *Advanced Scanning Electron Microscopy and X-Ray Microanalysis*. Plenum, New York.

Panish, M.B., Hayashi, I., and Sumski, S. (1970). "Double-Heterostructure Injection Lasers with Room Temperature Threshold as Low as 2300 A/cm^2," *Appl. Phys. Lett.* **16**, 326.

Parker, E.H.C., ed. (1985). *The Technology and Physics of Molecular Beam Epitaxy*. Plenum, New York.

Pucel, R.A., Haus, H.A., and Statz, H. (1975). "Signal and Noise Properties of GaAs Microwave Field-Effect Transistors," in L. Martin, ed., *Advances in Electronics and Electron Physics* **38**. Academic Press, New York, 195.

Read, W.T. (1958). "A Proposed High-Frequency Negative Resistance Diode," *Bell Syst. Tech. J.* **37**, 401.

Rees, H.D., and Gray, K.W. (1976). "Indium Phosphide: A Semiconductor for Microwave Devices," *Solid State Electron Devices* **1**, 1.

Ridley, B.K. (1977). "Anatomy of the Transferred-Electron Effect in III-V Semiconductors," *J. Appl. Phys.* **48**, 754.

Ridley, B.K., and Watkins, T.B. (1961). "The Possibility of Negative Resistance Effects in Semiconductors," *Proc. Phys. Soc. Lond.* **78**, 293.

Rosenberg, J.J., Benlamri, M., Kirchner, P.D., Woodal, J.M., and Pettit, G.D. 1985. "An $In_{0.15} Ga_{0.85}$ As/GaAs Pseudomorphic Simple Quantum Well HEMT," *IEEE Electron Device Lett.* **EDL-6**, 491.

Sah, C.T., Noyce, R.N., and Shockley, W. (1957). "Carrier Generation and Recombination in p-n Junction and p-n Junction Characteristics," *Proc. IRE* **45**, 1228.

Seeger, K. (1985). *Semiconductor Physics*, 3rd ed. Springer-Verlag, Berlin.

Shockley, W. (1949). "The Theory of p-n Junctions in Semiconductors and p-n Junction Transistors," *Bell. Syst. Tech. J.* **28**, 435.

Shockley, W. (1984). "The Path to the Conception of the Junction Transistor," *IEEE Trans. Electron Devices* **ED-31**, 1523.

Shockley, W., and Read, W.T. (1958). "Statistics of the Recombination of Holes and Electrons," *Phys. Rev.* **87,** 835.

Smith, R.A. (1979). *Semiconductors*, 2nd ed. Cambridge University Press, London.

Solymar L., and Walsh, D. (1988). *Lectures on the Electrical Properties of Materials*, 4th ed. Oxford University Press, Oxford, England.

"Special Issue on Solid State Microwave Millimeter-Wave Power Generation, Amplification, and Control" (May 1979). *IEEE Trans. Microwave Theory Tech.* **MTT-27**.

Sze, S.M. (1981). *Physics of Semiconductor Devices*, 2nd ed. Wiley, New York.

Thim, H.W. (1980). "Solid State Microwave Sources," in C. Hilsum, ed., *Handbook on Semiconductors* **4,** *Device Physics*. North-Holland, Amsterdam.

Thomson, G.H. B. (1980). *Physics of Semiconductor Laser Devices*. Wiley, New York.

Tsividis, Y.P. (1987). "Operation and Modeling of the MOS Transistor," McGraw-Hill, New York.

Wilson, J., and Hawkes, J.E.B. (1983). *Optoelectronics*. Prentice-Hall, New York.

Yariv, A. (1985). *Optical Electronics*, 3rd ed. Holt, Rinehart and Winston, New York.

Yariv, A., and Yeh, P. (1984). *Optical Waves in Crystals, Propagation and Control of Laser Radiation*. Wiley, New York.

CHAPTER 2

Electronic Wafer Probing Techniques

Hermann Schumacher

BELLCORE
RED BANK, NEW JERSEY

and

Eric W. Strid

CASCADE MICROTECH
BEAVERTON, OREGON

I. Introduction

In addition to the contactless techniques described elsewhere in this book, other, more conventional methods exist for the on-wafer characterization of high-speed electronic and optoelectronic circuits. Both high-speed probes (frequency > 1 GHz) and associated electronic instrumentation (high-speed sampling oscilloscopes, vector network analyzer) are available from a variety of sources. Optoelectronic measurement systems with bandwidths in excess of 20 GHz have been introduced into the market. Also, as we shall see, it is quite easy to modify microwave instrumentation for optoelectronic needs. Advantages and limitations of these techniques for characterizing devices and systems with a bandwidth in excess of one GHz (or waveform transitions of less than 300 ps) are discussed in this chapter.

Virtually all of the conventional characterization techniques require physical contact to the device under test. While contactless probing has the advantage of minimizing disturbance of the measurement by the instrumentation, functional testing of ICs requires that the signal inputs and outputs be properly terminated and that bias and control voltages be applied to the signal. This requires physical contact to the device under test.

One of the advantages of the more conventional characterization techniques is that their performance characteristics and sources of error have been investigated for quite some time. Error correction techniques are readily available to compensate for nonideal behavior in both the time and frequency domains. Some of these techniques will be outlined. The electrical properties of high-speed contact probes have been extensively characterized, and algorithms are available to correct measurement errors attributed to the probes, the instrumentation setup, and the interconnecting cables.

II. Electronic Characterization Methods

In the first part of this chapter, we consider some of the electronic instrumentation methods available for high-speed measurement needs and we discuss some hardware and software modifications that can significantly improve the performance of commercial equipment. The discussion begins with a description of some simple stimulus sources, moves on to a discussion of high-speed sampling oscilloscopes, and finally covers frequency-domain–characterization methods.

A. Stimulus Generators

1. Step Recovery Diode Generators

The "comb" waveform generator (e.g., Hamilton and Hall, 1967) using a step-recovery diode (SRD) is probably the most widely used source of narrow electrical pulses for a wide variety of applications, such as spectrum analyzer calibration, time-domain network analysis, or gain-switching modulation of laser diodes (Lin et al., 1980). The comb generator is a passive nonlinear network that transforms an input sine wave into a periodic train of narrow pulses with the same repetition rate. Figure 1 shows a practical example. Capacitors C_1 and C_2 and inductor L_1 form a matching network. Inductor L_2 and the diode junction capacitance C_j determine the width of the output pulse, together with the transit time associated with the depletion region of the diode. During the positive half-cycle of the sine wave, the SRD is forward-biased and in a low-impedance state, so no signal appears at the output. Once the diode becomes reverse-biased, it continues to conduct for a short while and then rapidly switches to the blocking state. This transition triggers a rapid transient in the resonator formed by L_2 and C_j. (C_2 can be considered to be

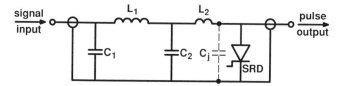

FIG. 1. "Comb" waveform generator using a step-recovery diode. C_1, L_1, and C_2 form a π-network that matches the impedance of the pulse-forming network (L_2 and SRD) to the source impedance at the comb generator's fundamental frequency.

of infinite capacitance here.) Only the first negative half-cycle of this transient will appear at the output because the SRD becomes conducting again at the onset of following positive half-cycle. The SRD comb generator can produce pulses of less than 100 ps FWHM. Its drawbacks are that the repetition rate is constant (given by the tuning circuit) and that the maximum voltage is limited by the reverse breakdown voltage of the SRD, typically 10–20 V for high-speed generators.

2. Avalanche Transistor Generators

Pulse generators using avalanche transistors are another simple means to obtain pulses with less than 400 ps FWHM. Consider the circuit in Fig. 2.

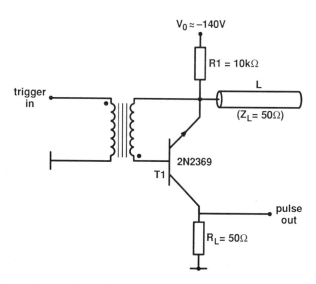

FIG. 2. Avalanche transistor pulse generator using a coaxial transmission line L as the timing element. The transformer polarity shown is for positive trigger pulses. Pulse-repetition rate is limited to a few MHz by thermal considerations for T_1.

While T_1 blocks (collector current $I_C = 0$), the open-ended transmission line L will be charged through R_1 to the voltage V. V is chosen just below the base-collector avalanche breakdown voltage of T_1. If a trigger signal is applied to the base of T_1, the transistor will become conducting, and the energy stored in L is discharged into the load impedance R_L in form of a pulse whose duration is determined by the length of L. Avalanche transistor generators are capable of delivering short pulses (<300 ps) of large amplitude (>30 V). The repetition rate, however, is limited to the low MHz range by thermal considerations for T_1.

3. Tunnel Diode Step Generators

Tunnel diode step function generators are widely used as stimuli in time-domain reflectometry (TDR) applications. Of the electronic stimuli discussed here, they are capable of delivering the sharpest transitions: risetimes (10–90%) of approximately 20 ps in commercially available models. Unlike the methods discussed previously, however, tunnel diode generators generate step-like waveforms with only one rapid transition.

4. Optical Stimuli-Pulsed Lasers

Semiconductor diode lasers are an important optical source for the stimulation of fast signals in the testing of high-speed optoelectronic circuits. They are quite inexpensive and small. Pulse generation is straightforward, and interfacing with electronic instrumentation does not pose a serious problem because laser diodes can be directly current modulated. Finally, laser diodes generate pulses at wavelengths of interest for fiber-optic communications, namely 800, 1300, and 1500 nm. Gain-switching and mode-locking are the most popular techniques for the generation of pulses with FWHM < 50 ps. Active and passive mode-locking techniques may generate pulses below one ps from semiconductor lasers (Yokoyama et al., 1982). The gain-switching technique has the advantage of great simplicity and the lack of external optical components.

Pulses of 30 ps or less may readily be obtained using gain-switched commercial laser diodes. The principle of gain-switching can be explained with the aid of Fig. 3. The laser is pulse-modulated without any prebiasing DC current. The electron density in the cavity increases when a brief (<300 ps) current pulse is injected. Once the electron density exceeds thresholds, a short photon pulse is emitted. Simultaneously, the carrier density decreases due to the stimulated recombination. If the injecting current pulse is sufficiently short, the threshold density will not be reached again; only a single light pulse is emitted. Note that the width of the emitted light pulse is mainly determined by the laser carrier dynamics and

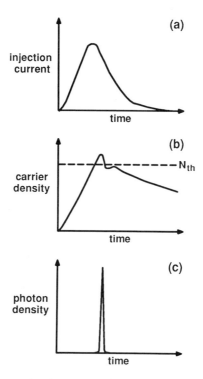

FIG. 3. Gain-switching modulation of a semiconductor laser diode showing (a) electrical injection current pulse shape applied to the laser, (b) electron density in the laser cavity, and (c) optical output pulse emitted from the laser.

can be considerably shorter than the injection current pulse. The width of the emitted light pulse decreases with increasing injection current amplitude. A more detailed study of gain-switching has been performed by Schoell et al. (1984).

Optically pumped, mode-locked dye lasers are the sources of choice in many experiments requiring sub-ps optical pulses. They have also been used to characterize photodetectors with bandwidths in excess of 100 GHz (Van Zeghbroeck et al., 1987). Most experiments are being performed around 600-nm wavelength, where suitable dyes and saturable absorbers are readily available. Results from these experiments are of limited validity at fiberoptic communications wavelengths, because the impulse response of high-speed photodetectors is generally wavelength-dependent. Dye lasers emitting in the infrared are gradually becoming available, however. Experiments using ps dye lasers are discussed extensively in other chapters in this book.

B. Time-Domain Methods

1. The Sampling Oscilloscope

The major advantage of high-speed real-time oscilloscopes is in the display of nonrecurrent waveforms. Oscilloscopes operating in continuous mode usually employ a travelling-wave structure cathode-ray tube where the vertical deflection signal travels with the same velocity as the deflected electron beam. While real-time oscilloscopes with less than 100-ps resolution have been reported (for a review, see Nahman, 1978), sampling methods are much more common in this speed range. We distinguish between equivalent-time sampling oscilloscopes, which need recurrent coherent signals to operate, and real-time sampling oscilloscopes, which do not have these restraints.

a. Real-Time Sampling

Theoretically, a real-time sampling oscilloscope multiplies the input signal s(t) with a periodic stream of impulses III(t) to yield an output function h(t) which contains the amplitude of the sampled signal at discrete times t:

$$h(t) = s(t)III(t). \tag{1}$$

In any practical system, h(t) will be distorted by the nonideal impulse response g(t) of the sampling gate:

$$h(t) = s(t)III(t) * g(t) \tag{2}$$

where $*$ is the convolution operator.

The limitation of this method lies in the fact that for a bandwidth-limited signal, where f_c is the highest frequency, the minimum sampling frequency would be $2f_c$. However, most observed waveforms like pulse responses are time-limited and hence not bandwidth-limited. (Bandwidth-limited signals have a cutoff frequency f_c above which the signal's spectral density is zero.) Sampling repeats the signal spectrum periodically in the frequency domain. For non–bandwidth-limited signals, parts of the repeated signal spectrum overlap, causing signal distortion which is called aliasing. The sampling frequency, which determines the periodic spacing of the spectra, has to be chosen as high as possible to minimize these effects. In their practical implementation, real-time sampling systems may follow quite diverse approaches. The most common technique is the use of a rapidly strobed electrical gate to sample the incoming waveform; one may instead guide the waveform along a transmission line and then "freeze" it by simultaneously strobing a set of sampling gates arranged periodically along the transmission line. (See Fig. 4.) Robin and Ramirez (1980) used a tech-

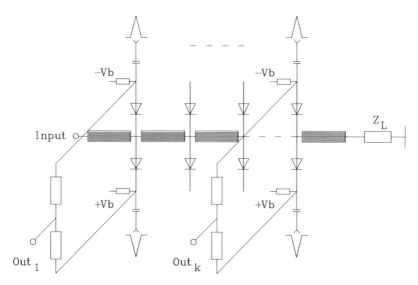

FIG. 4. Real-time sampling using a periodic structure of sampling gates along a transmission line (after Schwarte, 1972). The circuit contains k identical sampling elements strobed simultaneously (Λ). Z_L is the characteristic impedance of the transmission line. V_b is the bias voltage for the sampling diodes.

nique similar to a streak camera, where a ribbon-shaped electron beam is swept past a semiconductor detector array. Very short transitions, however, can only be measured using equivalent-time sampling concepts, which are discussed shortly. An excellent description of the development of time-domain ps measurements over the years can be found in the work of Nahman (1967, 1978, 1983).

b. Equivalent-Time Sampling

Consider, as in Fig. 5a, a recurrent train of coherent waveforms $s(t)$. We take a set of discrete amplitude samples at known delays τ after a fixed trigger point (Fig. 5b), only a single sample is taken during each occurrence of the waveform. If we allow τ to vary between 0 and T, T being the duration of the waveform, we can reconstruct $s(t)$ if we plot the amplitude value taken at delay τ as a function of τ (Fig. 5c). The time available for the measurement and display of the waveform has been expanded from T (for a real-time measurement) to NT, N being the number of amplitude samples or a multiple thereof. Hence, the equivalent-time image of a rapid waveform can be displayed on a slow oscilloscope. The high-speed circuitry needed is limited to the sampling gate and portions of the sampling time-base.

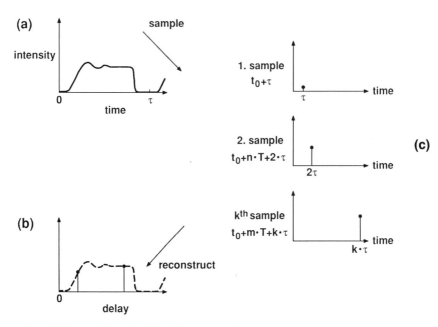

FIG. 5. The sampling principle: (a) reconstruction of a high-speed waveform (b) in equivalent time (c) through a set of samples. The waveform is periodic in T, and t_0 is a reference point with fixed delay after the trigger.

Equivalent-time sampling oscilloscopes may employ two different methods for performing the sampling process. In sequential-sampling oscilloscopes, the delay τ increases monotonically with every occurrence of the input waveform. Using a synchronized equivalent-time base, each amplitude sample is displayed as it is taken. One example of a sequential sampling time base is shown in Fig. 6. The trigger pulses with period T_{trig} are synchronized with the recurring waveform to be measured. Every occurrence of the trigger pulse during a single sweep increases the amplitude at the upper input of the comparator by a fixed amount. The resulting step function starts new with every sweep. Simultaneously, every trigger pulse starts a fast-rising ramp which is applied to the lower input of the comparator. When the voltages of the two waveforms are equal, a comparator generates the strobe pulse for the sampling gate, a high-speed semiconductor diode. The sampling interval is controlled by the step-height of the staircase function. Noise and drift in the amplitudes of both waveforms and in the timing of the fast-rising ramp translate into jitter and drift of the sampling time base. (See section II. B.1.c.)

In a random-sampling oscilloscope (Frye and Nahman, 1964) the signal

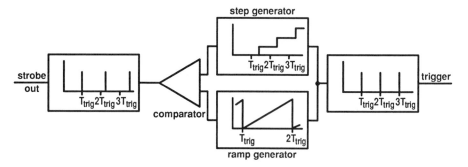

FIG. 6. Block diagram of a sequential-sampling time base. The trigger signal with period T_{trig} starts a fast ramp and increases the amplitude of the step generator on every occurrence of the trigger. The step generator is restarted at the beginning of each sweep. A strobe is supplied to the sampling gate when the amplitudes of step generator and ramp generator are equal.

triggers a linear ramp. The sampling pulse occurs at a random time, samples the signal amplitude at that point in time, and stops the ramp after a fixed delay. The amplitude value of the sample and the final value of the ramp are being stored. The final value of the ramp is related to the time position of the sample, so the signal shape can be reconstructed. While this method eliminates the need of synchronization between the trigger and the fast sampling pulse, it requires a memory for display.

The speed of sampling oscilloscopes can be characterized by an equivalent bandwidth or equally by the risetime the instrument would show when measuring an ideal step waveform. Commercially available sampling oscilloscopes currently have bandwidths of typically 20 GHz, equivalent to an intrinsic risetime of 18 ps. However, Marsland et al. (1988) have demonstrated a GaAs integrated sampling bridge with a bandwidth in excess of 100 GHz. A transient of 4 ps falltime has been experimentally shown. Two sampling diodes with a parasitic capacitance of 12 fF are monolithically integrated into a coplanar waveguide arrangement. The diodes are strobed by rapid electrical transients created in a nonlinear transmission line (Madden et al., 1988). The nonlinear transmission line distorts a 3–10 GHz input sinewave into a sawtooth waveform with 4-ps falltime. This sawtooth waveform is then differentiated to obtain the narrow strobe pulses for the sampling gates.

c. Time-Base Instabilities—Jitter and Drift

Drift is associated with long-term changes in the time base; jitter describes short-term fluctuations in the timing and trigger circuitry.

Jitter is usually random in nature. While jitter cannot be fully corrected, reduction techniques exist and are described shortly. A careful choice of the trigger signal form and amplitude can minimize the effect of jitter. The effect of jitter with time-invariant statistics on an averaged waveform is that of a low-pass filter whose cutoff decreases with an increasing amount of jitter (Elliott, 1970; Gans and Andrews, 1975; Gans, 1981). In the time domain, the averaged output waveform appears convoluted with the normalized probability density function of the jitter process. Jitter reduction, therefore, will increase the effective bandwidth of the instrument.

Humphreys (1988) has demonstrated a jitter-reduction technique that uses a recursive numerical treatment of the jitter-distorted data. Both the mean of the waveform and its variance must be recorded in this method. The variance $\sigma^2(t)$ of a signal s(t) distorted by jitter of probability density function $\phi(t)$ can be described as

$$\sigma^2(t) = \int_{-\infty}^{\infty} \phi(\tau)(f(t) - f(t - \tau))^2 \, d\tau. \tag{3}$$

Figure 7 shows that the variance of a pulse waveform with jitter has two maxima corresponding to the maximum slopes of the pulse: on the leading and on the trailing edges. A Gaussian shape is assumed for the jitter-probability-density function. The width of the Gaussian distribution is

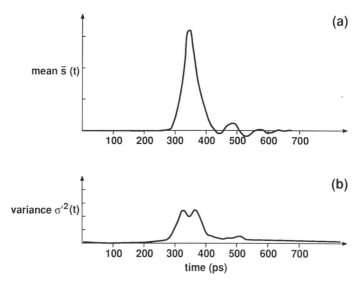

FIG. 7. High-speed pulse with jitter. (a) Mean signal. (b) Distribution function of the jitter contained in the signal. (After Humphreys, 1988)

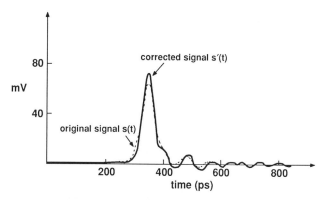

Fig. 8. Corrected pulse after numeric jitter reduction. The mean of the uncorrected signal is shown for comparison. (after Humphreys, 1988)

initially guessed and a corrected waveform is found by deconvolving the guess for the jitter-probability-density function $\phi'(t)$ from the mean of the data. A variance $\sigma'^2(t)$ is calculated from the corrected waveform $s'(t)$ and $\phi'(t)$. $\sigma'^2(t)$ is then compared with σ^2, the variance of the measured signal. This procedure is repeated with varying widths for the jitter-distribution function until the error is minimized, indicating that the best fit for the jitter distribution has been found. The associated $s'(t)$ is the best approximation for the corrected waveform, as shown in Fig. 8. Note the decreased rise and falltimes on the corrected signal.

Drift phenomena usually are not random in nature. They may seriously distort the displayed waveform if averaging is used to suppress noise without appropriate drift compensation.

Drift reduction is important if detection of small signal quantities by long-time averaging is to be attempted. Elliott (1976) has described a hardware method of drift elimination that compares the horizontal and vertical (equivalent time) outputs of a sampling oscilloscope in a phase-sensitive lock-in amplifier, using the horizontal deflection signal as the reference. In the absence of drift, the fundamental components of both waveforms must be in phase. If drift is present, the resulting slowly varying output signal of the lock-in amplifier is applied to the position control circuit of the sampling time base a negative feedback signal. (See Fig. 9.) Elliott was able to detect 5-μV signals in the presence of 8-mV system noise.

A software method of drift compensation has been described by Meckelburg and Matkey (1985). Their algorithm uses the fact that if $F(j\omega)$ is the Fourier transformation of the measured waveform $f(t)$, then the shifted

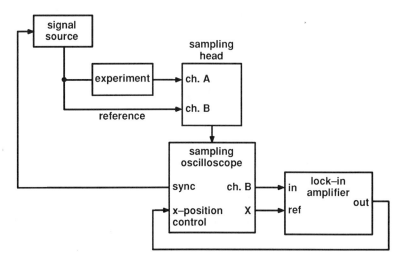

Fig. 9. Electrical time drift reduction (after Elliott, 1976). The fundamental frequency of the vertical oscilloscope output associated with the signal in channel B, and the fundamental frequency of the sweep time base (at the horizontal output) are compared in the lock-in amplifier. Any drift in the time base will produce a change in the lock-in output signal. This signal is fed back into the X-position control input.

signal $f(t - T)$ transforms as

$$f(t - T) -> F(j\omega) \exp(-j\omega T). \tag{4}$$

The drift error T can be determined at any frequency by comparison with the phase at that frequency obtained from the transform of a previous reference time sweep. Multiplying the Fourier transform of the drifted signal with $\exp(+j\omega T)$ compensates the drift completely. Figure 10 demonstrates that drift compensation significantly improves the ability to retrieve a noisy signal without distortion through averaging. As the residual uncertainty of time increases with decreasing signal-to-noise ratio, it can be advantageous to measure the drift of the time base using a reference signal of known amplitude in a second channel.

d. Time-Base and Amplifier Nonlinearities

While most authors find the linearity of vertical amplifiers to be sufficient, significant nonlinearities in the time-base of commercial sampling oscilloscopes have been observed. The magnitude of error was found to decrease as the delay between the trigger point and the start of the measurement window increases. One approach, therefore, is to increase this delay. One may, on the other hand, also recalibrate the time base to

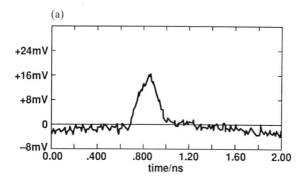

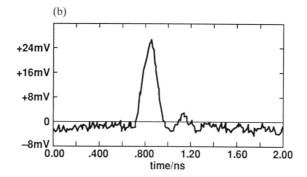

FIG. 10. Improvement in pulse shape in an averaged signal by numeric drift removal (after Meckelburg and Matkey, 1985). Phase change in subsequent sweeps is used to detect, quantify, and remove the drift. (a) Averaged without drift compensation. (b) Averaged with drift compensation.

remove the nonlinearity (Scott and Smith, 1986), a convenient method if the data are processed externally. For this purpose, a highly stable sine-wave signal with a frequency in the GHz range is fed into the sampling oscilloscope, and the times at which the waveform passes through zero are used to determine the measured period. In the presence of time-base nonlinearities, the measured period will not be constant during one sweep. As the theoretical period of the sine wave is accurately known, the error in linearity can be computed from the measured periods. Let t be the actual time and τ the time measured by the sampling oscilloscope. We can then express the actual time as a function of the measured time,

$$t = f(\tau) \tag{5}$$

The actual time at each of the sampled points is then approximately

$$t_j \approx t_{j-1} + \Delta\tau f'(j\Delta\tau) \tag{6}$$

where $\Delta\tau$ is the measured time interval between the samples. If further processing requires equally spaced time intervals, the samples can be repositioned in time using a spline interpolation through the corrected data.

e. Deconvolution

Several problems can mask the frequency response of a device under test during the measurement process. If a step or impulse response is desired, the limited risetime or finite pulse width of the stimulus is the first limit. Multiple reflections in the measurement setup present another source of error, as well as the bandwidth limitations of the sampling oscilloscope itself. Assuming that every component in the experimental setup can be described as a linear, time-invariant system, then these errors can be removed by deconvolution. Let us assume that the nonideal responses in our setup can be combined in a single transfer function $y(t)$. The dependency of the measured response $m(t)$ upon the true response $s(t)$ of our device under test can then be written as a convolution product:

$$m(t) = s(t)^*y(t) = \int_{-\infty}^{\infty} s(\tau)y(t - \tau)\,d\tau. \tag{7}$$

Solving this integral equation in the time domain is usually avoided by a transform to the frequency domain where the convolution product reduces to

$$M(f) = S(f)Y(F). \tag{8}$$

M, S, and Y are the Fourier transforms of m, s, and y, respectively. In the absence of noise, the deconvolution would reduce to a division operation:

$$S(f) = \frac{M(f)}{Y(f)} \tag{9}$$

followed by a transformation into the time domain. If noise is added to the system, its effect is amplified through the deconvolution process, especially if $Y(f)$ is small (for example, above the cut-off frequency of the setup). This severely limits the bandwidth increase that can be obtained by deconvolution. Similarly, errors in the observed waveform or the assumed system response are amplified by the deconvolution process. As a simple example, consider an observed risetime of 30 ps. We know the system response has a risetime of 25 ps, with an error margin of ±2 ps, or 16% of the mean value. Using an approximate rule-of-thumb deconvolution technique, the risetime of the deconvolved "true" signal response is anywhere between

$$t_{r,l} = \sqrt{30^2 - 27^2} = 13 \text{ ps and}$$
$$t_{r,u} = \sqrt{30^2 - 23^2} = 19 \text{ ps.}$$

The error margin has been increased from 4 ps (16%) in the system response to 6 ps (36%).

The signal-to-noise ratio of the deconvoluted signal can be significantly improved if a suitable filter in the frequency domain is added to the processing algorithm. Nahman et al. (1981) have shown that an optimum filter $R(f)$ can be implemented depending on the system response $Y(f)$:

$$R = \frac{|Y|^2}{|Y|^2 + \gamma |C_2|^2}. \tag{10}$$

Here C_2 is a multiplicative function that corresponds to the second derivative operator in the time domain. Humphreys (1988) has shown that

$$|C_2|^2 = \left(\frac{\pi i}{n}\right)^4 \tag{11}$$

is a sufficient approximation, where i is the harmonic number and n is the number of data points in the transform. The factor γ is operator-chosen and determines the amount of smoothing. Applying (9) to (8b), we get

$$S = \frac{M}{Y} R = \frac{ZY^*}{|Y|^2 + \gamma |C_2|^2} \tag{12}$$

for the deconvoluted response in the frequency domain. Y^* is the complex conjugate of Y.

The deconvolution not only corrects for bandwidth limitations to some extent, but also compensates for errors such as multiple reflections (as long as they are reproducible) and undesired relaxation oscillations from laser diode pulse generators in optoelectronic response measurements.

In summarizing the performance limitations of conventional sampling oscilloscopes, it is important to understand that it is not sufficient to improve solely the switching speed of the sampling gate. Time jitter and time-base drift also are important factors limiting the performance.

2. The Superconducting Sampling Oscilloscope

a. Principle

The performance of a sampling oscilloscope can be significantly improved if the electronic sampling gate and time base are replaced by a Josephson Junction sampling gate and time base. The Josephson Junction provides a higher switching speed than current electronic components, leading to a shorter instrument risetime. As critical elements of the time base are kept at cryogenic temperatures, timing instabilities due to noise are greatly reduced, suppressing instrument jitter. Furthermore, as this temperature is kept constant, temperature variations as an important source of long-term drift are eliminated.

The use of Josephson Junctions as amplitude discriminators in high-speed instrumentation has been demonstrated by Zappe (1975). In this technique, the signal to be measured plus a bias are applied to a Josephson detector that switches to the conductive state once its critical current is exceeded. By modifying the bias and observing the delay in the detector output, step-like signals can be mapped out. The technique does not work, however, for arbitrary pulse shapes. Hamilton et al. (1979) refined this technique and were able to measure rise times of 9 ps. Their result was limited by jitter and drift of approximately 5 ps in the room-temperature time base.

Faris (1980) introduced a Josephson sampler that is capable of characterizing arbitrary waveforms. His method extends the amplitude discrimina-

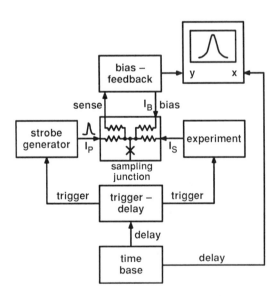

Fig. 11. Block diagram of a sampling oscilloscope for arbitrary waveforms using Josephson Junctions (after Faris, 1980). If the sum of signal current I_S (of unknown shape), strobe pulse current I_P, and D_C bias current I_B exceeds a threshold value, the sampling junction switches. The bias feedback circuit keeps the junction exactly at threshold. The feedback loop's control voltage maps the temporal shape of the signal current.

tor approach in that not only an adjustable bias, but also a very narrow pulse are superimposed on the signal under test. Adjusting the bias and delaying the probe pulse with respect to the beginning of the waveform permit sampling of the waveform. This is done in a feedback loop, as shown in Fig. 11. The bias is adjusted so that the sum of signal current I_S, pulse current I_P, and bias current I_B is exactly at the threshold of the sampling gate. Given a probe pulse of constant amplitude, adjusting the bias current as a function of delay time maps the signal under test. The signal level allowed at the input of the sampler has to be low enough not to switch the discriminator on its own. Harris et al. (1982) added an electronically adjustable delay using Josephson technology to provide low-jitter, low-drift timing.

Figure 12 depicts a Josephson sampler with integrated timing circuit which shows 9-ps (FWHM) impulse response and a maximum measurement window of 300 ps due to the internal delay circuit. Wolf et al. (1985) pushed the performance to 2.1-ps risetime and 0.8-μA sensitivity. An important trade-off between switching speed and tracking accuracy of the sampled signal toward the input signal has been pointed out by Van Zeghbroeck (1985): The shorter the risetime of the sampling gate, the more overshoot the measurement introduces to the signal under test. Therefore, it may be advisable to sacrifice speed and get a well-behaved step-response.

b. State-of-the-Art-Performance

Practical implementation of a superconducting oscilloscope has to deal with the problem of transferring signals from the room-temperature environment to the cryogenic environment of the Josephson sampler. A commercial sampler solves that problem by using a coplanar transmission line between the cryogenic sampling circuit and a coaxial connector mounted immediately at the end of the substrate. All cryogenic electronics are concentrated in a corner of the substrate (Fig. 13). This corner is spray-cooled with liquid helium and the rest of the substrate remains uncooled. The instrument has an intrinsic risetime of 5 ps, equivalent to a 70-GHz bandwidth. The maximum measurement window is 10 ns. The dynamic signal range is 50 μV to 10 mV. While the ability to detect small signals is definitely superior to that of conventional oscilloscopes, owing to the low-noise cryogenic electronics, one sees that the dynamic range is comparable (Weber, 1987).

c. Performance Limitations

Aside from the forementioned risetime-overshoot trade-off, the question of pulse dispersion on the transmission line between the signal source

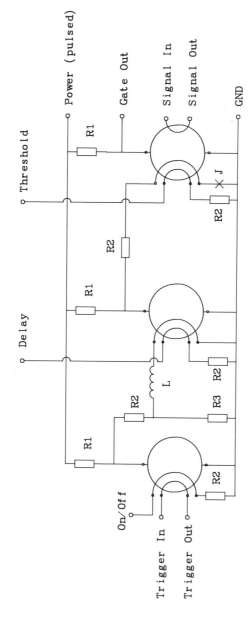

FIG. 12. Circuit diagram of a Josephson sampler with integrated delay circuit (after Harris et al., 1982). The left element is the strobe pulse generator, feeding into the delay circuit (center). The right element is the sampling gate. R1 = 100 Ω, R2 = 2.3 Ω, R3 = 0.13 Ω, L = 10 pH. J is a Josephson element (0.2 mA).

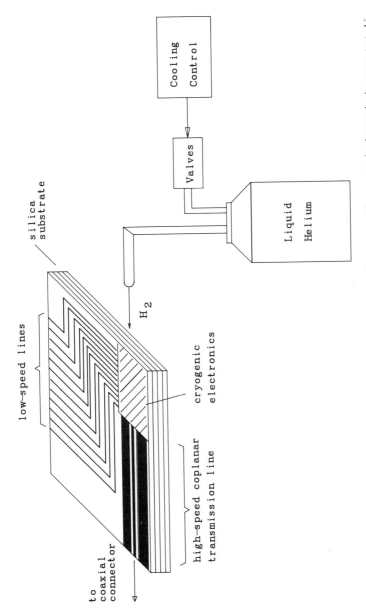

FIG. 13. Layout and cooling scheme of the Hypress Josephson sampling system. The cryogenic electronics is concentrated in the lower-right corner of the chip and spray-cooled with liquid helium.

and the sampling gate deserves further consideration. Having the transmission line at cryogenic temperatures, in a superconducting state, significantly reduces the series resistance of the transmission line. Unlike the DC resistance, however, AC resistance is not reduced to zero. At frequencies above the pair-breaking frequency, a significant loss is introduced, resulting in a strong dispersion as observed by Gallagher et al. (1987). For most practical applications, however, the series resistance does not significantly contribute to the pulse dispersion. In nonsuperconducting transmission lines, the series resistance increases with increasing frequency due to the skin effect. In planar transmission lines, waveguide dispersion is still present in the superconducting state. For waveguide dispersion in the important coplanar waveguide, see Hasnain et al. (1986). At frequencies above the cutoff for the next higher mode, the presence of several modes with different propagation velocities leads to modal dispersion. Depending on bandwidth and distance, pulse dispersion on transmission lines can still play a significant role in superconducting electronics.

Coaxial cables are usually used to connect the signal source at room temperature to the superconducting sampler. Here skin-effect induced dispersion and eventually modal dispersion will be the limiting factors. If for a given cable the attenuation a (dB/100 ft) is known at frequency f (GHz), then the risetime t_r of the step response of a cable of length l (ft) can be estimated by (Dreher, 1969):

$$t_r = 13.13 \ 10^{-6} \frac{a^2}{f} l^2 . \tag{13}$$

Consider a high-performance flexible cable that is specified with 85-dB attenuation per 100 ft at 26.5 GHz. A length of one meter of this cable will have an intrinsic risetime of 39 ps!

C. FREQUENCY-DOMAIN METHODS

1. Scalar Network Analyzer

The easiest way to characterize an electronic subsystem or device in the frequency domain is to compare the power levels at the input and output ports. When directional couplers are added, the magnitudes of input and output reflection coefficients can also be determined, as shown in Fig. 14. While simple errors, such as frequency-dependent cable losses and nonflat coupler responses, can be eliminated by calibration, a full error correction such as provided in the vector network analyzer described next is not possible.

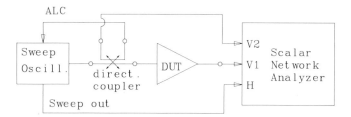

FIG. 14. Block diagram of a scalar network analyzer. The input power to the device under test (DUT) is kept constant by a power-leveling loop or automatic level control (ALC). Two channels, V1 and V2, are provided for analysis of the reflection coefficient and transmission, respectively.

2. Vector Network Analyzer

a. The S-Parameter Concept

At low frequencies, voltages and currents are the accepted way to describe input, output, and transfer parameters of electrical networks. As we move up in frequency, the direct measurement of voltages and currents becomes increasingly difficult, if not impossible, and parameters such as Z- and Y-parameters (which require a knowledge of voltages and currents) become of limited value. A more satisfactory approach is to use incident and reflected waves as variables, which are more easily measured at micro-wave frequencies. Scattering parameters, or S-parameters in short, are matrix elements relating these wave parameters.

The definition of S-parameters starts with the fact that at any point in a transmission line or waveguide, the total voltage is the sum of an incident and a reflected voltage wave. Consider the general twoport in Fig. 15. Here,

$$V_1 = V_{1i} + V_{1r} \quad \text{and} \tag{14a}$$

$$V_2 = V_{2i} + V_{2r} \tag{14b}$$

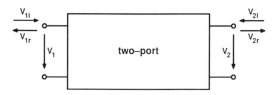

FIG. 15. Generic electrical twoport with incident ($V_{1i,2i}$) and reflected ($V_{1r,2r}$) voltage waves.

Using the characteristic impedance Z_0 of the transmission line or wave-guide as a normalizing factor, we define a set of wave variables:

$$a_1 = \frac{V_{1i}}{\sqrt{Z_0}}, \tag{15a}$$

$$b_1 = \frac{V_{1r}}{\sqrt{Z_0}}, \tag{15b}$$

$$a_2 = \frac{V_{2i}}{\sqrt{Z_0}}, \tag{15c}$$

$$b_2 = \frac{V_{2r}}{\sqrt{Z_0}}. \tag{15d}$$

Thus, a_k represents the incident wave, and b_k is the reflected wave at port k. Note that the a_k and b_k have the dimension of a square root of power. The S-parameters now relate the wave variables as follows:

$$b_1 = S_{11}a_1 + S_{12}a_2 \quad \text{and} \tag{16a}$$

$$b_2 = S_{21}a_1 + S_{22}a_2. \tag{16b}$$

As an example of how to use the S-parameters, consider the insertion gain of a twoport as the ratio of output power at port 2 and incident power at port 1:

$$G = \frac{P_{20}}{P_{1i}} = \frac{|b_2|^2}{|a_1|^2} = |S_{21}|^2. \tag{17}$$

Similarly, the S-parameter concept can be applied to systems with any number of ports.

b. Phase–Magnitude Characterization

To fully describe a linear twoport in the frequency domain, we need to know the twoport's complex S-parameters (or any other equivalent description) in magnitude and phase. If we insert two-branch directional couplers terminated with power meters at both the input and output ports, as in the scalar network analyzer, we obtain only the magnitude. In the vector network analyzer, the phases at the couplers' branches have to be compared as well. The most common way to achieve this is to downconvert the output signals and do the phase comparison at a much lower frequency, as Fig. 16 shows. A sweep oscillator can be connected either to the input (for S_{11}, S_{21}) or the output (for S_{22}, S_{12}) of the twoport under test. At both ports, the reference (or incident) signals are derived directly from the RF

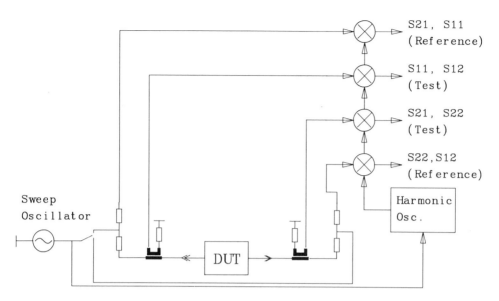

FIG. 16. Block diagram of a vector network analyzer using downconversion of the reference and test signals to a fixed intermediate frequency in a harmonic mixer. Amplitude and phase comparisons between the channels are done at the intermediate frequency. (IF processing is not shown.) DUT = device under test.

input via a power splitter, whereas the test (or reflected) signals are measured with two terminated reflectometers. The frequency of the local voltage-controlled oscillator tracks the frequency of the sweep oscillator in a way that a harmonic of the local oscillator (generated in a comb generator) will downconvert the reference and test signals (at the sweep oscillator frequency) to a fixed intermediate frequency to facilitate further processing. Phase and amplitude comparisons to determine the S-parameters are made at that intermediate frequency.

c. Error-Correction Techniques

Results represented by a network analyzer are subject to error. Sources of errors can be random errors such as noise, temperature fluctuations, and mechanical changes, but also systematic errors such as leakage in the setup, impedance mismatches, and the nonflat frequency response of the measurement system. Systematic errors can be detected by a calibration procedure and subsequently corrected.

The amount of effort necessary for an error correction largely depends on the application. Let us consider first the characterization of a one-port network. Figure 17 shows the corresponding signal flow graph. E_D, the

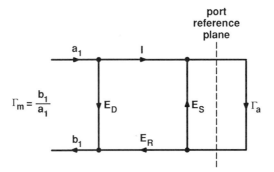

FIG. 17. Signal flow graph describing sources of measurement error in a one-port characterization: directivity error E_D, source match error E_S, and response error E_R.

directivity error, describes the imperfect properties of the directional coupler in the network analyzer as well as directivity errors due to cables, adapters, or probes. The source match error, E_s, accounts for any impedance mismatch at the interface between the device under test and the test setup. E_R, finally, models the analyzer frequency response as well as cable losses, etc. The measured reflection coefficient, Γ_m, is a function of the actual reflection coefficient, Γ_a, and the error coefficients:

$$\Gamma_m = E_D + \frac{\Gamma_a E_R}{1 - E_S \Gamma_a}. \tag{18}$$

Correction of the measurement error requires knowledge of the error vectors E_D, E_R, and E_S. Once they are known, the actual reflection coefficient can be calculated:

$$\Gamma_a = \frac{\Gamma_m - E_D}{E_S(\Gamma_m - E_D) + E_R}. \tag{19}$$

In case of the forward characterization of twoports, we add three more error vectors (Fig. 18). The additional interface at the output leads to a load match error, E_L. The frequency response of the load channel, including cables, probes, and adapters, is contained in the transmission tracking error E_T. Finally, the limited isolation between the channels is accounted for in an isolation error E_X. A full two-port characterization requires the same set of measurements in the forward and reverse directions, leading to a total of twelve error vectors that have to be determined by calibration. In cases where the device under test has high gain or isolation in both directions, the effects of load-match error and isolation error can be omitted, leading to a simpler eight-port error-correction procedure.

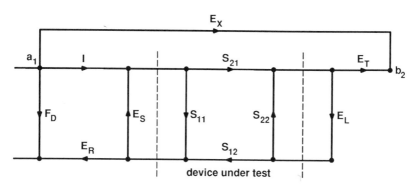

FIG. 18. Extended error flow chart for the forward characterization of twoports. The additional error vectors are the load match error E_L, the transmission tracking error E_T, and the isolation error E_X.

Several different procedures to determine error vectors have been suggested and will be reviewed briefly. All calibration procedures rely on simple, idealized standards for which the impedance is assumed to be known over the frequency range of interest. These procedures differ in the types of impedance standards used. Their relative advantages and disadvantages depend on the measurement circumstances. A comprehensive overview has been presented by Maury et al. (1987).

The OSL (open-short-load) technique (Fitzpatrick, 1978) is the most widely used calibration technique. For a full two-port calibration technique, an additional through-connection is required. An extremely low reflection termination will allow direct measurement of error vector $E_D(\Gamma_a = 0)$. The remaining two vectors can be found from the open $(\Gamma_a = 1)$ and short $(\Gamma_a = -1)$ measurements.

The through-short-delay (TSD) technique (Franzen and Speciale, 1976) has an advantage for sexless connector technologies such as the APC-7 connector. It uses shorts on both ports and a precision transmission line of known impedance. The exact delay of this line does not have to be known. It is the impedance of this transmission line that establishes the reference impedance. In an additional measurement, the two analyzer ports are directly connected together.

The through-reflect-line (TRL) technique introduced by Engen and Hoer (1979) differs from the TSD technique in not requiring a precision short. Reflection standards need only to be different from zero (perfect load) and equal.

A further generalization of the TRL technique is the line-reflect-line (LRL) technique, introduced by Hoer and Engen (1986). The zero-length

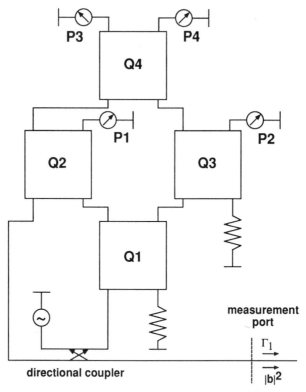

FIG. 19. Example of a six-port network analyzer (after Engen, 1977a). The six-port junction is formed by four quadrature hybrids, Q1–Q4. Two ports are connected to the measurement port through the directional coupler, the others terminated by the power meters P1–P4.

through-connection is replaced by a finite length of precision cable or waveguide. It is important to note that the two transmission lines need not be identical, enabling calibrations for devices that have a coaxial input and waveguide output, for example, without using additional adapters.

3. The Six-Port Network Analyzer

An existing alternative to the common automatic network analyzer will be mentioned briefly. The vector network analyzer requires, as we have seen, the determination of the magnitude and phase of incident and reflected signals at the input and output of a twoport. A system that would achieve the same goal of full S-parameter characterization and error correction by mere power measurements, without respect to phase, would substantially reduce the cost of such equipment.

This is what the six-port network analyzer (Engen and Hoer, 1972) attempts to do. The direct measurement of phase that requires a heterodyne downconverter in the vector network analyzer is replaced by additional power measurements at the ports of a six-port microwave junction. In the practical example of Fig. 19, the two-arm directional coupler is connected to a network of quadrature hybrids terminated by power meters. The complex reflection coefficient Γ of the network connected to the test port can be determined from the power readings and complex constants describing the six-port junction. The algorithm to calculate Γ is beyond the scope of this chapter, and a discussion has been presented by Engen (1977b). The key problem in the accurate use of the six-port network analyzer is its calibration. A review of the state-of-the-art in six-port network analyzer calibration, along with a new calibration approach, have been presented by Ghannouchi and Bosisio (1987).

III. Wafer Probing

The remaining section of this chapter will be devoted to wafer probing as a means of access to the device under test. The description of instrumentation methods in Sections I and II dealt with unknown signal sources or twoports in a very general way, without paying attention to the problem of exactly how to access the device. We shall limit the following to the direct probing of semiconductor devices and integrated circuits on wafer. Readers interested in measurements in a coaxial or waveguide environment can refer to the extensive literature (e.g., Maury et al., 1987).

We shall further confine ourselves to contact probing, which is the oldest—and by far the best—accepted probing method. Until recently, contact probes have been primarily used at frequencies below 50 MHz where the microwave characteristics of the probes are of no concern. Since 1986, probes with multi-GHz bandwidths have become available that extend the useful frequency range of contact probing well into the millimeter wave range. For most applications, the bandwidth of the probe matches or surpasses the bandwidth of the electrical characterization techniques previously discussed.

On-wafer contact probe measurements with a temporal resolution of better than 10 ps are feasible with commercial equipment. Power supply and control signals require physical contact to an integrated circuit under test, and these "low-frequency" probes require proper attention, as will be shown in this section. Finally, "end-to-end" performance tests of integrated circuits require the input and output to be terminated by the proper load impedances, where 50 Ω is the widely accepted standard. These terminations again require physical access to the chip. The probes provide a controlled impedance to the device under test that substitutes for the environment in which the device is normally operated. We refer to this application as substitutional probing. High-speed contact probes may also be used to supply bias and control signals to the device under test, as described in Section III.B.4.

Before the advent of high-speed wafer probes, it was common practice to dice up and package high-speed devices and then characterize the packaged device. This procedure not only required extensive de-embedding to eliminate the influence of the package on the results, but also was time-consuming and, therefore, expensive. Dicing up the processed wafer complicates the extraction of data relating yield to wafer position. A main goal of all high-speed wafer-probing techniques (not just contact probing) is to considerably reduce the effort of device characterization, increase test throughput, and decrease development and manufacturing costs. Wafer-probing results are also used to develop large databases for integrated-circuit design, the increased accuracy leading to improved device models. High-speed probes can be combined with autoprobers familiar from low-bandwidth IC testing to largely automate microwave measurements.

A. PROBE-PERFORMANCE PARAMETERS

Important probe features are repeatability, reflections, insertion loss, crosstalk, common-lead inductance, signal delay match, contact resistance, and bandwidth.

1. Repeatability

Repeatability describes the ability of the system to arrive at the same result when repeatedly measuring identical devices, and it is of foremost concern in probe evaluation. Repeatability may be adversely affected by the accuracy with which the probes can be placed, usually within 10 μm. Since electrical characteristics of the probe can be altered by the environment of the device under test (e.g., large metallized areas that couple to the probe geometry), repeatability is difficult to quantify.

2. Reflections

Reflections are introduced by an impedance mismatch between the probe and the feeding transmission line, which reflects energy back to the signal source and reduces the energy delivered to the device under test.

Reflections are usually specified in terms of the magnitude of the reflection coefficient Γ. How large a value of $|\Gamma|$ can be tolerated depends on the circumstances. If the signal source is reverse-terminated (that is, the source impedance corresponds closely to the transmission line impedance), a large reflection coefficient at the probe reduces the power delivered to the device under test, but does not distort the response. The same applies to the output port, if the following instrumentation is properly terminated. As power reflected back into the device under test by the output probe may severely interfere with the device's operation, the requirements for the output probe are usually more stringent than for the input counterpart. Qualitative measurements are possible in these cases, even with considerable mismatch at the probes. Quantitative measurements, however, require error correction (Section II.C.2.c). Moreover, if source and/or terminating instrumentation are mismatched, a large probe mismatch will result in multiple reflections in the time domain and pronounced resonances in the frequency domain. Even qualitative measurements will require error correction in this case to compensate for these signal distortions. Bear in mind, however, that strong mismatches attenuate the transmitted signal, decreasing the signal-to-noise ratio in the setup. Error correction cannot compensate for this. A wafer probe should have a return loss of larger than 10 dB.

3. Insertion Loss

The loss of signal due to a variety of factors such as reflection, radiation, dielectric, and resistive losses is called insertion loss. In well-designed probes using low-loss dielectrics, resistive losses dominate. Insertion loss in these probes increases monotonously with frequency due to the skin effect. This behavior can readily be compensated for, either by numeric error correction or by a suitable increase of signal power versus frequency in an otherwise "uncorrected" system. Faulty probes, on the other hand, may show high-Q series resonances at specific frequencies (sometimes dubbed "suckouts") that severely reduce the signal-to-noise ratio in the system at these frequencies. A specification of 3-dB maximum insertion loss is acceptable in most cases. Noise-figure measurements using a tuner to provide a specific impedance to the device may require tighter insertion-loss specifications, as will be shown. In this case, the insertion loss (expressed in terms of S_{21}) will decrease the effective tuning range Γ_{eff} of the tuner connected ahead of the probe:

$$\Gamma_{eff} = S_{11} + \frac{S_{12}S_{21}\,\Gamma_{tuner}}{1 - S_{11}\Gamma_{tuner}}. \tag{20}$$

Imagine an ideal tuner with $\Gamma_{tuner} = 1$. For a typical low-noise FET, the noise-optimum match corresponds to $\Gamma_{opt} = 0.8$. In order to tune close to the noise match, an insertion loss of less than 0.97 dB is required.

4. Crosstalk

Crosstalk is the unwanted coupling between two probes and is caused by radiation, capacitive coupling, and common lead inductance. As crosstalk can result in device oscillation, it must be minimized, regardless of whether or not error correction is used. While in principle crosstalk can be removed in an error-corrected measurement, this is rarely ever done. Part of the reason is that the removal of crosstalk by a 12-term error-correction model used commonly in network analyzers is accurate only when the impedance being measured is the same impedance used for the crosstalk calibration. A complete error correction capable of removing crosstalk for arbitrary impedances would require eight more calibration measurements. Crosstalk specification of better than -30 dB is sufficient in most cases.

5. Common-Lead Inductance

Common-lead inductance is the inductance in the ground path that is shared by two or more probes. Figure 20 shows the equivalent circuit of a probe assembly having two inputs sharing a common ground connection. The inductance in the ground lead is a major source of crosstalk between the two signal lines. In most probe designs having two or more signal lines, common-lead inductance is the main limit to their usability at high frequencies. It would be much more meaningful to specify such probes in terms of their inductance, not their bandwidth. Saying that a probe has, for example, a 5-GHz -3dB-bandwidth does not say whether the power is dissipated, reflected, radiated, or coupled to other lines. While many loss mechanisms can be compensated for, crosstalk due to common-lead inductance can jeopardize the functionality of the device under test.

6. Signal Delay Match

Digital ICs often require that the timing of signals applied to different ports of the circuit be kept within narrow tolerances. Certain analog applications require phase-matched input signals at different ports. When these signals are being applied by a multiple-line wafer probe, the delay time introduced by each signal transmission line is important. To be generally applicable, the individual delays should be equal. Probe cards with delays matched to within 5 ps are practical.

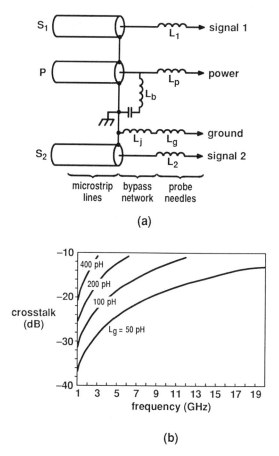

(a)

(b)

Fig. 20. (a) Equivalent circuit of a probe assembly with two signal lines sharing a common ground connection. (b) Resulting crosstalk versus frequency for various common-lead inductances.

7. Contact Resistance

Contact resistance depends on the area of the contact and the probe tip and also on the metallization on the device under test. Resistances of 0.1 Ω can be quite readily obtained with microwave contact probes, causing negligible error in high-speed measurements.

8. Bandwidth

The usable frequency range over which a probe can be used is called the probe's bandwidth. It can be expressed in terms of other parameters previously listed, such as reflection, transmission, and crosstalk. The obvious requirement that a probe have a usable bandwidth commensurate with the bandwidth of the measurement may not be sufficient. Consider, as an example, the characterization of a microwave transistor in the 2–18-GHz frequency range. If the probe provides a high output reflection coefficient at frequencies below 2 GHz, it may induce self-oscillations of the device below the measurement range which interfere with the measurement. Such problems can be insidious because the signal is not being monitored outside of the measurement range. It is therefore desirable that the usable frequency range of a probe extends to frequencies much lower than the measurement range.

Probe performance has been described so far in the frequency domain. The probe's limited frequency response also results in distortions of the measured signal in the time domain. One example is insertion loss, which increases with frequency and causes a pulse to ramp up slowly after an initial sharp rise. Another example is a series resonance which results in ringing in the pulse representation.

B. PROBES

1. Conventional Needle Probes

Needle probes have been used for many years for DC and low-frequency measurements up to 50 MHz. The disadvantage of needle probes lies in the fact that they are unshielded wires of uncontrolled impedance. Bringing coaxial cables instead of unshielded wire (typically 0.5–2 cm long) close to the needle offers some improvement. Probe inductance will only be considerably reduced if the coaxial shields are connected together close to the needles. The needle inductance is usually larger than 10 nH, introducing reflection and insertion losses that are high and that are a strong function of frequency. Furthermore, the needle probes act as fairly efficient antennae at higher frequencies, introducing crosstalk between virtually all other probes being used. A conventional needle probe has been measured to have a radiation resistance of approximately 600 Ω above 500 MHz. The resonant frequencies of the "antennae" appear as strong series resonances. The effect of the high needle impedance is more pronounced when probing a low-impedance node.

2. Improved Needle Probes

Reducing the length of the needle greatly improves probe performance. Limitations on this approach are mainly mechanical in nature. Some approaches that have been used include sharpening the center conductors

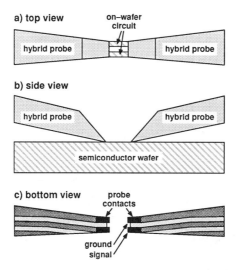

FIG. 21. Microwave probe using a microstructured coplanar transmission line instead of a needle. The transmission line has reinforced metal contact pads at the end and is supported by a ceramic substrate. A controlled impedance is maintained up to the device under test.

of semi-rigid coaxial lines of small diameter and attaching very short probe needles to microstrip lines.

Others have used microstrip transmission lines on a ceramic probe card that converge toward a hole in the ceramic substrate where short needles reach down to contact the device under test. A very low impedance bias port can be made by connecting a chip capacitor between the appropriate transmission line and a ground viahole in the subcarrier. These and similar approaches can be used to approximately 3 GHz. The main limitation is the self-inductance of the probe needles, especially in the ground and bias connections.

3. Microwave Probes

a. Principle: Microstructured Transmission Lines

A microwave probe design with vary large bandwidth is made by abandoning the needle concept alltogether and using lithographically defined transmission lines on a microwave substrate (Strid, 1986). As Fig. 21 shows, a coplanar transmission line geometry is maintained up to the end of the substrate, where small reinforced probe contacts are formed by an additional metallization deposited on the transmission line. The characteristic impedance of the transmission line can be kept to within close tolerances, even with geometries smaller than 100 μm. The advantage of this

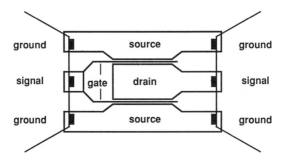

FIG. 22. Microprobe-compatible FET layout. The size and spacing of the source, drain, and gate contact pad match the dimensions of the probe contacts.

probe design is that it extends the controlled-impedance instrumentation environment (usually $Z = 50\ \Omega$) down to the chip level, eliminating the probe inductance. For very low impedance bias connections, the transmission-line impedance can be decreased further by changing the geometry of the coplanar waveguide and loading the line with multiple capacitors. An impedance of $4\ \Omega$ with 0.3 nH of residual inductance has been demonstrated. Unlike conventional needle probes, however, a microwave wafer probe requires a specific footprint of the contact pads on the device under test. Figure 22 shows, as an example, a probe-compatible layout for a microwave FET.

b. Calibration Process Using Microwave Probes

As discussed in detail in Section II.C.2.c, the accuracy of microwave measurements can be greatly enhanced by error-correction techniques. This also applies to microwave wafer probing. The calibration procedure is facilitated by a special impedance standard substrate containing loads, shorts, and through-connections for the calibration process, as well as other one- and twoports such as inductors, capacitors, and attenuators that can be used to verify the accuracy of a calibration by measurement of a device of known frequency response. The advantage of on-wafer calibration over calibrations in the coaxial environment is that the standards can be treated as lumped elements up to 50 GHz due to their small geometric size. Furthermore, the observation of the probing process under a microscope allows a better than 10-μm repeatability of the connections.

As the film thickness in the load standard resistors is much less than the skin depth, the real part of the load impedance standard remains constant with frequency. The main parasitic reactance on the thin-film loads is a series inductance of less than 20 pH. Additionally, unlike the coaxial en-

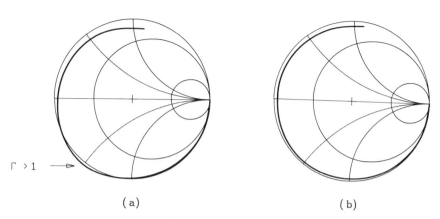

Γ > 1 →

(a)   (b)

FIG. 23. Verification measurement on a planar microwave capacitor showing: (a) End effect of the open probe uncompensated. The trace leaves the Smith chart (reflection coefficient larger than 1, which is not possible). (b) End effect compensated by the assumption of a negative capacitance for the open probe. The measured reflection stays within the Smith chart.

vironment, the probe-to-contact point region can be viewed directly, ensuring a geometric repeatability of the contact of about 5 μm. Some correction has to be applied to the open standard, if an OSL calibration is performed (which will be the rule). The open-circuit standard is simply formed by lifting the microwave probe at least 250 μm into the air. However, the electrical length of the open-ended probe in air will be different from the electrical length of a probe connected to a device. The effect of this error can be seen when measuring a planar capacitor, as shown in Fig. 23: the reflection coefficient moves out of the Smith chart (Fig. 23a). This effect can be eliminated by assuming a negative capacitance for the open, as shown in Fig. 23b, typically −10 to −20 fF. A polynomial expression for $C_{open} = f(f)$, as is common for coaxial shorts, it not necessary in this case.

The electrical length of the short-circuit standard can be shown to be sensitive to the distance the probe tip overlaps the short-circuit edge, as shown in Fig. 24. Consistently reproducible probe contact on the short-circuit standard is important in order to achieve repeatable measurements.

The electromagnetic fields of on-wafer elements are not sufficiently shielded to prevent coupling to nearby elements on the wafer. This applies not only to the devices to be tested, but also to impedance standards. The isolation error vector E_X in the 12-term error correction discussed in Section II.C.2.c is generally only a partial crosstalk correction, because it assumes the terminating impedances at the ports to be identical during

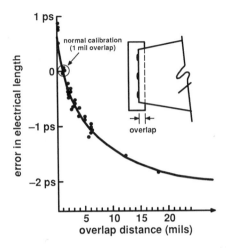

Fig. 24. Placement-sensitivity of the short-circuit standard. Resulting error in electrical length of the short standard versus overlap distance of the probe on the contact pad.

calibration and measurement, which is not always the case. A complete crosstalk error model requires, as mentioned, an additional array of calibration measurements. Minimization of the probe crosstalk is therefore desirable. The influence of probe crosstalk is examined with the aid of Fig. 25. Two similar coplanar waveguide structures with a slot of width d in the center conductor are being characterized. The probes are separated by 12 mils (305 μm) and 6 mils (152 μm), respectively. Except for the extreme case of slot width $d = 200$ μm, the added crosstalk by the probe heads proves to be negligible. A probe contribution to crosstalk of < -40 dB can be estimated from these measurements.

4. High-Impedance Probes

Measurements on the internal nodes of integrated circuits that do not disturb the device's operation require a very different kind of robe. Rather than provide a controlled, usually low impedance to the device, the "noninvasive" probes must present as high an impedance as possible to the probing point—a difficult task at high frequencies.

a. Passive Probes

The common approach to noninvasive probes at low frequencies is the compensated voltage divider probe which may load the probed circuit node with approximately 10 pF. This passive probe can be adapted to high-speed instrumentation by scaling down the probe size. A small thin-film chip or pellet resistor can be added to the tip of a probe to serve as a voltage

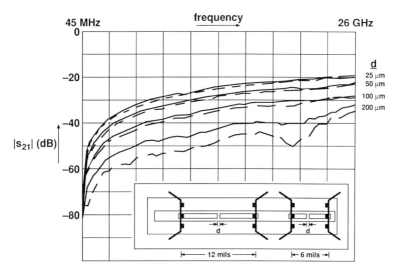

FIG. 25. Evaluation structures for probe crosstalk. The attenuation of a gap of width d in a coplanar transmission line, for gap widths of 25 to 200 μm, and probe spacings of 6 mils (solid curves) and 12 mils (dotted curves) coplanar transmission line is shown as a function of frequency.

divider, e.g., a series resistor of 450 Ω can serve as a 10:1 divider with the 50-Ω line impedance, as shown in Fig. 26. Adding a 50-Ω shunt resistor removes possible nonflatness of the frequency response which may be caused by an imperfect instrument input match, although at the expense of a higher signal attenuation.

b. Active Probes

Another method to raise the impedance of a transmission-line probe is to add a buffer amplifier at the probe tip, as in Fig. 27. Amplifier topologies using emitter or source followers at the input can provide a very high input impedance, low input capacitance, and a low output impedance.

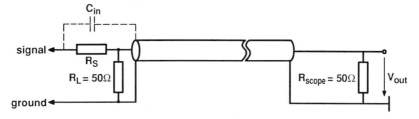

FIG. 26. Equivalent circuit diagram of a passive reverse-terminated 10:1 voltage divider probe.

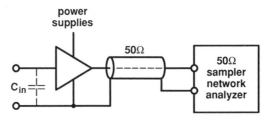

FIG. 27. Active high-impedance probe using a buffer amplifier. C_{in} is the input impedance of the probe presented to the circuit node under test.

Since the amplifier has gain, it gives a smaller attenuation of the tapped signal than a passive probe of similar input impedance. The active approach has, however, the disadvantage of added complexity and cost, as well as dynamic range limitations.

c. Accuracy of High-Impedance Probes

A major source of error when probing a high-impedance node is loading of the node with too low a probe impedance, which may significantly alter the performance of the device under test. In a digital circuit, loading may cause added delays, resulting in timing errors. Generally, loading errors cannot be corrected by characterization of the probe. Inductance in the ground reference lead of any probe creates a frequency-dependent voltage divider with the probe input impedance. The sensitivity of the probe to this inductance depends on the input impedance of the probe. The higher the probe impedance, the lower the effect of ground inductance. For very high impedance probes, the radiation resistance of the probe's ground conductor is low enough to provide a proximity ground reference. The DC attenuation accuracy and frequency response of the probe cause pulse distortion. As in the case of the microwave probes previously discussed, frequency-dependent losses on the connecting cables as well as mismatches can significantly influence the measurement. These errors, however, are generally correctable. Figure 28 summarizes the parasitic elements of an active high-impedance probe.

5. DC Bias Supply Probes

The bias current drawn by digital and some other devices exhibits rapid fluctuations of considerable magnitude as the device state changes. The probes delivering the bias to such devices must have a low impedance to prevent large bias voltage swings at the device pads. These voltage swings are very effective in propagating crosstalk to other functional blocks on the chip.

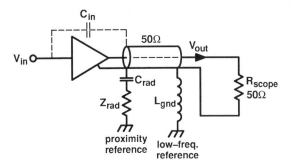

FIG. 28. Major parasitic elements for an active high-impedance probe.

This consideration also applies to circuits characterized by contactless probing techniques. Very low impedance bias probes, which have been developed for high-speed contact probing, thefore benefit accurate high-speed measurements in general.

IV. Conclusions

Electronic sampling oscilloscopes operating at room temperature are available with bandwidths of 20 GHz or 18-ps intrinsic risetime. In many applications, such as time-domain network analysis, the setup can be carefully characterized with respect to the combined frequency response of stimulus and sampler, and deconvolution techniques may yield an additional 30–50% improvement in usable bandwidth with satisfactory error margins. Depending upon the type of application, sampling oscilloscopes with up to 30-GHz bandwidth are therefore considered state-of-the-art, and future improvements can be expected from emerging high-speed III-V semiconductor device technologies. The dynamic range is currently limited to approximately one mV to one V where the lower limit can be extended considerably by signal processing, provided jitter and drift have been eliminated as described in Section II.B.

The state-of-the-art of sampling oscilloscopes using superconducting Josephson electronics is currently at 70-GHz bandwidth, or 5-ps intrinsic risetime. At these speeds, the interconnect technology (cables and connectors) that join oscilloscope and device under test at room temperature largely determines whether this bandwidth can actually be used. The dynamic range is 50 μV to 10 mV and can, as for the conventional sampling oscilloscopes, be shifted to higher values using input attenuators. The signal-processing capabilities are enhanced by the fact that the cryogenic temperatures reduce noise, jitter, and drift.

Figure 29 compares the progress in developing higher-bandwidth wafer probes with improvements in the frequency limits of solid-state amplifiers. Wafer-probe bandwidths are seen to be approaching the maximum frequencies of three-terminal device operation. 50-GHz wafer probes became available in 1988 and 100-GHz devices may follow, permitting designers to measure microwave-device S-parameters and noise parameters with contact-type 50-Ω probes to at least 100 GHz, at the die or hybrid level. The accuracy of these measurements is expected to improve as more is learned about calibration methods and planar calibration standards. Contact-type high-impedance probes will continue to achieve better accuracies, higher bandwidths, and lower capacitances than are currently available. The frequency range of solid-state devices appropriate for active probe buffers will continue to increase in step with the devices to be measured, providing the probe input capacitance can be decreased while wideband performance is maintained. In general, the need to mini-

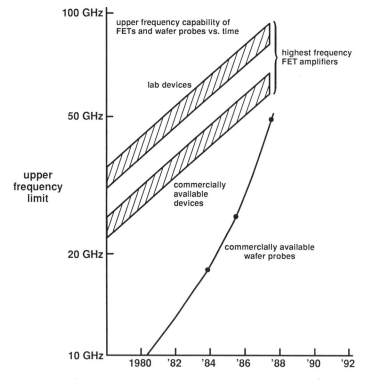

FIG. 29. Evolution of water-probe bandwidth and comparison with maximum operating frequencies of three-terminal devices.

mize losses and phase shift will dictate more signal processing closer to the probe tip. On-probe high-speed signal processing will ultimately offer increased performance and measurement throughput by minimizing high-speed interconnects.

Physical contact to the device under test cannot be avoided in most cases. For on-wafer end-to-end testing of microwave and high-speed digital circuits, contact probing is preferred as it provides the inputs and outputs of the device under test with a controlled impedance that substitutes for the "real-world" operating environment. The bandwidth available from contact probes combined with commercial equipment exceeds 50 GHz and is sufficient for the large majority of today's demands.

The diagnostic access to internal nodes of a high-speed circuit, however, requires a different approach because it has to be noninvasive. In many cases, the size of a contact probe prevents access to an internal node. At frequencies above approximately 5 GHz, the additional node loading introduced by even a well-designed contact probe's parasitics can no longer be neglected. In these circumstances, contactless probing, as described in the remainder of this book, has an undisputable advantage.

The application of low frequency bias and control signals forces a combined contact–contactless probing approach. It must be kept in mind that inductance and crosstalk on seemingly "low-frequency" probe leads can severely deteriorate the tested circuit's performance, and proper contact-probe design thus remains of importance even in contactless probing.

References

Dreher, T. (1969). "Cabling Fast Pulses," *Electronic Engineer* (August), 71–75.

Elliott, B.J. (1970). "System for Precise Observations of Repetitive Picosecond Pulse Waveforms," *IEEE Trans. Instrum. Meas.* **IM-19** 391–395.

Elliott, B.J. (1976). "High-Sensitivity Picosecond Time-Domain Reflectometry," *IEEE Trans. Instrum. Meas.* **IM-25**, 376–379.

Engen, G.F. (1977a). "An Improved Circuit for Implementing the Six-Port Technique of Microwave Measurements," *IEEE Trans. Microwave Theory Tech.* **MTT-25**, 1080–1085.

Engen, G.F. (1977b). "The Six-Port Reflectometer: An Alternative Network Analyzer," *IEEE Trans. Microwave Theory Tech.* **MTT-25**, 1075–1080.

Engen, G.F., and Hoer, C.A. (1972). "Application of Arbitrary Six-Port Junction to Power Measurement Problems," *IEEE Trans. Instrum. Meas.* **IM-21**, 470–474.

Engen, G.F., and Hoer, C.A. (1979). "Thru-Reflect-Line: An Improved Technique for Calibrating the Dual Six-Port Automatic Network Analyzer," *IEEE Trans. Microwave Theory Tech.* **MTT-27**, 987–993.

Faris, S.M. (1980). "Generation and Measurement of Ultrashort Current Pulses with Josephson Devices," *Appl. Phys. Lett.* **36**, 1005–1007.

Fitzpatrick, J. (1978). "Error Models for Systems Measurements," *Microwave Journal* (May), 63–66.

Franzen, N.R., and Speciale, R.A. (1976). "Accurate Scattering Parameter Measurements on Non-Connectable Microwave Networks," *Proc. IEEE 6th European Microwave Conference*, 210–214.

Gallagher, W.J., Chi, C.-C., Duling, I.N., Grischkowsky, D., Halas, N.J., Ketchen, M.B., and Kleinsasser, A.W. (1987). "Subpicosecond Optoelectronic study of Resistive and Superconducting Transmission Lines," *Appl. Phys. Lett.* **50**, 350–352.

Gans, W.L. (1981). "The Measurement and Deconvolution of Time Jitter in Equivalent-Time Waveform Samplers," *IEEE Trans. Instrum. Meas.* **IM-32**, 126–133.

Gans, W.L., and Andrews, J.R. (1975). "Time-Domain Automayed Network Analyzer for Measurement of RF and Microwave Components." National Bureau of Standards, Tech. Note 672. Boulder, CO.

Ghannouchi, F.M., and Bosisio, R.G. (1987). A New Six-Port Calibration Method Using Four Standards and Avoiding Singularities," *IEEE Trans. Instrum. Meas.* **IM-36**, 1022–1027.

Hamilton, S., and Hall, R. (1967). Shunt-Mode Harmonic Generation Using Step-Recovery Diodes," *Microwave Journal* **10**, 69.

Hamilton, C.A., Lloyd, F.L., Peterson, R.L., and Andrews, J.R. (1979). "A Superconducting Sampler for Josephson Logic Circuits," *Appl. Phys. Lett.* **35**, 718–719.

Harris, R.E., Wolf, P., and Moore, D.F. (1982). "Electronically Adjustable Delay for Josephson Technology," *IEEE Electron Dev. Lett.* EDL-3, 361–363.

Hasnain, G., Dienes, A., and Whinnery, J.R. (1986). "Dispersion of Picosecond Pulses in Coplanar Transmission Lines," *IEEE Trans. Microwave Theory Tech.* **MTT-36**, 738–741.

Hoer, C.A., and Engen, G.F. (1986). "Calibrating a Dual Six-Port or Four-Port Network Analyzer for Measuring Two-Ports with Any Connectors," *1986 IEEE Microwave Theory and Techniques Symposium, Baltimore, MD Technical Digest*, 665–668.

Humphreys, D.A. (1988). Personal communication.

Lin, C., Liu, P.L., Damen, T.C., Eilenberger, D.J., and Hartmann, R.L. (1980). "Simpler Picosecond Pulse Generation Schema for Injection Lasers," *Electron. Lett.* **16**, 600–601.

Madden, C.J., Rodwell, M.J.W., Marsland, R.A., Bloom, D.M., and Pao, Y.C. (1988). "Generation of 3.5-ps Fall-Time Shock Waves on a Monolithic GaAs Nonlinear Transmission Line," *IEEE Electron Dev. Lett.*, 303–305.

Marsland, R.A., Valdivia, V., Madden, C.J., Rodwell, M.J.W., and Bloom, D.M. (1988). "100 GHz GaAs MMIC Sampling Head," *1988 IEEE IEDM, Technical Digest*, San Francisco (postdeadline paper).

Maury, M.A. Jr., March, S.L., and Simpson, G.R. (1987). "LRL Calibration of Vector Automatic Network Analyzers," *Microwave Journal* **30** (5), 387–392.

Meckelburg, H.-J., and Matkey, K. (1985). "A High-Sensitivity Microwave Sampling System with Adaptive Time-Drift Compensation," *IEEE Trans. Instrum. Meas.* **IM-34**, 427–430.

Nahman, N.S. (1967). "The Measurement of Baseband Pulse Rise Times of Less than 10^{-9} Second," *Proc. IEEE* **55**, 855–864.

Nahman, N.S. (1978). "Picosecond-Domain Waveform Measurements," *Proc. IEEE* **66**, 441–454.

Nahman, N.S. (1983). "Picosecond-Domain Waveform Measurement: Status and Future Directions," *IEEE Trans. Instrum. Meas.* **IM-32**, 117–124.

Nahman, N.S. (1981). "Deconvolution of Time-Domain Waveforms in the Presence of Noise," NBS Technical Note 1047, Boulder, CO.

Robin, N.A., and Ramirez, R. (1980). "Capture Fast Waveforms Accurately with a 2-Channel Programmable Digitizer," *Electron. Des.* **3**, 50–55.

Schoell, E., Bimberg, D., Schumacher, H., and Landsberg, P.T. (1984). "Kinetics of

Picosecond Pulse Generation in Semiconductor Lasers with Bimolecular Recombination at High Current Injection," *IEEE J. Quantum Electron.* **QE-20,** 394.

Schwarte, R. (1972). "New Results of an Experimental Sampling System for Recording Single Events," *Electron. Lett.* **8,** 95–96.

Scott, W.R. Jr., and Smith, G.S. (1986). "Error Corrections for an Automated Time-Domain Network Analyzer," *IEEE Trans. Instrum. Meas.* **IM-35,** 300–303.

Strid, E.W. (1986). "26 GHz Wafer Probing for MMIC Development and Manufacture," *Microwave Journal* (August).

Weber, S. (1987). "Found—A Practical Way to Turn Out Josephson Junction Chips," *Electronics* (February 19), 49–52.

Wolf, P., Van Zeghbroeck, B.J., and Deutsch, U. (1985). "A Josephson Sampler with 2.1 ps Resolution," *IEEE Trans. Magn.* **MAG-21,** 226.

Yokoyama, H., Ito, H., and Inaba, H. (1982). "Generation of Subpicosecond Coherent Optical Pulses by Passive Modelocking of an AlGaAs Diode Laser," *Appl. Phys. Lett.* **40,** 105.

Zappe, H.H. (1975). "A Subnanosecond Josephson Tunneling Memory Cell with Nondestructive Readout," *IEEE J. Solid State Cir.*, **SC-10,** 12–19.

Van Zeghbroeck, B.J. (1985). "Model for a Josephson Sampling Gate," *J. Appl. Phys.* **57,** 2593–2596.

Van Zeghbroeck, B.J., Harder, C., Halbout, J.M., Jaeckel, H., Meier, H., Patrick, W., Vettiger, P., and Wolf, P. (1987). "5.2 GHz Monolithic GaAs Optoelectronic Receiver," IEDM 1987, Washington, D. C., December 6–9, 1987, Technical Digest, 229–232.

CHAPTER 3

Picosecond Photoconductivity: High-Speed Measurements of Devices and Materials

D.H. Auston

DEPARTMENT OF ELECTRICAL ENGINEERING
COLUMBIA UNIVERSITY
NEW YORK, NEW YORK

I. Introduction

Steady progress in semiconductor-electronics technology, spurred by new material techniques such as molecular beam epitaxy and metal-organic chemical-vapor deposition, has led to new electronic devices that operate at speeds approaching a few picoseconds. For example, heterostructure field effect transistors have demonstrated switching speeds as fast as 6 picoseconds (Shah et al., 1986). These new devices have outpaced the ability of conventional instruments to measure their speed of response.

New techniques are needed to characterize both the structures and materials used to make them.

This chapter reviews an extremely effective technique for making high-speed measurements of materials and devices. The technique uses sub-picosecond optical technology and high-speed photoconducting materials to generate and measure extremely fast electrical pulses. It has an advantage over conventional electronics because optics is inherently faster than electronics. There is a substantial gap of one to three orders of magnitude between optical and electronic measurement capabilities. It is in this gap, optoelectronic technology resides and plays its unique role in combining the speed and flexibility of optics to develop new and faster electronic devices and measurement systems.

This chapter also reviews research in picosecond photoconductivity with particular emphasis on novel device and measurement concepts. First, a brief overview of basic device concepts illustrates the wide range of novel applications of ultrafast optoelectronics for the generation and measurement of ultrafast electrical transients, including microwave, mm-wave, and far-infrared generation and detection. Next, there is a discussion of photoconducting materials for ultrafast optoelectronics. Specific details of these applications are described in subsequent sections. Many of these applications are advancing rapidly. For this reason, discussion of many of these topics may already be outdated. The chapter concludes with a brief discussion of current challenges and future trends.

The scope of this chapter is not exhaustive and all-inclusive, it merely highlights the key concepts of ultrafast optoelectronics and illustrates them with specific examples of measurements and devices. Literature referenced in the text is listed at the end of this chapter. For additional reading, refer to the articles in (Lee, 1984) as well as the proceedings of the 1985 and 1987 Picosecond Electronics and Optoelectronics Conferences (Mourou et al., 1985; Leonberger et al., 1987) and the 1984 and 1986 Ultrafast Phenomena Conferences (Auston and Eisenthal, 1984; Siegman and Fleming and Siegman, 1986). An extensive account of current research in ultrashort optical pulses and their applications can be found in (1988).

Some generic optoelectronic devices based on the use of photoconducting materials are illustrated in Figs. 1 and 2. Although each performs a different electronic function, all are based on variations of the central concept of a light-pulse–producing conductivity modulation by electron-hole injection in a semiconductor. With moderate optical pulse energy, it is possible to produce a photoresistance that is relatively low compared to the characteristic impedance of the transmission line. This results in a switching action that permits a fast optical pulse to initiate a high-speed electrical signal. Figure 1a is a schematic illustration of the basic photoconducting

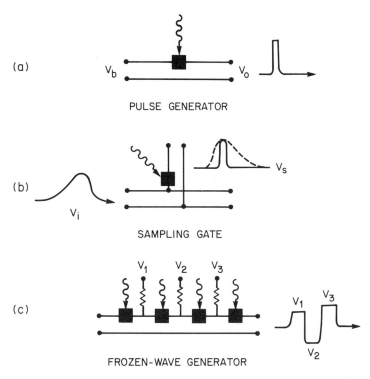

FIG. 1. Schematic illustration of basic optoelectronic device concepts: (a) photoconducting electrical-pulse generator, (b) photoconducting electrical-sampling gate, and (c) frozen-wave generator. In each case, an ultrashort optical pulse illuminates a photoconductor to produce a fast conducting transient in a transmission line.

electronic pulse generator. It consists of a photoconducting material mounted in a high-speed transmission line. As described in detail in Section II, the amplitude and shape of the electrical pulse depends on the details of the device geometry and the materials as well as the optical pulse. With proper choice of these parameters, extremely fast electrical pulses having relatively large amplitudes can be generated by this technique. Figure 1b illustrates a photoconducting sampling gate. In this case, the input electrical signal is a time-varying waveform that is sampled by the photoconductor by diverting a small portion of the signal to a sampling electrode. By varying the relative timing between the incoming electrical waveform and the optical pulse, the amplitude, v_s, of the sampled pulse gives a stroboscopic replica of the desired waveform. As in other sampling measurement systems, it is not necessary to time-resolve the sampled

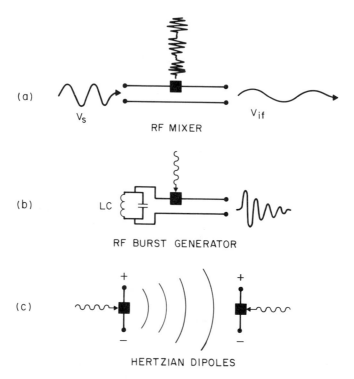

(a) RF MIXER

(b) RF BURST GENERATOR

(c) HERTZIAN DIPOLES

FIG. 2. Other ultrafast optoelectronic devices that use high-speed photoconductors. (a) A radio-frequency mixer, which mixes a local oscillator signal, V_s, with an envelope encoded optical signal, which illuminates a high-speed photoconductor, to produce an intermediate frequency, V_{if}. (b) A radio-frequency burst generator, which produces a damped oscillation at the frequency of the tank circuit in which the photoconductor is mounted. (c) Photoconductors in which a rapid current transient radiates into free space and is detected by a second receiving photoconducting antenna.

signal v_s, so that highly sensitive low-frequency electronics can be used. A novel variation on the pulse generator uses multiple switches to form a frozen wave generator as shown in Fig 1c. Simultaneous illumination of the photoconductors releases the "frozen" waveform, which can have an arbitrary shape determined by the number of photoconductors and the DC bias voltages applied to them. This case requires a long photoconductivity falltime to enable each section to discharge in tandem. For the same reason, a relatively high energy optical pulse is required to ensure low resistance in each photoconductor. The amplitudes and durations of each segment of the waveform can, in principle, be adjusted arbitrary by changing the bias voltages and lengths of the transmission lines separating each photoconductor.

Figure 2a shows a radio-frequency mixer in which the input signal is a high-frequency sine wave with amplitude modulated by the optical pulse. The optical signal envelope modulation frequency is equal to the local oscillator so that the output signal at the intermediate frequency is proportional to the product of the electrical and optical input signals.

Figure 2b illustrates the use of photoconductors for producing short bursts of radio frequencies. In this case, a resonant circuit controls the frequency of the generated waveform. Coupling the circuit to an external transmission line causes the signal to decay rapidly, thus producing short bursts of radio frequencies. By using high bias voltages and large optical pulse energies, this method can produce relatively high power signals.

When the photocurrent risetime is extremely rapid, the photoconductor can directly radiate an electromagnetic signal into free space as illustrated in Fig. 2c. Photoconductors can also be used as receiving antennas by sampling a radiated electromagnetic pulse, when illuminated by an optical pulse. The combination of optically triggered transmitting and receiving antennas forms a measurement system that is phase-coherent and has extremely good time resolution.

II. Materials for Picosecond Photoconductors

Many materials are used for picosecond photoconductors. A summary of the key properties of some of the more important semiconductor materials is given in Table I. They can be divided into the following classes: intrinsic, impurity-dominated, radiation-damaged (d), polycrystalline (p), and amorphous (a) semiconductors. Before discussing their relative merits, some important intrinsic properties of semiconductor materials that are of interest for applications to picosecond photoconductors are summarized. For a review of the basic properties of photoconductors, see Rose (1963).

A. Intrinsic Speed of Response of Semiconductors

The absorption of a photon by a semiconductor and the subsequent generation of an electron–hole pair is intrinsically very fast. This process is limited by the uncertainty principle and the requirement that the frequency spectrum of the optical pulse falls within the absorption bands corresponding to electronic transitions from bound to free states. Because the width of these bands is a few eV, this time can be as short as 10^{-15} s, or one optical cycle, and consequently does not limit the photocurrent risetime.

The quantum efficiency of the initial photocurrent is determined by the probability that the electron–hole pair will escape its mutual coulomb field. In most high-mobility semiconductors, this probability is very close to

TABLE I

MATERIALS FOR PICOSECOND PHOTOCONDUCTORS

Material	Band gap E_g(eV)	Resistivity $\rho(\Omega cm)$	Mobility $\mu(cm^2/vs)$	Decay time τ_c(ps)	References
Si	1.12	4×10^4	1950	10^7–10^3	Auston (1975)
GaAs:Cr	1.43	10^6–10^7	~2000	300	Lee (1977)
InP:Fe	1.29	2×10^8	2200	150–1000	Leonberger and Moulton (1979)
CdS.$_5$Se.$_5$	2.0	10^7	400	2×10^4	Mak et al. (1980)
GaP	2.24	10^8	240	60–500	Margulis and Sibbett (1981)
Diamond	5.5	10^{13}–10^{15}	1800	50–300	Vermeulen et al. (1981)
d-SOS	1.12	10^5	~10–100	1–300	Smith et al. (1981a)
d-GaAs	1.43	10^7	~100	<5	Hammond et al. (1984)
d-InP	1.29	10^6	100–1000	1–100	Foyt et al. (1982), Downey and Iell (1984)
d-InGaAs	0.75	~1	800–4000	40–800	Downey and Schwartz (1984)
p-Si	~1.1	10^5		2–50	Bowman et al. (1985)
p-Ge	.85		~3	~50	DeFonzo (1983)
p-CdTe	~1.5	3×10^7	60	4	Johnson et al. (1985)
a-Si	1.4	10^5 10^7	~10	3 30	Johnson et al. (1981)

unity. However, in some low-mobility photoconductors, such as amorphous selenium, the situation is quite different; geminate recombination of electron–hole pairs can significantly lower quantum efficiency.

Although the onset of photoconduction is very rapid, there are a number of effects that can influence the subsequent evolution of the current following excitation by a very short optical pulse. These are illustrated schematically in Fig. 3, where the photocurrent caused by an optical pulse of infinitesimal duration is plotted versus time on a logarithmic scale. The initial risetime is limited by the elastic scattering rate of the photogenerated electrons and holes (momentum relaxation rate). Energy relaxation then occurs and can either speed up or slow down the response, depending on the electric field and photon energy. The decay of the photocurrent is dominated by recombination, but may be speeded up by the introduction of defects to produce rapid trapping of free carriers. If the photon energy exceeds the threshold for photoconduction (the band gap in an intrinsic semiconductor) by an amount greater than the thermal energy, kT, the excess energy is given to the electron–hole pair as kinetic energy. Because most scattering mechanisms (Nag, 1980; Seeger, 1982) are stronger at

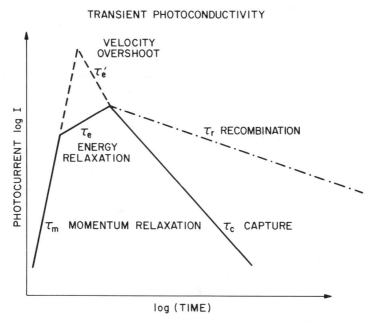

FIG. 3. Schematic illustration of the temporal response of a semiconductor to a fast optical pulse (transient photoconductivity).

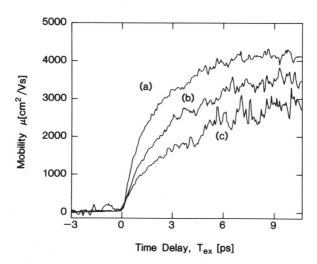

FIG. 4. The rise of the low field mobility of a sample of GaAs due to the optical injection of carriers at 2 eV by an ultrashort optical pulse. The three curves illustrate the dependence on injection density and are for densities of (a) 5×10^{17} cm^{-3}, (b) 5×10^{18} cm^{-3}, and (c) 1.2×10^{19} cm^{-3}. (Nuss and Auston, 1987)

higher energies (an important exception being ionized impurity scattering), the initial mobility of the free carriers is smaller than in equilibrium conditions. The transfer of this excess energy to the lattice by phonon emission produces an additional mobility transient with a risetime ranging from a fraction of a picosecond to hundreds of picoseconds, depending on the material, the excitation density, the excess energy, and the lattice temperature. These effects have been studied extensively using picosecond and femtosecond optical techniques (Shah, 1981, 1986; Oudar et al., 1985; Knox et al., 1986; Block et al., 1986). At high excitation densities, these rates can be influenced by screening of the electron–phonon interaction, which tends to slow the cooling of the electrons and holes.

Measurements in GaAs (Nuss and Auston, 1987) show that these effects produce a mobility transient that rises from an initial value of approximately 500 cm^2/Vs to 4000 cm^2/Vs in a time that varies from 2 to 5 picoseconds depending on the excitation density. Figure 4 illustrates this result. The photon energy in this experiment (2 eV) caused an appreciable population of the satellite valleys of GaAs, accounting for the relatively "slow" risetime of the mobility. The mobility at longer times depended on the excitation density, an effect that is expected from electron–hole scattering. The density and temperature dependence of electron–hole scattering is relatively complicated due to the screening of the coulomb cross section at high densities. This tends to produce a minimum in the carrier mobility versus density dependence. Theoretical calculations of electron–hole scattering have been done by McLean and Paige (1960, 1961), Paige (1960), Appel (1961, 1962), and Meyer and Glicksman (1978). Measurements have been made by Vaitkus et al. (1976), Auston and Johnson (1977), and Grivitskas et al. (1984). In the experiment of Auston and Johnson, the effective mobility ($\mu_n + \mu_h$) of silicon was measured in a picosecond photoconductor over a wide range of densities and two different temperatures. At room temperature, they found that the mobility was depressed by a factor of approximately two due to electron–hole scattering at densities in the range of 10^{19} cm^{-3}. However, at 80 K, the mobility was depressed almost an order of magnitude.

B. HIGH ELECTRIC FIELD EFFECTS

If the bias voltage applied to the photoconductor is very large, the electric field heats the photogenerated carriers. This produces a non-equilibrium distribution that can result in a number of important effects that strongly influence the dynamic and steady-state properties of photoconductors (Reggiani, 1985). On a short time scale (less than 10 ps), this can produce a transient in the current waveform due to velocity overshoot effect (Rees, 1969; Reggiani, 1985). A number of experiments have been directed at measuring this effect (Shank et al., 1981; Laval et al., 1980; Hammond, 1985; Mourou, 1987). Although many of these measurements demonstrate that velocity overshoot can influence the transient photocurrent under high field conditions, the detailed features of velocity overshoot remain to be elucidated. The situation is more complicated when electron heating is caused by both optical injection and high electric fields, as in the case of photoconductors with large bias voltages.

On longer time scales, once an equilibrium carrier distribution has been established, high electric fields produce substantially lower mobility due to saturation of the drift velocity. For example, in silicon, at fields in excess of 10^4 V/cm, the electron and hole drift velocities saturate at approximately 10^7 cm/s. In GaAs and other materials with satellite valleys in their conduction bands, the steady-state drift velocity reaches a maximum at a field of approximately 3×10^3 V/cm and then decreases at higher fields. This negative resistance property can lead to electric-field instabilities due to the Gunn effect and can limit the maximum DC bias that can be placed across a photoconductor.

C. TRAPPING AND RECOMBINATION

In an intrinsic semiconductor, recombination rates are determined by radiative transitions. These are relatively slow, producing a long-lasting current waveform which is a major problem for applications to picosecond photoconductors. Direct-gap semiconductors have radiative lifetimes that are a few nanoseconds. Indirect-gap materials can have lifetimes that are much longer, sometimes extending into the millisecond range. Alternative materials and techniques must be used to produce the fast current decays necessary for picosecond applications.

In p–n junction diodes, carrier sweep-out determines the speed of response, which can be substantially faster than the intrinsic recombination time. However, drift velocity saturation limits the sweep-out rate (e.g., 10 ps for a saturation velocity of 10^7 cm/s in a one-μm path length). As a result, sweep-out is primarily used in semiconductor devices where extremely high speeds are not required. Also, the photovoltaic behavior of p–n junctions and the need for a large reverse bias make them unsuitable for some picosecond devices such as sampling gates.

An effective method to reduce the free carrier lifetime is to introduce a moderate density of defects into the semiconductor. The defects act as traps and recombination centers. This can be done by radiation damage, impurities, or the use of materials with large densities of naturally occurring defects such as polycrystalline and amorphous semiconductors. Table I lists some examples of materials in these classes. The much faster decay times relative to the intrinsic materials are evident.

The capture time, τ_c can be estimated from the expression

$$\tau_c = \frac{1}{N_t \sigma_c \langle v_{th} \rangle} \tag{1}$$

where Z_t is the trap density, σ_c the capture cross section, and $\langle v_{th} \rangle$ the mean thermal carrier velocity. According to Eq. (1), trap densities of 10^{18} to 10^{20} cm^{-3} produce free carrier lifetimes of approximately one picosecond if the capture cross sections are between 10^{-13} and 10^{-15} cm^2.

D. RADIATION DAMAGE

The specific nature of the defects produced by radiation damage is a subject that has been extensively studied. (For a review, see Vavilov and Ukhin, 1977). The type, density, and stability of the defects depend on the material, type of radiation, background impurities, and temperature. In the case of damage produced by light-and medium-weight ions, it is thought that clusters of defects, possibly as large as 10 to 20 Å are also produced (Stoneham, 1980) as well as elementary displacements such as Frenkel pairs (vacancy plus interstitial). Capture cross sections have been measured by deep-level transient spectroscopy (DLTS) for a variety of materials and defects. Although a wide range of levels exist, each having

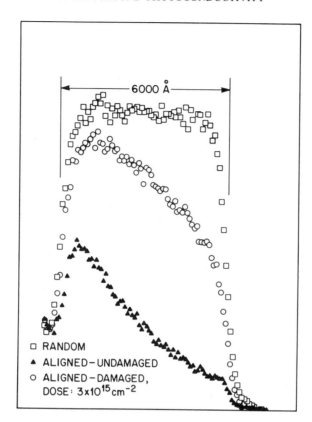

FIG. 5. Composite plot of Rutherford backscattering and channeling measurements of a silicon-on-sapphire sample bombarded with 2-MeV argon ions at a fluence of approximately 3×10^{15} cm^{-2}. Spectra are shown of the damaged and unirradiated samples in both the random and $\langle 100 \rangle$ aligned crystallographic directions. The estimated density of damage sites for the dose shown is 2.56×10^{22} cm^{-3}; one half that of fully damaged material. (Smith, Auston and Nuss, 1988).

different energies and capture cross sections, certain general trends can be identified. For example, in silicon, a commonly occurring defect is the electron trap at -0.53 eV having a capture cross section of approximately 2×10^{-15} cm^2 at room temperature (Chen and Milnes, 1977). A similar confluence of hole traps occurs at an energy of 0.30 eV above the valence band with a cross section of approximately 10^{-15} cm^2.

Experiments with picossecond photoconductors showed that extremely fast photocurrent decay times can be produced with moderate levels of radiation damage. (See Table I for specific references.) Figure 5 illustrates the damage profile of a sample of silicon-on-sapphire measured by

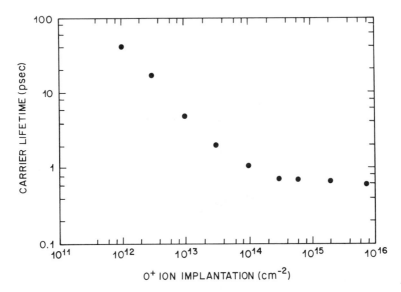

FIG. 6. The relaxation of optically injected carriers in radiation-damaged silicon-on-sapphire, as measured by time-resolved optical reflectivity with femtosecond opticaal pulses. (Doany et al., 1987)

Rutherford backscattering and channeling (Smith, Auston and Nuss, 1988). In this example, defects were produced by bombardment of the sample with argon ions of 200 MeV energy at a fluence of 3×10^{15} cm^{-2}. Although very high, the defect density is less than a fully amorphous sample. Measurements of the carrier mobility in this sample show that it has the value of 25 cm^2/Vs, which is greater than would be expected for a fully amorphous sample (Smith and Auston, 1988). The photocurrent decay time was estimated to be 0.6 ps (Smith and Auston, 1988).

In Fig. 6, the carrier lifetime of a radiation-damaged silicon-on-sapphire film is plotted as a function of radiation dose. This measurement used fematosecond optical pulses to probe the free carrier absorption following excitation. The maximum decay rate of 0.7 ps was produced with damage levels that are close to the point of rendering the material completely amorphous.

In GaAs, many deep levels exist having capture cross sections in the range of 10^{-14} to 10^{-13} cm^2 (Martin et al., 1977; Mitonneau et al., 1977). For example, bombardment of GaAs produces electron and hole traps at -0.71 eV with cross sections of 1.3×10^{-13} and 2.3×10^{-13} cm^2 (Li et al., 1980).

The introduction of moderately large densities of defects into crystalline semiconductors has a number of additional effects. Some of these are advantageous and some are not. If the traps lie deep within the band gap, free carriers due to dopants are removed from the transport bands, moving the Fermi level close to midgap and greatly increasing the resistivity. For most applications to photoconductors, this a desirable property since it reduces the dark current. A second feature of high-defect densities is an enhanced optical absorption in the spectral region below the edge for direct transitions due to the introduction of new states and the relaxation of selection rules for indirect transitions. This effect is more pronounced in indirect-gap materials such as silicon.

Another advantage of high-defect-density materials is the ease of fabrication of ohmic contacts. It is generally very difficult to make low resistance, nonrectifying contacts to semiinsulating materials. A Schottky barrier usually forms at the metal–semiconductor interface. This produces non-ohmic behavior together with an internal electrostatic field, which results in a photovoltaic response in addition to a photoconductive signal (Rhoderick, 1978). When the injected carrier density exceeds the background carrier density, as usually occurs in picosecond applications, it is even more difficult to make ohmic contacts. It is important to have contacts that are ohmic for both electrons and holes, a condition that is extremely difficult to achieve. Although not fully understood, an empirical property of metallic contacts to semiconductors with moderate-to-high defect densities is their tendency to exhibit ohmic behavior without the need for special processing. The most probable explanation for this phenomenon is the extremely short depletion layers in these materials which permit efficient tunneling currents to pass through the contacts, analogous to heavily doped materials.

E. CARRIER MOBILITIES

The major disadvantage of high-defect-density materials is their lower carrier mobilities caused by increased elastic scattering from defects. The lack of a strong temperature dependence of the mobilities suggests that most of the defects are neutral rather than charged if relatively high doses of radiation damage are used. Estimates of the influence of elastic scattering from neutral defects were made by Erginsoy (1950). He estimated a mobility:

$$\mu = \frac{1.4 \times 10^{22}}{N_t} \frac{m^*}{m_0} \cdot \frac{\varepsilon}{\varepsilon_0} \text{ cm}^2/\text{Vs} \tag{2}$$

where the defect density, N_t, is in units of cm^{-3}. Using the previously estimated numbers for N_t required to produce capture times of one ps in silicon and GaAs, the corresponding mobilities from Eq. (2) would be 350 cm^2/Vs for GaAs and 5 cm^2/Vs for silicon (using $\sigma = 10^{-13}$ cm^2 for GaAs and 10^{-15} cm^2 for Si). The clear advantage of GaAs relative to silicon arises primarily from the apparently larger capture cross sections observed in GaAs. This conclusion is substantiated by experiments.

Although the introduction of relatively high densities of defects dramatically affects the free-carrier lifetime, it is also important to provide recombination paths to prevent thermal emission of trapped carriers. Thermal emission can produce sustained photoconduction that forms long tails on the current waveforms. If the defect density is sufficiently high, trapped carriers can tunnel between defect, sites, and recombination will occur without emission. It is usually best to use a defect density somewhat higher than necessary to achieve short capture times. Thermal emission is usually exacerbated by the presence of shallow defects which can have relatively fast emission rates. If emission occurs more rapidly than recombination, the effective mobility will be reduced by the inelastic capture and fast release of free carriers.

In general, a particular method of introducing defects results in a variety of traps, each having different energies and capture cross sections. Some of these will be deep levels with fast capture times and slow emission rates. Others may be too shallow for picosecond applications. Different combinations of materials and methods of introducing defects should be explored to optimize this approach. The use of annealing to remove shallow defects, but retain deep traps, is potentially useful for improving the mobility without loss of speed (Foyt et al., 1982).

F. OTHER PHOTOCONDUCTING MATERIALS

Other methods of achieving fast photocurrent decay include the use of compensating impurities such as Fe in InP (Leonberger and Moulton, 1979; Hammond et al., 1981), polycrystalline materials such as p-Ge (De-Fonzo, 1983), and amorphous semiconductors (Auston et al., 1980a,b). In the latter case, the mobilities tend to be very small (1–10 cm^2/vs) and, although the response times can be very fast, radiation damage appears to provide a better compromise between speed and sensitivity than the use of completely disordered (i.e., amorphous) materials.

III. Generation and Detection of Ultrafast Electrical Pulses

A. PICOSECOND PHOTOCONDUCTORS

Many different materials and geometric configurations have been used for picosecond photoconductors. However, most have a common design that optimizes the speed of response by directly coupling the photocurrent

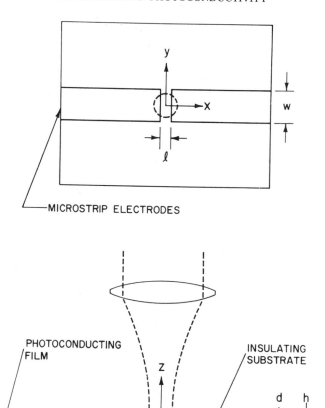

FIG. 7. Schematic diagram of a thin-film photoconductor as an integral component of a high-speed microstrip transmission line. The active region is the gap in the top electrode at which the light is focused.

to a high-speed transmission-line structure. This is usually accomplished by making the electrodes of the photoconductor a part of the transmission line. Figure 7 illustrates a typical example: a photoconducting film on an insulating substrate on which metallic electrodes were deposited to form a microstrip transmission line. A picosecond light pulse is focused on the active region of the photoconductor, which consists of a small gap in the top microstrip electrode. Other transmission lines, such as coaxial and coplanar waveguides, have also been used. (The next section describes the high-speed properties of different transmission-line structures.) The photo-conductor may be used as the substrate material if it is semi-insulating. A rigorous analysis of the voltage waveform produced in Fig. 7 is difficult. It requires the solution of the time-varying electromagnetic equations for the electric field produced by the radiating currents and charges in the gap. A more expedient method is to disregard retardation effects and represent the photoconductor by time-varying macroscopic circuit elements. These approximations are valid if the dimensions of the gap and the transmission-line cross section are small relative to the distance an electromagnetic signal travels in a time equal to the electrical pulse duration.

Following this approach, the photoconductor is represented as a time-varying conductant $G(t)$, in parallel with a capacitance C_g, embedded in a transmission line. Theoretical estimates of the static capacitance across this gap have been made for microwave applications. These estimates have values of one to 100 fF for the specific geometries of interest (Maeda, 1972). The capacitance increases logarithmically as the gap becomes smaller. The sensitivity decreases as the square of the gap length, so that it is possible to use very small gap lengths to make sensitive photoconductors without appreciable loss of speed. Although a small shunt capacitance may exist in some configurations, its value is usually small and will be omitted here.

A general expression for the time-varying conductance $G(t)$ of a photo-conductor can be derived from the rate of dissipation of electrical energy,

$$Gv_g^2 = \int_V dx \mathscr{E} \cdot \mathbf{J}, \tag{3}$$

where v_g is the voltage across the gap, $\mathbf{E}$ is the electric field, and $\mathbf{J}$ is the current density. The integration extends over the entire active volume of the photoconductor. When Ohm's law can be used to relate the current density and electric field, the conductance is

$$G(t) = \frac{1}{v_g^2} \int_V dx \, \sigma |\mathscr{E}|^2 \tag{4}$$

where σ is the conductivity. G can be separated into two components: (1) a constant G_0, representing the dark conductivity, and (2) a time-varying term photoconductance

$$g(t) = \frac{1}{v_g^2} \int_V dx (n_e e \mu_n + n_h e \mu_p) |\mathscr{E}|^2, \tag{5}$$

where n_e and n_h, and μ_n and μ_p are the electron and hole densities and mobilities, respectively. Although $n = p$ initially, their evolution may differ due to drift and different capture rates by defect sites. Assuming the optical pulse to be negligibly short in duration, the initial carrier densities are

$$n_e(t = 0^+) = n_h(0^+) = \frac{(1 - R)\alpha W(x, y)e^{-\alpha z}}{\hbar\omega} \tag{6}$$

where R is the reflectivity of the photoconductor surface, α is the optical absorption constant, W is the incident optical pulse energy per unit area, and $\hbar\omega$ is the energy of one photon.

The electric-field distribution in the gap, $\mathscr{E}(x)$, depends both on the geometric configuration of the photoconductor and on the distribution of the photoexcited carriers. The latter effect is particularly important in the high-injection regime typical of picosecond photoconductors. If the electrical contacts to the photoconductor are not ohmic, they can also influence the electric-field distribution. For the radiation-damaged and amorphous semiconductors, however, good contacts can usually be made using relatively simple methods.

1. Response of an Ideal Photoconductor in a Transmission Line

This section describes some features of incorporating a high-speed photoconductor into a transmission line (Auston, 1983). The most important feature is that the bias voltage can be a fast electrical transient that need not be a constant voltage. The effective load seen by the photoconductor is the characteristic impedance, Z_0, of the transmission line. Also, the travelling wave nature of the electrical signals necessitates the description of the response of the photoconductor in terms of incident, reflected, and transmitted waves. To determine the response for an arbitrary incident, $v_i(t)$, and conductance, $G(t)$, it is convenient to first focus attention on the instantaneous voltage, $v_g(t)$, across the gap. By straightforward application of circuit laws, the following equation for $v_g(t)$ is derived:

$$Z_0 C_g \frac{dv_g}{dt} + \frac{1}{2}[1 + 2Z_0 G(t)]v_g(t) = v_i(t), \tag{7}$$

from which the reflected and transmitted waves can be determined:

$$v_r(t) = \frac{1}{2} v_g(t) \tag{8}$$

and

$$v_t(t) = v_i(t) - \frac{1}{2} v_g(t) \tag{9}$$

2. Photoconducting Electrical-Pulse Generators

To illustrate some basic features of the response of an ideal photoconductor in a transmission line, consider the case of a step-function conductance with a constant bias, i.e.,

$$G(t) = \begin{cases} 0; & t < 0 \\ G_1; & t \geq 0 \end{cases}$$

$$v_i(t) = \frac{1}{2} V_b = \text{constant.}$$

From equations (7) and (8), the transmitted signal is

$$v_t(t) = \left(\frac{V_b}{2}\right)\left[\frac{2Z_0 G_1}{1 + 2Z_0 G_1}\right]\left\{1 - \exp\left[-\left(\frac{1}{2Z_0 C_g} + \frac{G_1}{C_g}\right)t\right]\right\}. \tag{10}$$

It is important to distinguish two different regimes, determined by the magnitude of the photoconductance, G_1. For small values of G_1, $v_t(t)$ has a rise time of $2Z_0 C_g$ and a steady-state value of $Z_0 G_1 V_b$ which is linearly proportional to G_1. For large values of G_1 (i.e., $Z_0 G_1 \gg 1$), the risetime decreases and the steady-state transmitted signal saturates at one-half the constant bias voltage. For example, if $Z_0 G_1 = 5$, the risetime is 11 times faster than the small signal case and the incident wave is transmitted with approximately 90% efficiency. The response is very different in the small signal or linear response regime, and the large signal or saturated regime.

Picosecond photoconductors were first demonstrated by Auston (1975) and Lawton and Scavannec (1975). When the optical illumination is large, picosecond photoconductors act as light-activated electric switches, as demonstrated experimentally in crystalline semiconductors by Auston (1975). In this case, illustrated in Fig. 8, optical pulses from a mode-locked Nd:glass laser were used to inject high densities of free carriers into a silicon microstrip photoconductor. A visible pulse at 0.53 μm obtained by second harmonic generation was used to initiate the photocurrent. Because the material was a high-quality crystal, the free carrier lifetime produced a current that persisted for a relatively long time. However, it could be terminated by absorbing a second optical pulse at the fundamental wavelength of 1.06 μm. The pulse at this wavelength penetrated through to the ground plane, making a short circuit with the top electrode. This "turned off" the switch by reflecting all further incoming signals. By varying the delay between the two pulses, the duration of the current pulse could be continuously varied from approximately 15 ps to many nanoseconds. The same device also functioned as a sampling gate when an electrical pulse was used as a bias rather than a constant voltage.

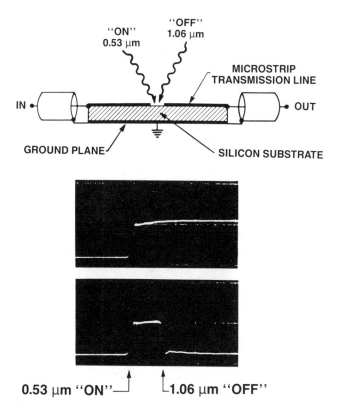

Fig. 8. A picosecond photoconductor that uses a visible (0.53-μm) picosecond pulse of approximately 5 μJ to initiate conduction and an infrared pulse of approximately 5 μJ to terminate the photocurrent by introducing a short circuit between top electrodes and ground plane. The oscilloscope photograph (2 ns/div.) illustrates the variable aperture feature of the device, which can be controlled by the time delay between the two optical pulses. Peak electrical transmission of the gate was 95%. (Auston, 1975)

3. Photoconducting Electronic Sampling

Picosecond photoconductors can also be used as electronic gates to measure the waveform of a high-speed electrical signal (Auston, 1975; Lawton and Andrews, 1975, 1976). In this case, a pulsed signal is the bias on the photoconductor rather than a constant voltage. The active photoconducting area is usually a gap between a main transmission line and a secondary line. When a light pulse strikes the sampling gap, small sample of charge is transferred from the signal on the main transmission line to the sampling line. The response of this sampling gate can be analyzed similarly to the single gap in a main line considered in the last section. Because the response time of the photoconductor determines the speed of the sampling gate, it is sufficient to measure the total charge generated in the sampling transmission line. This is usually accomplished by measuring the average current in the sampling line while the laser is operating at a high repetition rate. Time-resolving the sampled voltage waveform is not necessary. The signal that is capacitively-coupled across the sampling gap does not contribute any net charge to the sampled signal. The sampled charge, $Q_s(\tau)$, can be written in the form of a convolution of a sampling function, $f_s(t)$, with the incident voltage waveform, $v_i(t)$:

$$Q_s(\tau) = Q_0 + \int_{-\infty}^{+\infty} dt' \, f_s(t' - \tau) v_i(t'), \tag{11}$$

where

$$f_s(t) = \frac{2}{3Z_0 C_s} \left(\frac{1}{1 + \frac{3}{2} Z_0 G_0} \right) \int_t^{\infty} dt' \, g(t') e^{-\gamma(t' - t)} \quad \text{and} \tag{12}$$

$$\gamma = \frac{2}{3Z_0 C_s} + \frac{G_0}{C_s}$$

where C_s is the capacitance of the sampling gap.

The sensitivity and signal-to-noise ratio of photoconductive sampling is extremely good, permitting fast electrical signals having amplitudes of approximately one μV to be measured with realistic integration times. The limiting noise properties of photoconductive sampling gates have not been studied in detail, nor have they been optimized. In the ideal case, the limiting noise level would be determined by generation-recombination noise and Johnson noise (Forrest, 1985) in the photoconductor. For typical sampling gates, this would predict a limiting sensitivity of a few tens of nV/$\sqrt{\text{Hz}}$. In practice, however, imperfect ohmic contacts give rise to a photovoltaic signal that tends to dominate the noise properties. This produces a noise background signal due to laser amplitude fluctuations. Although the quantum efficiency of the photovoltaic signal is extremely small, it is often sufficient to dominate the noise properties and results in sensitivity limits that are more typically a few μV/$\sqrt{\text{Hz}}$. Development of better ohmic contacts would greatly improve the minimum signal measurable by photoconductive sampling.

4. Electronic Autocorrelation Measurements

A simple and effective technique for evaluating the response of photoconducting pulse generators and sampling gates is to directly connect them together. This allows the output of the pulse generator to be the input to the sampling gate (Auston, 1975; Auston et al., 1980b). This approach,

UHV EV a-Si   AUTOCORRELATION

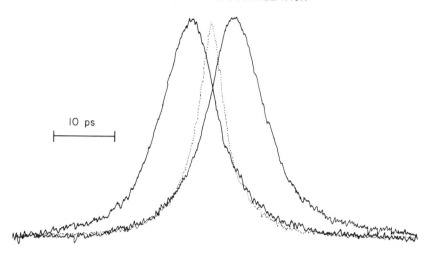

FIG. 9. The picosecond electronic autocorrelation of two amorphous silicon photoconductors using the configuration shown in the inset of the figure. The nonlinear optical autocorrelation of the optical pulses is shown for comparison (dotted line). When the bias and sampling connections are reversed, a mirror image of the electronic autocorrelation function is produced. (From Auston et al., 1980b)

called electronic autocorrelation, accurately measures system response. It has important applications for characterizing specific photoconducting materials, electrode geometries, transmission-line structures, and mounting configurations. The system response approximately equals the convolution of the individual responses of the generator and the sampling gate. (For a detailed discussion, see Auston, 1983.)

This approach has been used to measure the photocurrent decay rates in amorphous silicon (Auston et al., 1980b; Johnson et al., 1981), silicon-on-sapphire (Grivitskas et al., 1984), radiation-damaged silicon-on-sapphire (Smith et al., 1981a); Ketchen et al., 1986), radiation-damaged indium phosphide (Downey et al., 1983, Karin et al., 1986; Hammond et al., 1984), gallium arsenide (Lee et al., 1977), radiation-damaged gallium arsenide (Hammond et al., 1984; Downey et al. 1983), and polycrystalline silicon (Hammond et al., 1985). In many cases, the capacitance of the electrode geometry limits the speed of response. For example, in microstrip structures having widths of 0.2 mm, gaps of 25 μm, and substrate thicknesses of 0.25 mm, the limiting circuit response has a full-width at half maximum of approximately 10 ps. Figure 9 illustrates an electronic autocorrelation circuit used by Auston et al. to estimate the photocurrent decay time in amorphous silicon. In this experiment, the biased electrode injected a pulse of charge into the main transmission line when a picosecond optical pulse illuminated the top gap. The main transmission line served as a load, permitting the charge to propagate away quickly. Before dissipating, however, it was sampled by the second optical pulse striking the lower gap. The measured quantity was the variation of the average current in the sampling electrode as the relative time delay changed between the two optical pulses illuminating the gaps. Figure 9 shows two autocorrelation traces. The first trace corresponds to normal biasing. The second trace has the bias and signal electrodes interchanged. This is equivalent to a time reversal and permits an accurate determination of the zero time delay (i.e., simultaneous arrival of the two optical pulses at the gaps). A detailed analysis of this experiment permits the determination of the relative contributions of the circuit and material time constants (Auston, 1983).

Ketchen and coworkers (1986) scaled down an autocorrelation circuit using 5-μm coplanar strip transmission lines and obtained a response with a full width at half maximum of only 1.6 ps. They used a novel "sliding-contact" technique involving the illumination of regions between the conductors. They applied this technique to the measurement of dispersion on coplanar transmission lines by sampling the generated electrical pulse at different distances along the coplanar line. By using superconducting coplanar lines, they observed the contribution to the propagation constant due to the finite superconducting bandwidth (Gallagher et al., 1987).

B. Ultrafast Electrical Transmission Lines

Ultrafast electrical pulses require transmission lines that are extremely broadband, have low dispersion and loss, and can be readily interconnected to optoelectronic and other devices. Unlike microwave and millimeter-wave signals, the frequency spectrum of ultrafast electrical pulses ranges from zero to many hundreds of gigahertz and, in some cases, even terahertz. For this reason, waveguides are unsuitable because of their large dispersion and low frequency cutoff. Some transmission line structures that have been used for ultrafast optoelectronics are coaxial, microstrip, coplanar, and coplanar microstrip. Figure 10 illustrates the geometric crosssections of these different configurations. For details of the electrical properties of these structures, the reader is referred to the standard texts on this subject such as Ramo et al. (1984), Gupta et al. (1979), and Edwards (1981). Coaxial lines have good bandwidth, low loss, and low dispersion. They have been used extensively for high-power optoelectronic switching with photoconductors mounted in a gap in the center conductor. They are used less for high-speed, low-power, optoelectronic applications

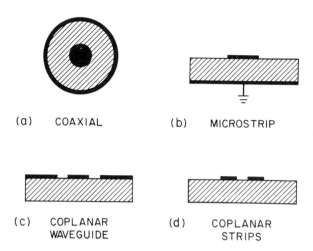

(a)   COAXIAL   (b)   MICROSTRIP

(c)   COPLANAR   (d)   COPLANAR
      WAVEGUIDE   STRIPS

FIG. 10. Schematic illustration of the cross section of different transmission-line structures used for picosecond optoelectronics.

because they are difficult to interconnect to other structures such as microstrip lines and discrete devices (although schemes for accomplishing this have been developed). The microstrip, coplanar waveguide, and coplanar microstrip structures are all compatible with semiconductor microelectronic fabrication technology. Photoconductors can be integrated into the transmission lines at a gap in an electrode or between two electrodes. The ability to use microelectronic processing technology means these structures can be scaled to extremely small sizes, thereby increasing the useful bandwidth.

Dispersion is a major problem in transmission lines for ultrafast electrical pulses because it produces loss of speed due to pulse broadening. Theoretical calculations of the dispersive broadening of pulses on microstrip and coplanar lines were made by Li et al. (1982) and Hasnain et al. (1986). They found good agreement with experimental measurements of Mourou and Meyer (1984). Measurements of pulse dispersion on microstrip lines by Cooper (1985) indicated that microstrip is much less dispersive than predicted by theory. Goossen and Hammond (1985) made time-domain calculations of the dispersion and loss of electrical pulses on microstrip lines on silicon substrates. Their results show the strong influence of substrate losses and demonstrate the need for high-resistivity substrates. Mourou (1986) made extensive measurements of pulse dispersion on different lines. Their results, which are illustrated in Fig. 11, demonstrate that ultrashort pulses can propagate on suitably designed lines with negligible broadening due to dispersion.

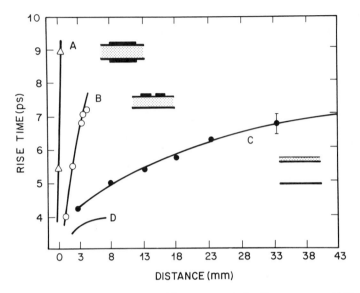

Fig. 11. Measured pulse dispersion on different transmission-line structures: (A) balanced stripline on $LiTaO_3$ with 500-μm electrode separation; (B) coplanar strips on $LiTaO_3$ with 50-μm electrode separation; (C) inverted air-spaced stripline with the top electrode supported by a thin glass superstrate 500 μm above the ground plane; (D) superconducting coplanar transmission line with line dimensions of 50 μm. (From Mourou, 1986)

The principal advantage of the coplanar geometries is that the size can be reduced to very small dimensions. Although microstrip can also be scaled down, it requires the use of very thin substrates, which is often impractical.

A problem with the coplanar-strip and coplanar-waveguide geometries for short pulse propagation is attenuation due to radiation into the dielectric substrate. Rutledge et al. (1983) have estimated this effect and find it can be extremely important for frequencies having wavelengths comparable to and smaller than the cross-section dimensions of the transmission line. Figure 12 plots the attenuation due to radiation into the dielectric substrate. The attenuation coefficient varies as the cube of the frequency and is a more serious problem in coplanar waveguide (two slots) than coplanar strips. To avoid this problem, the dimensions of the line should be kept small. However, if normal (not superconducting) metals are used, skin-effect losses increase as the line becomes smaller. For a given bandwidth and conductor material, there is an optimum size for the cross-section dimension of the line.

The geometric properties of various transmission-line structures have

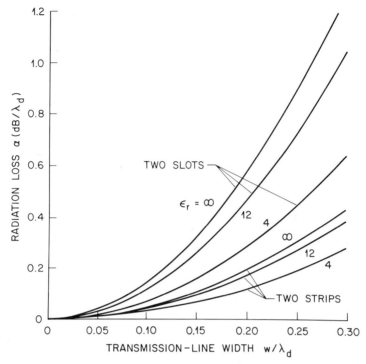

FIG. 12. Calculated attenuation coefficient for coplanar waveguide (two slots) and coplanar strips (two strips) with the indicated dielectric constants as a function of the width of the transmission line relative to the wavelength in the dielectric medium. (From Rutledge et al., 1983)

been exploited for pulse-shaping applications. For example, Buck et al. (1980) showed how a stub microstrip transmission line can be used to shape the electrical pulse from a photoconductor. Margulis and Persson (1986) used a coaxial differentiator to accomplish similar results. Chang et al. (1982) mounted a picosecond photoconductor in a coaxial Blumlein structure for pulse-shaping applications. Li et al. (1984) reviewed the topic of pulse forming with optoelectronic switches.

A number of theoretical papers have examined the properties of nonlinear transmission lines. For example, Landauer (1960) and Khokhlov (1961) showed that shock waves are expected to develop for high-voltage pulses propagating on transmission lines with nonlinear dielectrics. Paulus et al. (1984) suggested that soliton-like behavior should be possible in transmission lines having a quadratic nonlinearity and a cubic dispersion. The implementation of nonlinear transmission lines to the picosecond time domain could result in the development of electrical-pulse–compression techniques analogous to the soliton-like techniques developed for optical pulses.

IV. Optoelectronic Measurement Systems and Their Applications

A. OPTOELECTRONIC MEASUREMENT SYSTEMS

The use of picosecond and femtosecond optical pulses for control and measurement of electronic devices and circuits substantially improves performance and flexibility compared to conventional electronic-measurement systems such as sampling oscilloscopes. In addition, an entirely new class of devices has evolved from the use of this approach. These devices have properties and applications that go beyond conventional electronic devices.

The previous section described different optoelectronic devices that use picosecond photoconductivity. There are basically two types: sources of electrical transients and electrical sampling gates. The combination of an optoelectronic signal generator and an optoelectronic sampling gate, as illustrated in Fig. 13, is a flexible high-speed measurement system that can be used to measure a wide range of electronic devices and materials. An important feature of this system is the use of the same source of optical pulses for triggering both the optoelectronic pulse generator and the sampling gate (Smith et al., 1981b). An optical beam splitter is used to split each pulse from the laser into two pulses. One pulse goes to the generator

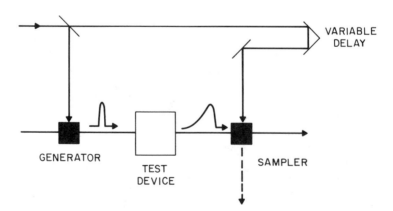

FIG. 13. An ultrafast optoelectronics measurement system for measuring the response of high-speed devices and materials. Both the generator and sampler can be photoconducting or electro-optic devices. The stroboscopic measurement is accomplished by varying the optical delay between the two optical pulses used to time the generator and sampler.

and the other pulse goes to the sampling gate after passing through a variable delay line. The variation of this delay permits accurate stroboscopic measurements with negligible jitter.

Speed is the most important advantage of optoelectronics. It is the main incentive for the extensive research activity in this field. This speed advantage comes from the availability of short pulse lasers. In addition to speed, there are other equally important advantages derived from the use of optics, which the remaining part of this section summarizes and illustrates with examples.

1. Sensitivity

The high sensitivity of optoelectronic measurement systems is primarily due to the high repetition rate of pulsed lasers, which permits extensive signal averaging. For a single event, these measurement techniques are relatively insensitive. The signal-to-noise ratio (S/N) can be readily enhanced however by averaging multiple events at high repetition rates. At a repetition rate of 100 MHz, this results in an improvement of $10^4:1$ in S/N for a one-second integration time, making possible microvolt-sensitivity for photoconducting sampling gates.

2. Timing Accuracy

The absence of trigger jitter is important. Without it, obtaining high timing precision would not be possible. In a conventional sampling oscilloscope, the jitter between the triggering of the sampling gate and the measured signal can limit the useful time resolution. With optoelectronic sampling, the timing of the test electrical pulse and the sampling is controlled by the same optical pulse. (See Fig. 13.)

3. Low-Temperature Environments

The picosecond photoconducting devices described in the previous sections are relatively rugged and perform well at low temperatures as well as room temperature. Picosecond photoconductors have been used at 125°K (Auston and Smith, 1982; Johnson et al., 1981) and at 2°K (Ketchen et al., 1986; Sprik et al., 1987).

4. High-Power

The high-power scaling of photoconductors is extremely important for generating multikilovolt electrical transients. It is possible to switch electrical signals that have peak powers that exceed the incoming optical pulse. (See Section IV. E.).

5. Simultaneous Time and Frequency Measurements

The bandwidth of picosecond optoelectronic measurement systems must be extremely broad to avoid loss of speed due to pulse broadening. For picosecond systems, a flat baseband response from zero to hundreds of gigahertz is required. This places some demanding constraints on the design of transmission structures and interconnections between devices. Once achieved, however, an important benefit results from broad bandwidth. Because the measurement system is phase-coherent (i.e., amplitudes, not intensities, are measured), the spectral response of a device can be inferred directly from its temporal response (Auston and Cheung, 1985). This concept of time-domain metrology is equivalent to variable frequency measurements of both phase and amplitude over terahertz bandwidths.

6. Electrical Isolation

The use of optical pulses for triggering optoelectronic devices results in isolation between the control signal (optical) and the generated or measured signal (electrical). The ability to accurately focus and position optical beams also results in good isolation between different optoelectronic devices. Because the triggering and sampling information is conveyed on optical beams, the device or material under test can be in remote environments such as low-temperature cryostats or high magnetic fields, and can be interfaced to an external environment with optical beams and perhaps one or two DC or low-frequency electrical connections. This avoids the problems of requiring high-speed coaxial or other electrical transmission systems to bring out (or send in) electrical transients.

B. CHARACTERIZATION OF HIGH-SPEED DISCRETE DEVICES

1. Photoconductive Sampling of High-Speed Photodetectors

Radiation-damaged GaAs photoconductors have been used as sampling gates to measure the response of a high-speed silicon p–i–n photodiode (Auston and Smith, 1982). The p–i–n structure had an intrinsic drift region of 3 μm and an active area of 25 μm. The diode chip, which measured 250 μm square, was mounted on a microstrip in a configuration that permitted the photocurrent pulses to be injected into a 50 Ω transmission line terminated at each end. The sampling photoconductor was a small chip of semi-insulating GaAs that had been irradiated with 3×10^{15} protons/cm^2 at 300 keV. Using an identical GaAs photoconductor as a pulse generator, an electronic autocorrelation measurement of the sampling gate gave a response time of approximately 12 ps for the effective sampling aperture. When used with the photodiode, the FWHM was observed to be 36 ps at full bias at room temperature. Compared to a measurement of the same photodiode with a sampling oscilloscope, the photoconductive sampling technique was superior with regard to time resolution, trigger, jitter, signal-to-noise level, and ringing. Measurements of the photodiode were also made at 125°K and with different bias voltages. Because the dominant contribution to the speed of response was due to the electron transit time in the intrinsic layer of the photodiode, it was possible to estimate the electron drift velocities from the results. It was found that the saturated drift velocity of electrons in silicon was 8.8×10^6 and 1.0×10^7 cm/s at 125 and 293°K, respectively.

2. Optical Mixing in Photodetectors

A very different approach to characterizing the response of high-speed photodetectors has been demonstrated by Carruthers and Weller (1986). They have mixed two picosecond optical pulses in a GaAs Schottky-barrier photodiode and observed the variation in the average detector current as a function of the delay between the optical pulses. A nonlinearity with respect to the excitation amplitude produces an autocorrelation signal that can be used to extract information about the intrinsic speed of response of the detector. They suggest the probable physical mechanism for the nonlinearity is the perturbation of the electric field by the injected carriers and the consequent change in carrier collection efficiency.

3. Impulse Response of High-Speed Transistors

An early example of the application of a picosecond optoelectronic-instrumentation system is the measurement of the impulse response of a GaAs FET (Smith et al., 1981b).

The configuration used for these measurements is illustrated in Fig. 14. Picosecond photoconductors were used for electrical pulse generators and sampling gates. The circuit consisted of two radiation-damaged silicon-on-sapphire wafers with microstrip transmission lines in the pulse-injection and sampling goemetries that were previously described. A "floating" main-transmission line, onto which pulses are injected and sampled, provides flexibility for making various types of measurements.

The DC bias for the gate and drain were supplied through the lines to set the operating point of the FET without disturbing the photoconducting sampling gates and pulse generators. Two photoconductors on the input (left) side of the figure allowed calibration of the input signal by using one as a pulse generator (with constant bias) and the second as a sampling gate

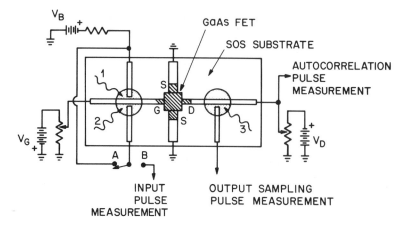

F$_{\text{IG}}$. 14. Schematic of picosecond optical electronic circuit used to measure electronic response of GaAs FET. Picosecond optical pulses, indicated by wavy lines 1, 2, and 3, are focused on gaps in the microstrip transmission lines where very high speed photoconductors have been made by radiation damage to the silicon-on-sapphire wafers. Biased photoconductors (V_B) act as electrical-pulse generators by injecting charge onto the main transmission line. When the electrical pulses of interest (e.g., the output drain current pulse) are used as bias signals, the photoconductors act as sampling gates (e.g., photoconductor 3). DC gate-source and drain-source bias voltages are applied through the main microstrip line. (Smith et al., 1981b)

to make an autocorrelation measurement as described in Section II.D. This same configuration was also used in a time-domain reflectometer mode. In that case, the optical delay on pulse 2 was lengthened to observe the reflected electrical signal returning from the gate. This approach was used to evaluate the response of a high-speed GaAs MESFET. The drain signal due to excitation of the gate with a fast electrical pulse was broadened and delayed by approximately 20 ps. By comparing the response of the system with and without the FET, it was possible to make absolute measurements of gain and propagation delay.

A third type of measurement that provides information about the non-linear response of the device uses both input photoconductors as pulse generators. The total average current in the drain (without sampling pulse 3) was measured as the relative delay between optical pulses 1 and 2 was changed. Using this technique, a nonlinear electronic autocorrelation measurement was made of a GaAs MESFET. Using this technique, the response time was approximately 20 ps (Smith et al., 1981b). Because the system was free of jitter, very accurate measurements of the delay between gate and drain signals were possible.

Similar measurement techniques were used by Cooper and Moss (1986a) to determine the high-frequency scattering parameters of a GaAs FET. By Fourier analysis of time-domain waveforms of the reflected and transmitted drain and gate signals, they were able to determine the scattering parameters to frequencies greater than 60 GHz. They have also used direct optical stimulation of FETs (Cooper and Moss, 1986b).

C. OPTOELECTRONIC MEASUREMENTS OF INTEGRATED CIRCUITS

An important application of picosecond optoelectronic techniques is the measurement of high-speed integrated circuits. Conventional wafer probers and network analyzers have limited speed and bandwidth and they cannot probe internal nodes of a complex circuit without perturbing its performance. A number of approaches that use picosecond optoelectronics will be summarized in this section.

1. Direct Optical Stimulation of Integrated Circuits

Jain and coworkers (Jain and Snyder, 1983a,b; Jain et al., 1984) have developed a method to characterize high-speed integrated circuits by direct illumination of logic gates with picosecond optical pulses. Direct injection of optical pulses into the channel of a GaAs FET produced fast logic level switching. The output of the stimulated FET initiates a sequence of logic operations in subsequent logic elements. Accurate relative timing was obtained using two or more stimulating optical pulses to trigger different logic gates in the circuit. A NOR or AND circuit was used to determine the coincidence of logic signals produced by different optical pulses. The basic concept of this approach is illustrated in Fig. 15. In this case, the propagation delay of an intervening set of logic elements can be measured with high precision. An advantage of this approach is that it is noninvasive. It does not require physical contact to the IC and can in principle be used with silicon ICs as well as gallium arsenide ICs. Experiments on conventional GaAs D-flip-flop logic circuits demonstrated latching times of 475 ps and transition times of less than 100 ps (Jain and Suyder, 1983a,b). More recent measurements in higher-speed GaAs ICs have demonstrated switching speeds of less than 10 picoseconds (Jain et al., 1987; Zhang and Jain, 1987).

2. Integrated Photoconductors

Hammond and coworkers developed an approach to characterizing silicon integrated circuits that involves the fabrication of polysilicon photoconductors on the IC for pulse generators and sampling gates (Hammond et al., 1984; Bowman et al., 1985; Eisenstadt et al., 1985). The polysilicon

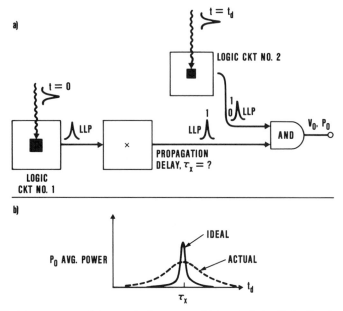

FIG. 15. Measurement of the speed of response of integrated circuits by the direct optical stimulation of logic gates. The pulses designated "LLP" are logic-level pulses stimulated by the optical pulses. (Jain et al., 1983, 1984)

photoconductors are made with standard silicon integrated-circuit processing techniques to ensure full compatibility with standard VLSI processes. Photoresist-masked, ion-beam irradiation was used to generate trapping sites in the polysilicon to give fast response times. Test measurements of dispersion in silicon-based stripline show the response times of the photoconductors to be less than 3 ps FWHM.

D. Microwave, Millimeter-Wave, and Far-Infrared Applications

The ability to generate and measure extremely fast electrical signals by optoelectronic techniques has extremely important applications for the generation and detection of high-frequency electromagnetic waves. The frequency range spanned by optoelectronic devices is very large. For example, an electrical signal having a risetime of one ps has a base bandwidth that extends from zero to approximately 300 GHz. Faster electrical transients such as those produced by photoconducting antennas (Smith, Auston and Nuss 1988) have spectral components as high as 2 THz. The reciprocal property of photoconducting devices enables the phase-coherent detection of these high-frequency signals with good sensitivity.

1. Photoconductive Switching and Gating

Using photoconductive switches to gate continuous microwave sources is an effective method of producing short bursts of radiation with precise timing and variable durations. Because photoconductive switches have large bandwidths, a microwave bias can be applied to the photoconductor instead of a DC bias voltage. When two optical pulses are used to control the opening and closing times of the switch, a variable duration burst of microwaves is produced (Johnson and Auston, 1975; Platte and Appelhaus, 1976; Platte, 1977, 1978). A disadvantage of this approach is the poor on–off ratio of the switch. This results from the capacitive coupling of the microwave bias across the switch. Also, the peak microwave power is limited by the CW microwave source. Platte (1978) reviewed the use of photoconductive gating of microwave signals.

A second method of producing short bursts of microwaves is the impulse excitation of DC-biased photoconductors that are mounted in tuned circuits or waveguides. This approach can produce much higher peak powers. The basic concept uses the tuned circuit or waveguide as a frequency-selective load so that a particular frequency component of the current pulse is extracted. Figure 17 illustrates this approach (from Mourou, et al., 1981a). In the experiment of Mourou et al. (1981a), the current pulse from a semi-insulating GaAs photoconductor was coupled into an X-band wave-guide to produce short bursts of X-band radiation. They measured the duration of these pulses to be approximately 50 ps by observing the time-resolved reflection of the microwave signals from an optically pumped sample of germanium. The technique was also applied to a radar ranging

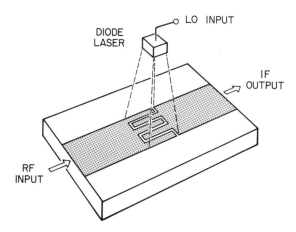

FIG. 16. A radio-frequency mixer using a photoconductor illuminated by a semiconductor laser diode. (Foyt et al., 1981)

experiment in which microwave echoes from targets were resolved within a few centimeters at a distance of 4 m.

Auston and Smith (1983) used small radiation-damaged silicon-on-sapphire photoconductors that were resonant at 55 GHz to produce short bursts of millimeter waves. In this experiment, the reciprocal property of the photoconductors made possible phase-coherent detection of the signals with an excellent signal-to-noise ratio. Lee et al. (1985) used impulse excitation of GaAs:Cr and InP:Fe photoconductors in coaxial resonators to produce bursts of RF energy of 300 MHz. The DC-to-RF energy-conversion efficiency was as high as 50 %.

Proud and Norman (1978) applied the concept of a frozen-wave generator to picosecond optoelectronics. They used a sequence of photoconductive switches mounted in tandem in a microstrip line. Line segments between the switches were independently biased. When the photoconductors were simultaneously illuminated, the frozen wave, determined by the bias conditions, was launched down the line. The result was a short RF burst. The waveshape was determined by the static voltage profile established in the device by the bias voltages. Lee et al. (1985) extended this approach to high-powers by generating an RF pulse of two-and-one-half cycles in duration at 250 MHz having a peak-to-peak amplitude of 850 volts.

Mooradian (1984) demonstrated a novel form of microwave generation by using multiple pulses to illuminate either a GaAs avalanche photodiode or an InP:Fe photoconductive switch. An optical circuit consisting of beam splitters and multiple path length delays produced an optical pulse having a variable pulse-repetition rate. A microwave signal of variable frequency could be produced by this technique.

The use of optical pulses to modulate and control the operation of conventional semiconductor devices was explored by Kiehl (1978, 1979, 1980) and Carruthers et al. (1981) and Carruthers (1984). Kiehl (1979) used optical excitation of an avalanche photodiode to gate RF signals with excellent on–off ratios and reverse isolation. He also used optical injection to quench the output of an Impact Avalanche Transit Time (IMPATT) oscillator. This produced short bursts of microwaves (Kiehl, 1980). Phase control of a TRAPATT oscillator was produced by injection of a modulated optical signal (Kiehl, 1978, 1980) Carruthers et al. (1981) and Carruthers (1984) made extensive measurements of the effects of injecting picosecond optical pulses into transferred electron devices.

2. *Phase Modulation of Microwave and Millimeter Waves*

Lee et al. (1980) used picosecond optical pulses to inject electron-hole plasmas in dielectric waveguides. This produced phase modulation of the

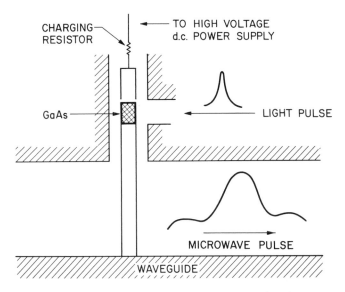

FIG. 17. Schematic illustration of a picosecond optoelectronic microwave generator. (From Mourou et al., 1981b)

microwave and millimeter-wave signals in the waveguides. The basic concept of their millimeter-wave phase modulator is illustrated in Fig. 18. The phase shift arises from the contribution to the real part of the dielectric response due to the free electrons and holes. It can be estimated from the expression

$$\varepsilon(\omega) = (m + ik)^2$$
$$= \varepsilon_s - \sum_\alpha \frac{\omega_{p\alpha}^2}{\omega^2 + \gamma_\alpha^2} + i \sum \frac{\gamma\alpha}{\omega} \frac{\omega_{p\alpha}^2}{\omega^2 + \gamma_\alpha^2} \tag{13}$$

where the plasma frequencies, $\omega_{p\alpha}$ are (in MKS units)

$$\omega_{p\alpha}^2 = \frac{n_\alpha e^2}{m_\alpha} \tag{14}$$

and the summation is over all species of free carriers (i.e., electrons and light and heavy holes). The static dielectric constant is ε_s, and n_α, γ_α, and m_α are the carrier densities, damping rates, and effective masses. Useful phase shifts at 94 GHz were produced with this device without large attenuation from losses produced by the imaginary part of the dielectric response.

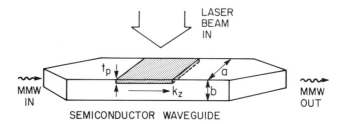

F<small>IG</small>. 18. A photoconducting microwave phase modulator. The phase velocity of the dielectric waveguide is modified by the optical injection of free carriers. (Lee et al., 1980)

3. Radio-Frequency Mixing

Foyt et al. (1981) demonstrated a radio-frequency mixer that used an interdigitated InP optoelectronic switch (Leonberger and Moulton, 1979). This device, which is illustrated in Fig. 16, used a semiconductor laser diode as a light source. It was modulated at the local oscillator frequency and the switch was operated in a bilinear mode so that the output signal at the intermediate frequency was proportional to the product of the input RF signal and the light signal. This produced a high quality IF signal with very low third-order intermodulation. This device and related applications of InP optoelectronic devices were summarized in Foyt and Leonberger (1984).

4. Photoconducting Antennas

The concept of a photoconducting antenna is based on the radiation of a photocurrent into free space, as opposed to being coupled into a transmission line or other guiding structure. This approach has the advantage of greater speed because it overcomes the limitations imposed by dispersion and losses characteristic of conventional transmission systems.

Photoconductors were used to drive a dipole antenna by Mourou et al. (1918b). A GaAs:Cr photoconductor was illuminated with subpicosecond pulses to produce a short current pulse. It was then fed to a dipole antenna. The duration of the radiated signal from the antenna was measured by gated transmission through a thin slab of germanium. It was estimated to be approximately 3 ps. In a similar experiment, Heidemann et al. (1983) used a photoconductive switch to drive an exponentially tapered slot-line antenna. They produced a radiated pulse of a single cycle having a duration of approximately one ns.

Auston and Cheung (1984) have used photoconductors for both transmitting and receiving antennas. The reciprocal property of photoconducting antennas enables them to be used as high-speed, sensitive detectors of radiated electrical pulses. In this case, the radiation field induced the bias signal in the photoconductor of the receiving antenna, rather than a DC signal. The average current at the receiving dipole was measured as a function of the delay between the optical pulses illuminating the transmitting and receiving dipoles. This produced autocorrelation measurements of the system response. Response times of approximately 1.6 ps were measured for radiation-damaged silicon-on-sapphire Hertzian dipoles.

The properties of the radiated field from elementary dipoles excited by short current pulses from photoconductors can be estimated from the classical field of a Hertzian dipole. The photocurrent signal produces a time-varying dipole moment whose radiation field, $E_\theta(\theta)$ at a distance r and angle θ is:

$$E_\theta(r, \theta) = \frac{1}{4\pi\varepsilon} \left(\frac{p}{r^3} + \frac{n}{cr^2} \frac{\partial p}{\partial t} + \frac{n^2}{c^2 r} \frac{\partial^2 p}{\partial t^2} \right) \sin \theta. \tag{15}$$

The photocurrent, $i(t)$, is equal to the first time-derivative of the dipole moment, $p(t)$, divided by the length of the dipole. (This dimension is assumed to be small relative to the shortest radiated wavelength.) The three terms in the expression for the electric field are identified as the static, near-, and far-field components, varying respectively by r^{-3}, r^{-2}, and r^{-1}. Each of these has a different temporal variation. The far-field term is proportional to the second derivative of the photocurrent. A short unipolar current pulse radiates a far field that is bipolar. For short pulses, the distinction between the far and near fields is given by the simple relation $r \gg \tau_p c/n$, where τ_p is the pulse duration. The specific geometry of the photoconducting antenna can substantially modify the radiation field. For example, a dipole of finite length introduces resonance effects at frequencies where the dipole is an odd multiple of a half wavelength.

Karin et al. (1986) observed response times of less than one ps in He^+ ion–bombarded InP:Fe Hertzian dipoles. They used a geometry in which the photocurrent was orthogonal to the photoconductor electrodes to suppress the slower radiation signal from currents in the electrodes.

DeFonzo et al. (1987) and DeFonzo and Lutz (1987) used tapered exponential waveguide structures as anetnnas to produce signals with strong unipolar components. They measured correlations having full widths at half maximum of 10 picoseconds.

Smith et al. (1988) made resonant halfwave photoconducting dipole antennas having a measured frequency response of up to 2 THz. These antennas were dipoles having lengths as short as 50 μm. They were excited by radiation-damaged silicon-on-sapphire photoconductors having estimated photocurrent decay times of approximately 0.6 ps. Figure 19 shows an experimental measurement of the electronic correlation signal from two

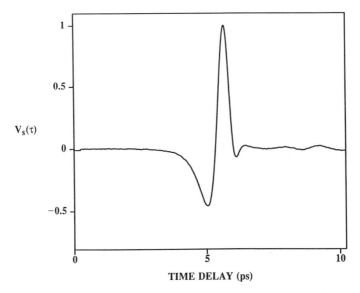

FIG. 19. Electronic correlation measurement of two identical 50-μm photoconducting dipole antennas separated by 500 μm of sapphire. (From Smith, Auston and Nuss 1988)

50-μm antennas separated by 0.5 mm of sapphire. The response is extremely fast and has a frequency spectrum that peaks at 0.6 THz, has a FWHM of one THz, and rolls off by 10 db at 2 THz.

E. High-Power Optoelectronic Switching

Photoconductors use the bulk properties of materials and consequently can be readily scaled to switch high voltages and currents. This has led to a number of important applications where extremely fast dv/dt transients are required for switching high-voltage instruments. In principal, it is possible to have power gain in a photoconductor (i.e., to switch an electrical power that exceeds the optical power).

The design of photoconductors for high-power applications requires a substantially different approach than for low powers. Electric-field breakdown and thermal dissipation are two effects that greatly influence the choice of materials and geometry. Since the size of high-power photoconductors scales with the switching voltage, careful attention to geometry and mounting is important to optimize the speed versus power trade-off.

As discussed in Section II, electronic transport in semiconductors under high-electric-field conditions is substantially different than for low fields. In silicon, for example, the electron velocity saturates at fields above 3 to

5 kV/cm at a value of approximately 10^7 cm/s. The behavior of holes is similar. In GaAs and related III-V direct-gap semiconductors, the electron velocity reaches a maximum at a few kV/cm and then decreases at higher fields. This behavior produces a negative differential mobility that can lead to Gunn-type instabilities. Transient high-fields effects such as velocity overshoot can lead to improved speed of rresponse in high-voltage photoconductors.

Departures from linear ohmic behavior can influence the efficiency and stability of high-power photoconductors. Additional effects, such as inductive (Nunnally and Hammond, 1984) and electromagnetic transit times (Kerstan and Schubert, 1984) may also play a role in determining the speed of response of high-power photoconductors.

High-power photoconductors use higher optical intensities and consequently the carriers' densities can be extremely high and can produce results that differ from low injection conditions. For example, in silicon at room temperature, electron densities above 10^{19} cm^{-3} will produce a degenerate Fermi distribution. The conductivity is determined by elastic scattering at the Fermi energy and generally results in a lower effective mobility. An additional effect of high carrier densities is the reduced mobility due to electron-hole scattering. (See Section II.)

Breakdown by high electric fields constrains the size of a high-power picosecond photoconductor. The gap size which scales linearly with voltage. Because the photoconductance is proportional to the optical energy and inversely proportional to the square of the gap length, the required optical energy increases as the square of the bias voltage.

Thermal effects are also important considerations in high-power photoconductors. Thermal runaway can limit the hold-off voltage. This effect occurs when the temperature rise due to the dark current is sufficient to create additional carriers by thermal generation of e–h pairs. Practical solutions to this problem include the use of high-resistivity materials, low temperatures, and pulsed bias voltages. For a more detailed discussion of these and related aspects of high-power photoconductors, see Mourou et al. (1984) and Nunnally and Hammond (1984).

1. Applications of High-Power Photoconductors

Picosecond photoconductive switching above one kV was first demonstrated by LeFur and Auston (1976). They used a silicon microstrip photoconductor with a pulsed bias. The output voltage drove a travelling-wave Pockel's cell of $LiTaO_3$. Transverse probing of the Pockel's cell by a second light pulse showed the voltage wave had a risetime of less than 25 ps.

Antonetti et al. (1977) used a silicon photoconductor mounted in a coaxial transmission line to switch voltages up to 10 kV. These were used to drive a travelling wave Kerr cell and a fast commercial Pockel's cell. Rise-times of less than 50 ps were reported.

Agostinelli et al. (1979) used a GaAs photoconductive switch to generate electrical pulses of 3 kV. They used them to drive a Pockel's cell and measured rise-and falltimes as fast as 40 ps. A streak camera was used to observe the transmitted optical signals that were sliced out of a long optical pulse by the Pockel's cell. Improved switching of high voltages was reported by Mourou and Knox (1979), Stavola et al. (1979), and Koo et al. (1984).

The highest-voltage photoconductor reported as of early 1989 was demonstrated by Nunnally and Hammond (1984) to produce 1.8 kA into a 25-Ω load for a peak voltage of 45 kV and a peak power of 80 MW. Although the risetime was relatively slow (5 ns), it is expected that further improvements in high-power switching technology will produce results of less than one nanosecond.

Mourou and Knox (1980) used a high-voltage silicon photoconductor to synchronize a streak camera to optical pulses from a mode-locked Nd:YAG laser. Their experiment, which is illustrated in Fig. 20, used the output of the silicon phottoconductor to drive the deflection plate of the streak camera. Greatly reduced jitter enabled extensive shot-to-shot averaging for improved time resolution and signal-to-noise ratios. Similar work was reported by Yen et al. (1984), who used an InP photoconductor to synchronize a proximity focused streak camera.

F. APPLICATIONS TO MATERIALS CHARACTERIZATION

The speed of response of conventional electronic devices is currently limited to approximately 5 ps. Optoelectronics, however, has a capability of making measurements with time resolution below one picosecond. In this range, the most important applications are the characterization of the properties of materials. Because optoelectronics gives jitter-free phase-coherent measurements, the frequency response can be directly deduced from the temporal response. For characterization of materials, optoelectronics provides measurements over an extremely broad range of frequencies from DC to submillimeter waves. In this section, some specific examples illustrate this capability.

1. Measurements of Superconducting Transmission Lines

Dykaar et al. (1986) measured the response of superconducting transmission lines to fast-rising electrical pulses from a GaAs photoconductor.

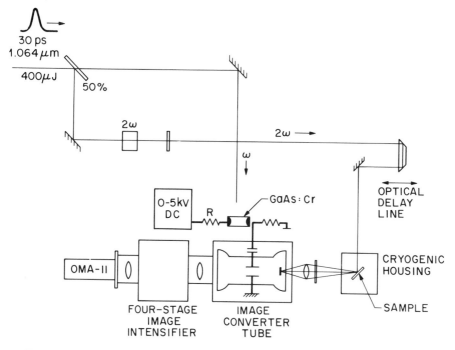

FIG. 20. Optoelectronic synchronization of a streak camera by a high-voltage photoconductor. (From Mourou and Knox, 1980)

By sampling the waveform at different distances down the line with an electro-optic crystal, they determined the attenuation and dispersion.

Measurements by Sprik et al. (1987) of the propagation of short pulses on niobium transmission lines have clearly illustrated the influence of the finite superconducting band gap on the attenuation and dispersion of pulses. They made measurements on 5-μm coplanar strips of niobium. Excitation and sampling were done with radiation-damaged silicon-on-sapphire photoconductors having estimated speeds of response of 0.6 ps. Figure 21 illustrates some of their results. The finite superconducting band gap of the niobium superconductor produced a ringing on the tail of the pulse when the temperature was well below the transition temperature of 9.4 K.

2. Picosecond Photoconductivity in Amorphous Semiconductors

Picosecond photoconductivity has been very effective for measurements of mobilities and carrier lifetimes in amorphous semiconductors. (See Johnson, et al., 1981.) In the case of the hydrogenated amorphous silicon,

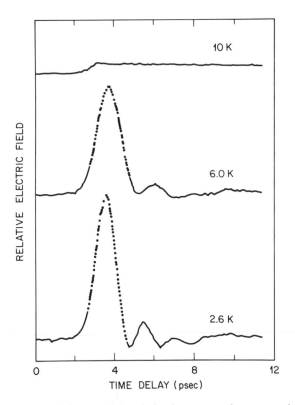

Fig. 21. Propagation of picosecond electrical pulses on a coplanar transmission line composed of niobium metal at different temperatures. The niobium becomes superconducting below 9.4 K. The ringing on the tail of the pulse is due to the finite bandwidth of the superconducting niobium. (From Gallagher et al., 1987)

measurements were made over a wide range of temperatures, as illustrated in Fig. 22. This was accomplished by mounting a-Si:H and RD-SOS photoconductors on the tip of the cold-finger of a variable-temperature cryostat. The radiation-damaged SOS photoconductor was used to generate a pulsed bias signal for the amorphous silicon sample. This enabled correlation measurements to be made by varying the time delay between the optical pulses illuminating the two photoconductors. The use of a pulsed bias voltage substantially reduced the problems associated with non-ohmic contacts in the a-Si sample.

The results showed a dramatic increase in the speed of the decay of the photocurrent as the temperature was reduced. This was expected due to the reduction in the thermal emission rate of captured carriers in localized

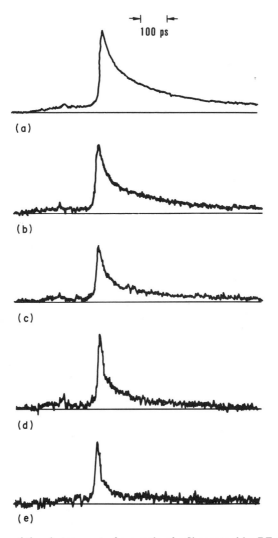

FIG. 22. Decay of the photocurrent of a sample of a-Si prepared by RF glow discharge. These measurements were made at different temperatures by cross correlation using a fast (<10 ps) radiation-damaged silicon-on-sapphire photoconductor as a bias pulse generator. The temperature-dependent non-exponential decay is a characteristic of a multiple trapping relaxation of the photoexcited carriers by a continuum of localized states.(From Johnson, et al., 1981)

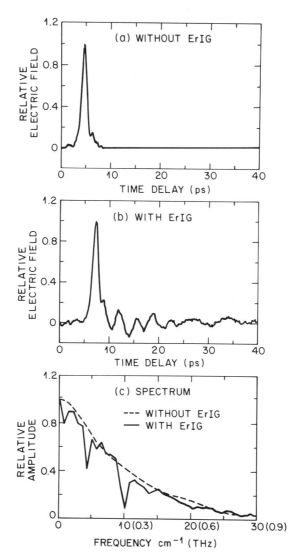

FIG. 23. Frequency spectrum of a picosecond electrical pulse after propagating through a sample of erbium iron garnet showing distinct absorption lines in the far-infrared. The sample temperature was 5 K. (From Sprik et al., 1987)

states back to the extended states. The nonexponential shape of the photo-current decay is also a characteristic feature of this phenomenon.

The initial mobility was determined from the peak amplitude of the photocurrent. It was expected to increase as T^{-1} when the temperature was lowered. However, the experiment gave the opposite result. It decreased approximately by a factor of 10 from room temperature to 100°K. It appears that at low temperatures, the photocurrent transient consists of capture without reemission. Thermalization of the hot free carriers within the continuum of extended states may also be important.

3. Picosecond Photoconductivity in Polydiacetylene

Moses et al. (1987) measured the response of PDA-TS (i.e., polydiacetylene bis p-toluene sulfonate) photoconductors using picosecond optical ilumination. Small gap spacings permitted them to observe carrier sweep-out. They determined the carrier mobility to be approximately 5 cm^2/Vs at room temperature. Their results suggest that geminate recombination does not play an important role, as had previously been thought to be true.

4. Submillimeter Wave Spectroscopy of Magnetic Garnets

Sprik et al. (1987) applied picosecond optoelectronics to measurements of the absorption spectra of magnetic garnets in the frequency range from DC to 0.9 THz. They placed a powdered sample of erbium iron garnet over a coplanar strip transmission line on which they measured the propagation of subpicosecond electrical pulses. Figure 23 shows a spectrum taken at 5 K. The difference between the spectrum of the transmitted pulse with and without the ErIG sample clearly shows the sharp absorption lines due to magnetic resonances.

V. Discussion and Future Trends

It is clear from the diversity and volume of work summarized in this chapter that picosecond photoconductivity is a well-developed and rapidly growing field of research. The use of optical pulses to generate and detect picosecond and subpicosecond electrical pulses is now relatively well developed. However, we can expect to see substantial improvements with regard to speed and sensitivity. Optical pulses are now available with durations as short as 6 fs. The application of these extremely short pulses to optoelectronics presents an exciting challenge for future work. The available optical bandwidth extends as hig as 10^{14} Hz, well into the far-infrared region of the spectrum. If techniques can be developed that can utilize most of this bandwidth, it will make possible measurements over an enormous range of the electromagnetic spectrum. This will fill the gaps in the spectrum where coherent sources do not exist today.

New materials for optoelectronics will have an important influence on future trends. Although the traditional bulk properties of the better known semiconductors established the foundation for picosecond-optoelectronic-device concepts, new materials are required to produce faster speeds and better sensitivities. For example, semiconductors grown by molecular beam epitaxy have already proven to have valuable properties for high-speed optical modulators and bistable optical logic elements. The "self-electro-optic" effect and the quantum-confined stark effect will probably find applications in high-speed optoelectronic measurement systems as well. As the technology of the new high-temperature oxide superconductors advances, we can expect to see these materials used in picosecond optoelctronic circuits for low-loss, high-speed transmission lines and possibly as new switching devices.

In the area of applications, ultrafast optoelectronics is now experiencing its most rapid growth. As a precision tool for testing high-speed discrete devices and integrated circuits, optoelectronic measurement systems are now being widely implemented and are at the stage where useful design information can be obtained about device performance. Measurements of novel high-speed electronic devices such as resonant tunneling transistorss and "ballistic" transistors will be an important guide to research and development of new high-speed technologies.

In the very high speed area, the trend toward applying these techniques to materials characterization is expected to continue. For example, measurements of the detailed kinetics of electronic transport on the sub-picosecond time scale over a wide range of density, electric field, and temperature are critically important to the development and understanding of electronic and optical materials.

The continued development of techniques for generating and detecting large-amplitude electrical pulses will have important applications for the study of the transient nonlinear response of materials and devices. For example, we might expect to observe electrical solitons on nonlinear transmission lines or in nonlinear dielectric materials, analogous to the optical solitons that have been produced in optical fibers. This could lead to pulse compression and result in even shorter electrical pulses than can be produced today.

References

Agostinelli, J., Mourou, G., and Gavel, C.W. (1979). *Appl. Phys. Lett.* **35**, 731–733.

Antonetti, A., Malley, M.M., Mourou, G., and Orszag, A. (1977). *Opt. Comm.* **23**, 435–438.

Appel, J. (1961). *Phys. Rev.* **122**, 1760.

Appel, J. (1962). *Phys. Rev.* **125**, 1815.

Auston, D.H. (1975). *Appl. Phys. Lett.* **26**, 101–103.

Auston, D.H. (1983). *IEEE J. Quant. El.* **QE-19**, 639–648.

Auston, D.H. (1984). In *Picosecond Optoelectronic Devices*, C.H. Lee, ed. Academic Press, New York, Chapter 4.

Auston, D.H., and Cheung, K.P. (1984). *Appl. Phys. Lett.* **45**, 284–286.

Auston, D.H., and Cheung, K.P. (1985). *J. Opt. Soc. Amer.* **B2**, 606–612.

Auston, D.H., and Eisenthal, K.B., eds. (1984). *Ultrafast Phenomena* **IV**, Springer-Verlag Series in Chemical Physics **38**, Springer-Verlag, Berlin.

Auston, D.H., and Glass, A.M. (1972). *Appl. Phys. Lett.* **20**, 398–399.

Auston, D.H., and Johnson, A.M. (1977). In *Ultrashort Light Pulses*, S.L. Shapiro, ed. Springer-Verlag, Berlin, 245–247.

Auston, D.H., and Smith, P.R. (1982). *Appl. Phys. Lett.* **41**, 599–601.

Auston, D.H., and Smith, P.R. (1983). *Appl. Phys. Lett.* **43**, 631–633.

Auston, D.H., Glass, A.M., and Ballman, A.A. (1972). *Phys. Rev. Lett.* **28**, 897–900.

Auston, D.H., Glass, A.M., and LeFur, P. (1973). *Appl. Phys. Lett.* **23**, 47–48.

Auston, D.H., Lavallard, P., Sol, N., and Kaplan, D. (1980a). *Appl. Phys. Lett.* **35**, 66–68.

Auston, D.H., Johnson, A.M., Smith, P.R., and Bean, J.C. (1980b). *Appl. Phys. Lett.* **37**, 371–373.

Auston, D.H., Cheung, K.P., and Smith, P.R. (1983). In *Proc. SPIE Symp. on Picosecond Optoelectronics*, G.A. Mourou, ed.

Block, D., Shah, J., and Gossard, A.C. (1986). *Solid St. Comm.* **59**, 527.

Bowman, D.R., Hammond, R.B., and Dutton, R.W. (1985). *IEEE Electr. Dev. Lett.* **EDL-6**, 502–504.

Buck, J.A., Li, K.K., and Whinnery, J.R. (1980). *J. Appl. Phys.* **51**, 769.

Carruthers, T.F. (1984). In *Picosecond Optoelectronic Devices*, C.H. Lee, ed. Academic Press, New York, 339–371.

Carruthers, T.F., and Weller, J.F. (1986). *Appl. Phys. Lett.* **48**, 460–462.

Carruthers, T.F., Weller J.F., Taylor, H.F., and Mills, T.G. (1981). *Appl. Phys. Lett.* **38**, 202–204.

Chang, C.S., Mathur, V.K., Rhee, M.J., and Lee, C.H. (1982). *Appl. Phys. Lett.* **41**, 392–394.

Chang, C.S., Jeng, M.C., Rhee, M.J., Lee, C.H., Rosen, A., and Davis, H. (1984). *IEEE MTT-S International Microwave Symposium Digest*, 540.

Chen, J.W., and Milnes, A.G. (1977). *Sol. St. Electr.* **22**, 684–686.

Cheung, K.P., and Auston, D.H. (1986a). *Phys. Rev. Lett.* **55**, 2152.

Cheung, K.P., and Auston, D.H. (1986b). *Infrared Phys.* **26**, 23–27.

Cooper, D.E. (1985). *Appl. Phys. Lett.* **47**, 33–35.

Cooper, D.E., and Moss, S.C. (1986a). *IEEE J. Quant. El.* **QE-22**, 94–100.

Cooper, D.E., and Moss, S.C. (1986b). In *Ultrafast Phenomena* **V**, A.E. Siegman and G.R. Fleming, eds. Springer-Verlag series in Chemical Physics **46**, Springer-Verlag, Berlin, 117.

DeFonzo, A.P. (1983). *Appl. Phys. Lett.* **39**, 480–481.

DeFonzo, A.P., and Lutz, C.R. (1987). *Appl. Phys. Lett.* **51**, 212.

DeFonzo, A.P., Jarwala, M., and Lutz, C. (1987). *Appl Phys. Lett.* **50**, 1155.

Doany, F.E., Grischkowsky, D., and Chi, C.C., eds. (1987). *Digest of the Second Topical Conference on Picosecond Electronics and Optoelectronics*. Optical Society of America, Washington, D.C.

Downey, P.M., and Tell, B. (1984). *J. Appl. Phys.* **56**, 2672–2674.

Downey, P.M., and Schwartz, B. (1984). *Appl. Phys. Lett.* **44**, 207.

Downey, P.M., Auston, D.H., and Smith, P.R. (1983). *Appl. Phys. Lett.* **42**, 215–217.

Downey, P.M., Martin, R.J., Nahory, R.E., and Lorimer, O.G. (1985). *Appl. Phys. Lett.* **46**, 396–398.

Dykaar, D.R., Sobelewski, R., Whitaker, J.F., Hsiang, T.Y., Mourou, G.A., Hollis, M.A., Clifton, B.J., Nichols, K.B., Bozler, C.O., and Murphy, R.A. (1986). In *Ultrafast Phenomena* **V**, G.R. Fleming and A.E. Siegman, eds. Springer-Verlag Series in Chemical Physics **46**, Springer-Verlag, Berlin, 103.

Edwards, T.C. (1981). *Foundations for Microstrip Circuit Design*, Wiley, New York.

Eisenstadt, W.R., Hammond, R.B., and Dutton, R.W. (1985). *IEEE Trans. Electr. Dev.* **ED-32**, 364–369.

Erginsoy, C. (1950). *Phys. Rev.* **79**, 1013–1014.

Fleming, G.R., and Siegman, A.E., eds. (1986). *Ultrafast Phenomena* **IV**, Springer-Verlag, Berlin.

Forrest, S.R. (1985). *IEEE J. Lightwave Tech.* **LT-3**, 347.

Foyt, A.G., and Leonberger, F.J. (1984). In *Picosecond Optoelectronic Devices*, C.H. Lee, ed. Academic Press, New York, 271–311.

Foyt, A.G., Leonberger, F.J., and Williamson, R.C. (1981). *Proc. SPIE* **269**, 109–114.

Foyt, A.G., Leonberger, F.J., and Williamson, R.C. (1982). *Appl. Phys. Lett.* **40**, 447–449.

Gallagher, W.J., Chi, C.C., Duling, I.N., Grischkowsky, D. and Halas, N.J. (1987). *Appl. Phys. Lett.* **50**, 350.

Gobel, E.O., Kuhl, J., and Veith, G. (1984). *J. Appl. Phys.* **56**, 862–864.

Goossen, K.W., and Hammond, R.B. (1985). *IEEE Trans. Micr. Th. Tech.* **MTT-33**.

Grivitskas, V., Willander, M., and Vaitkus, J. (1984). *Sol. St. Electr.* **27**, 565–572.

Gupta, K.C., Garg, R., and Bahl, I.J. (1979). *Microstrip Lines and Slotlines.* Artech House, Dedham, Massachusetts.

Hasnain, G., Arjavalingam, G., Dienes, A., and Whinnery, J.R. (1983). *Proc. SPIE Conf. on Picosecond Electro-optics*, SPIE Press, vol. 439, Bellingham, WA.

Hammond, R.B. (1985). In *Proceedings of the 1987 International Conference on Hot Electrons*, E. Gornick, ed., Innsbruck, Austria.

Hammond, R.B., Paulter, N.G., Iverson, A.E., and Smith, R.C. (1981). *Tech. Dig. of IEEE 1981 Int. Electr. Dev. Mtg.*, 157–159.

Hammond, R.B., Paulter, N.G., Wagner, R.S., and Eisenstadt, W.R. (1984). *Appl. Phys. Lett.* **45**, 404.

Heidemann, R., Pfeiffer, T., and Jager, D. (1982). *Electr. Lett.* **18**, 783–784.

Heidemann, R., Pfeiffer, T., and Jager, D. (1983). *Electr. Lett.* **19**, 316–317.

Jain, R.K., and Snyder, D.E. (1983a). *IEEE J. Quant. El.* **QE-19**, 658.

Jain, R.K., and Snyder, D.E. (1983b). *Opt. Lett.* **8**, 85.

Jain, R.K., Synder, D.E., and Stenersen, K. (1984). *IEEE Electr. Dev. Lett.* **EDL-5**, 371.

Johnson, A.M., and Auston, D.H. (1975). *IEEE J. Quant. El.* **QE-11**, 283–287.

Johnson, A.M., Auston, D.H., Smith, P.R., Bean, J.C., Harbison, J.P., and Adams, A.C. (1981). *Phys. Rev.* **B23**, 6816–6819.

Johnson, A.M., Kisker, D.W., Simpson, W.M., and Feldman, R.D. (1985). In *Picosecond Electronics and Optoelectronics*, G.A. Mourou, D.M. Bloom, and C.H. Lee, eds. Springer-Verlag Series in Electro-physics **21**, Springer-Verlag Berlin, 188.

Kaiser, W. ed. (1988). *Ultrashort Light Pulses*, Springer-Verlag, Berlin.

Karin, J.R., Downey, P.M., and Martin, R.J. (1986). *IEEE J. Quant. El.* **QE-27**, 677.

Kerstan, F., and Schubert, M. (1984). *Optical and Quant. El.* **16**, 477–486.

Ketchen, M.B., Grischkowsky, D., Chen, T.C., Chi, C.C., Duling, I.N., and Halas, N.J. (1986). *Appl. Phys. Lett.* **48**, 751–753.

Khokhlov, R.V. (1961). *Radio Eng. Electr. Phys.* (USSR) **6**, 817–824.

Kiehl, R.A. (1978). *IEEE Trans. Electr. Dev.* **ED-25**, 703–710.

Kiehl, R.A. (1979). *IEEE Trans. Micr. Th. Tech.* **MTT-27**, 533–539.

Kiehl, R.A. (1980). *IEEE Trans. Micr. Th. Tech.* **MTT-28**, 409–413.

Knox, W.H., Hirliman, C., Miller, D.A.B., Shah, J., Chemla, D.S., and Shank, C.V. (1986). *Phys. Rev. Lett.* **56**, 1191.

Koo, J.C., McWright, G.M., Pocha, M.D., and Wilcox, R.B. (1984). *Appl. Phys. Lett.* **45**, 1130–1131.

Landauer, R. (1960). *IBM J. Res. Develop.* **4**, 391–401.

Laval, S., Bru, C., Arnodo, C., and Castagne, R. (1980). *Digest of the 1980 IEDM Conference.* IEEE Press, New York, 626–627.

Lawton, R. A., and Andrews, J.R. (1975). *Electr. Lett.* **11**, 138.

Lawton, R.A., and Andrews, J.R. (1976). *IEEE Trans. Instr. Meas.* **25**, 56–60.

Lawton, R.A., and Scavannec, A. (1975). *Electr. Lett.* **11**, 74–75.

LeFur, P., and Auston, D.H. (1976). *Appl. Phys. Lett.* **28**, 21–23.

Lee, C.H. (1977). *Appl. Phys. Lett.* **30**, 84–86.

Lee, C.H., ed. (1984). *Picosecond Optoelectronic Devices*, Academic Press, New York.

Lee, C.H., Antonetti, A., and Mourou, G.A. (1977). *Optics Comm.* **21**, 158.

Lee, C.H., Mak, S., and DeFonzo, A.P. (1978). *Electr. Lett.* **14**, 733.

Lee, C.H., Mak, P.S., and De Fonzo, A.P. (1980). *IEEE J. Quant. El.* **QE-16**, 217–288.

Lee, C.H., Li, M.G., Chang, C.S., Yurek, A.M., Rhee, M.J., Chauchard, E., Fischer, R.P., Rosen, A., and Davis, H. (1985). *Proceedings of IEEE-MTT-S Intl. Micr. Symp., St. Louis*, 178–191, *IEEE Press*, New York.

Leonberger, F.J., ed. (1987). *Picosecond Electronics and Optoelectronics* **II**, Springer-Verlag Series in Electronics and Photonics, Springer-Verlag, Berlin.

Leonberger, F.J., and Moulton, P.F. (1979). *Appl. Phys. Lett.* **35**, 712–714.

Li, K.K., Arjavalingam, G., Dienes, A., and Whinnery, J.R. (1982). *IEEE Trans. Micr. Theor. Tech.* **MTT-30**, 1270–1273.

Li, K.K., Whinnery, J.R., and Dienes, A. (1984). In *Picosecond Optoelectronic Devices*, C.H. Lee, ed. Academic Press, New York, Chapter 6.

Li, S.S., Wang, W.L., Lai, P.W., and Owen, R.T. (1980). *J. Electr. Mat.* **9**, 335–354.

Low, A.J., and Carroll, J.E. (1978). *Sol. St & Electr. Dev.* **2**, 185–190.

Maeda, M. (1972). *IEEE Trans.* **MTT-20**, 390–396.

Mak, P.S., Mathur, V.K., and Lett, C.H. (1980). *Opt. Comm.* **32**, 485–488.

Margulis, W., and Persson, R. (1986). *Rev. Sci. Instr.* **56**, 1586–1588.

Margulis, W. and Sibbett, W. (1981). *Opt. Comm.* **37**, 224–228.

Martin, G.M., Mitonneau, A., and Mircea, A. (1977). *Electr. Lett.* **13**, 191–193.

McLean, J.P., and Paige, E.G.S. (1960). *J. Phys. Chem. Sol.* **16**, 220.

McLean, J.P., and Paige, E.G.S. (1961). *J. Phys. Chem. Sol.* **18**, 139.

Meyer, J.R., and Glicksman, M. (1978). *Phys. Rev.* **B17**, 3227–3238.

Mitonneau, A., Martin, G.M., and Mircea, A. (1977). *Electr. Lett.* **13**, 666–667.

Mooradian, A. (1984). *Appl. Phys. Lett.* **45**, 494.

Moses, D., Sinclair, M., and Heeger, A. J. (1987). *Phys. Rev. Lett.* **58**, 2710.

Mourou, G., and Knox, W. (1979). *Appl. Phys. Lett.* **35**, 492–495.

Mourou, G., and Knox, W. (1980). *Appl. Phys. Lett.* **36**, 623–626.

Mourou, G., Stancampiano, C.V., and Blumenthal, D. (1981a). *Appl. Phys. Lett.* **38**, 470.

Mourou, G., Stancampiano, C.V., Antonetti, A., and Orszag, A. (1981b). *Appl. Phys. Lett.* **39**, 295.

Mourou, G., Knox, W.H., and Williamson, S. (1984). In *Picosecond Optoelectronic Devices*, C.H. Lee, ed. Academic Press, New York, 219–248.

Mourou, G.A. (1986). In *High Speed Electronics*, Kallback and Beneking, eds. Springer-Verlag Series in Electronics and Photonics **22**, Springer-Verlag, Berlin, 191.

Mourou, G.A. (1987). *Digest of the second Second Topical Meeting on Picosecond Electronics and Optoelectronics*, Optical Society of America, Washington, D.C., 186–187.

Mourou, G.A., and Meyer, K.E. (1984). *Appl. Phys. Lett.* **45**, 492–494.

Mourou, G.A., Bloom, D.M., and Lee, C.H., ed. (1985). *Picosecond Electronics and Optoelectronics*, Springer-Verlag Series in Electro-physics **21**, Springer-Verlag, Berlin.

Nag, B.R. (1980). *Electronic Transport in Compound Semiconductors*, Springer-Verlag, Berlin.

Nunnally, W.C., and Hammond, R.B. (1984). In *Picosecond Optoelectronic Devices*, C.H. Lee, ed. Academic Press, New York, 373–398.

Nuss, M.C., and Auston, D.H. (1987). *Phys. Rev. Lett.* **58**, 2355–2358.

Oudar, J.L., Hulin, D., Migus, A., Antonetti, A., and Alexandre, F. (1985). *Phys. Rev. Lett.* **55**, 1191.

Panchhi, P.S., and van Driel, H. (1986). *IEEE J. Quant. El.* **QE-22**, 101.

Paige, E.G.S. (1960). *J. Phys. Chem. Sol.* **16**, 207.

Paulus, P., Wedding, B., Gasch, A., and Jager, D. (1984). *Phys. Lett.* **102A**, 89–92.

Paulus, P., Brinker, W., and Jager, D. (1986). *IEEE J. Quant. El.* **QE-22**, 108–111.

Platte, W. (1977). *Electr. Lett.* **12**, 437–438.

Platte, W. (1978). *Optics and Laser Technology* (February), 40–42.

Platte, W., and Appelhaus, G. (1976). *Electr. Lett.* **12**, 270.

Proud, J.M. Jr., and Norman, S.L. (1978). *IEEE Trans. Micr. Th. Tech.* **MTT-26**, 137.

Ramo, S., Whinnery, J.R., and van Duzer, T. (1984). *Fields and Waves in Communication Electronics*. Wiley, New York.

Rees, H.D. (1969). *J. Phys. Chem. Sol.* **30**, 643–655.

Reggiani, L., (ed.) (1985). *Hot-Electron Transport in Semiconductors*, Solid State Sciences **58**. Springer-Verlag, Heidelberg.

Rhoderick, E.H. (1978). "Metal-Semiconductor Contacts," Oxford University Press, New York.

Rose, A. (1963). *Concepts in Photoconductivity and Allied Problems*. Kreiger, New York.

Rutledge, D.B., Neikirk, D.P., and Kasilingham, D.P. (1983). *In Infrared and Millimeter Waves*. **10**, Part II, K.J. Button, ed. Academic Press, Chapter 1.

Seeger, K. (1982). *Semiconductor Physics*, Solid State Sciences **44**. Springer-Verlag, Heidelberg.

Shah, J. (1981). *J. Phys.* 42 (Suppl. 10), C7 445–C7462.

Shah J. (1986). *IEEE J. Quant. El.* **QE-22**, 1728.

Shah, N.J., Pei, S.S., Tu, C.W., and Tiberio, R.C. (1986). *IEEE Trans. Electr. Dev.* **ED-33**, 543–547.

Shank, C.V., Fork, R.L., and Greene, B.I. (1981). *Appl. Phys. Lett.* **38**, 104.

Shapiro, S., ed. (1977). *Ultrashort Light Pulses*. Springer-Verlag Topics in Applied Physics **18**, Springer-Verlag, Berlin.

Smith, P.R., Auston, D.H., Johnson, A.M., and Augustyniak, W.M. (1981a). *Appl. Phys. Lett.* **38**, 47–50.

Smith, P.R., Auston, D.H., and Augustyniak, W.M. (1981b). *Appl. Phys. Lett.* **39**, 739–741.

Smith, P.R., Auston, D.H., and Nuss, M.C. (1988). *IEEE J. Quant. El.* **24**, 255–260.

Sprik, R., Duling, I.N., Chi, C.-C., and Grischkowsky, D. (1987). *Appl. Phys. Lett.* **51**, 548.

Stavola, M., Agostinelli, J.A., and Sceats, M.G. (1979). *Appl. Opt.* **18**, 4101–4105.

Stoneham, A.M. (1980). *Physics Today* (January), 34–42.

Vaitkus, J., Grivitskas, V., and Storasta, J. (1976). *Sov. Phys. Semicond.* **9**, 883.

Vavilov, V.S., and Ukhin, N.A. (1977). *Radiation Effects in Semiconductors and Semiconductor Devices*. Plenum, New York.

Vermeulen, L.A., Young, J.F., Gallant, M.I.A., and van Driel, H.M. (1981). *Sol. St. Comm.* **38**, 1223–1225.

Yen, R., Downey, P.M., Shank, C.V., and Auston, D.H. (1984). *Appl. Phys. Lett.* **44**, 718–720.

Zhang, X.-C., and Jain, R.K. (1987). *Digest of the Conference on Lasers ans Electro-Optics*. Optical Society of America, Washington.

CHAPTER 4

Electro-Optic Measurement Techniques for Picosecond Materials, Devices, and Integrated Circuits

J. A. Valdmanis

ULTRAFAST SCIENCE LABORATORY
UNIVERSITY OF MICHIGAN
ANN ARBOR, MICHIGAN

I. Introduction

A. MOTIVATION

Until 1980, high-speed electrical signals were measured almost exclusively by purely electronic techniques, most notably the venerable cathode ray tube oscilloscope. "Ultrafast" measurements were considered those under one nanosecond, the bulk of which were performed with sampling oscilloscopes. During the 1970s, the workhorses of the ultrafast regime were sampling oscilloscopes. The fastest of these were based on step-recovery diode technology and offered a temporal resolution of approximately 25 ps.

Since 1980, development of ultrafast electronic and optoelectronic devices, which will be the fundamental elements of the next generation of computers and communications systems, has progressed rapidly. Many new devices that are the building blocks for more complicated integrated circuits have response times below 25 ps and many electronic materials show temporal characteristics at or below 1 ps (Mourou et al., 1985; Leonberger et al., 1987; Bloom and Sollner, 1989).

These advances in high-speed devices resulted from novel semiconducting-material growth and processing techniques, such as molecular beam epitaxy (MBE) and ion implantation, as well as the development of high-spatial-resolution (submicron) fabrication techniques. MBE is directly responsible for the invention of a new breed of ultrafast transistors known as modulation-doped field-effect transistors (MODFETs) (Berenz, 1987). These devices attained switching speeds of nearly 10 ps; permeable base transistors as fast as 5.4 ps (Bozler et al., 1980). Epitaxial growth techniques also resulted in development of new graded band-gap phototransistors and superlattice photodetectors with response times of only tens of picoseconds. Heterostructure transistors have achieved propagation delays below 6 ps (Shah et al., 1986). Gas-source MBE has now made possible the invention on InP/GaInAs bipolar transistors that operate at frequencies up to 140 GHz (Chen et al., 1989). Even faster are detectors based on photoconductive ion-bombarded materials such as silicon and indium phosphide, the former having a response time below one ps (Ketchen et al., 1986). Significant advances have also been made in superconducting Josephson-junction technology, demonstrating switching speeds of only a few picosecond (Wolf, 1985).

To sustain advances in device performance, suitable ultrafast instruments must be developed to accurately evaluate the performance of these new devices. However, within this new time regime, the constraints placed on the measurement system and the way it accesses the signals of interest are stringent. This book presents several new solutions to the measurement dilemma, some of which are advanced and refined versions of existing technology. Since about 1980, however, picosecond pulse lasers have spawned a new class of measurement technique, one that is optically based (Ketchen et al., 1986; Leonberger et al.; Mourou et al., 1985). Several of the techniques described in this book now use picosecond laser pulses as an exciting new tool for making high-speed electrical measurements. What differentiates these new optical techniques is the type of ultrafast physical mechanism each uses for effecting the optical interaction with electrical signals.

This chapter deals with the fastest of the current optical techniques available, electro-optic sampling (Valdmanis and Mourou, 1986; Valdmanis et al., 1982; 1983). It uses electro-optic crystals as the electrical signal sensor. In this system, optical pulses interact directly with the electrical signal via the electro-optic effect. As such, it is an all-optical technique and is one of the simplest ways to exploit the availability of picosecond optical pulses directly as sampling gates. To better appreciate the advantages and limits of this technique, this chapter and Chapter 5 consider some crucial aspects of measurements in the picosecond domain (from one GHz up to and over one THz).

B. Interacting with Ultrafast Electrical Signals

In addition to the standard considerations involved with electrical measurements such as loading and sensitivity, picosecond electrical signals offer the added complexity of extremely broad bandwidth requirements. Signals with risetimes on the order of 10 ps contain frequencies in excess of 35 GHz: a 10-ps sinusoid has a fundamental frequency of 100 GHz. Signals with characteristics near one ps can contain frequencies approaching one THz. These signals require (1) a measurement system with similar bandwidth capabilities and (2) a means by which to couple to the signal without distorting it. An additional complication is that most high-speed signals are generated by devices that have physical dimensions on the micron scale. These constraints require major rethinking of how to make electrical measurements in this temporal regime. No longer can we just roll up the scope and hook up the probe—or, for that matter, even the sampling head—to find out what's going on.

Consider two general classes of electrical devices that generate signals in the picosecond domain: discrete devices and integrated circuits. Discrete devices are simple electrical elements that may be used alone or incorporated into an integrated circuit after individual testing and evaluation. Integrated circuits consist of many devices (often thousands) fabricated on a common substrate.

These two classes have different measurement constraints. Discrete devices can be connected directly to the measurement device because the fast signal of interest is usually an output signal or is accessible. However, this possibility also necessitates extremely careful design of the interconnection to faithfully preserve signal characteristics. At these frequencies, where the characteristic wavelengths that correspond to the signal frequencies approach the physical dimensions of the connectors, impedance matching alone is not sufficient. Geometric matching of the electrical lines as well as dispersive effects due to the line dimensions and the dielectrics involved must be considered. Conductor skin effects can also contribute to signal deterioration.

Integrated circuits introduce yet another complication, accessibility to the signal itself. There is great interest in monitoring signals at points internal to an integrated circuit. However, with line dimensions in the micron regime, a probe cannot be "hung" on the line of interest. Radically new techniques are required not only to access the signal of interest, but also to preserve the integrity of the signal under these conditions.

C. Advantages of Electro-Optic Interaction

The most significant advantages of the electro-optic measuring techniques are high temporal resolution and flexibility. These techniques have been used to measure a wide variety of electrical devices in many different forms (Valdmanis and Mourou, 1986). Applications include measurement of single devices such as the 5.4-ps permeable base transistor (PBT) (Bozler et al., 1979), characterization of transmission lines (Valdmanis and Mourou, 1984; Meyer and Mourou, 1985; Kryzak et al., 1985; Hsiang et al., 1987), internal probing of GaAs integrated circuits (Kolner and Bloom, 1984; Freeman et al., 1985; Weingarten et al., 1985; Taylor et al., 1986a), and InP circuits (Wiesenfeld et al., 1987), cryogenic measurement of superconducting devices and materials (Dykaar et al., 1985), (including the latest high-T_c materials), external probing of devices and integrated circuits (Valdmanis, 1987; Valdmanis and Pei, 1987; Nees and Mourou, 1986), and even the detection of freely propagating terahertz radiation pulses (Auston et al., 1984).

The flexibility is attributed to the simple mechanism on which the technique is based: the electro-optic or Pockels effect (Pockels, 1906; Yariv and Yeh, 1984), where the optical properties (birefringence) of a crystal change when an electric field is applied across it. By shining light through the crystal and measuring the change of polarization, the amplitude of the applied electrical signal can be determined. If short optical pulses are used, repetitive electrical signals can be sampled with a temporal resolution limited basically by the optical pulse duration. Today's picosecond and subpicosecond lasers thus make high-resolution electro-optic sampling straightforward. The only limits encountered are those of physically incorporating the electro-optic crystal into the measurement situation. However, this can be accomplished in many ways, as discussed in Section III.C. (Modulator Geometries).

Electro-optic techniques also provide good sensitivity and high dynamic range with excellent linearity. Although typical electro-optic modulators require kV/mm electric fields for large-amplitude modulation, in device measurements, fields as low as one V/m can be detected with high-repetition-rate sampling, high-frequency–tuned detection techniques (lock-in amplifiers), and signal averaging. Because of this sensitivity, the dynamic range can extend over seven orders of magnitude.

Electro-optic systems offer another advantage. They do not require that charge be removed from the circuit under study because they rely on the applied electric field only. Thus loading considerations are limited to those of capacitive and geometric effects. This chapter shows how these situations are dealt with in real systems.

Other advantages of electro-optic techniques include the fact that the optical system is simple, many different lasers can be used, and the system works at room temperature and pressure, unlike electron-beam–based techniques that require a vacuum, or superconducting electronics that require a cryogenic environment.

The electro-optic technique has matured to the point where a commercial instrument based on this technology is available. In 1987, EG&G Princeton Applied Research introduced its Model 8300 electro-optic sampling unit for the characterization of discrete, ultrahigh-speed devices. It is used in conjunction with a picosecond laser system and achieves 4-ps temporal resolution. It is described in Section III.C on hybrid modulator geometries.

Also, electro-optic–based techniques have been developed for nonsampling systems where novel high-speed detection schemes have been introduced. Two that are discussed here are the real-time optical oscilloscope (Valdmanis, 1986) and the electron electro-optic oscilloscope (Williamson and Mourou, 1985). Both have mechanisms that faintly resemble a conventional oscilloscope, but obtain ps resolution by using electro-optic modulators.

D. REAL-TIME VERSUS SAMPLING SYSTEMS

The electro-optic interaction naturally lends itself to high-speed measurements because of, among other advantages, the inherent frequency response of the mechanism involved. However, to complete the measurement, the high-speed modulation of the optical beam must be detected. This detection can be in real time or by sampling. In real-time measurements, the temporal resolution is achieved simply by optical detection that is fast enough to resolve the high-speed optical modulation as it happens (real time). In this detection system, the optical beam can be a continuous wave (CW) relative to the speed of the electrical signal. However, the speed of conventional real-time optical detection systems is at most only a few GHz, as of the late 1980s. To significantly increase temporal resolution, the initial response is to turn to sampling schemes.

In sampling systems, the optical beam is the source of temporal resolution. Instead of a CW beam, short pulses of light, synchronized with the electrical signal, sample the modulation at a particular instant. Thus, the temporal resolution in a sampling system is determined basically by the duration of the optical pulse. By making the signal repetitive and delaying the optical pulse with respect to the electrical signal, the electrical waveform can be reconstructed in an equivalent time frame determined by the delaying mechanism. Thus, the detector does not limit the bandwidth of the system.

In summary, real-time systems use a "slow" laser with fast detection, while sampling systems reverse the roles by using a "fast" laser with a slow detector. Because lasers producing optical pulses of <100-fs duration are readily available, electro-optic sampling systems with frequency responses above one THz are now possible.

This chapter and the following chapter on internal (or "direct") electro-optic sampling show that the bulk of electro-optic measurements to date are performed with sampling-type systems, due to the ease of obtaining high temporal resolution with picosecond pulse lasers and conventional electronics. However, two real-time techniques that use electro-optic modulation, together with novel, ultrafast optical detection schemes are also discussed.

E. Chapter Outline

The remainder of this chapter will describe the electro-optic–measurement technique and a variety of applications in detail.

Section II explains the mechanism of the electro-optic effect, tells how it is implemented to achieve intensity modulation, and then describes several popular electro-optic materials and their properties.

Section III addresses the principles of electro-optic sampling. It considers the optical and electronic systems, the modulator geometries commonly encountered, and the performance limits of temporal resolution and sensitivity. Techniques for synchronizing the optical system and the electrical circuit under study are also presented.

Section IV presents a wide variety of applications of electro-optic sampling with hybrid modulator structures. Both devices and materials were studied as well as the most recent (late 1980s) state-of-the-art investigations of high-T_c superconductors.

Section V describes the latest configuration of electro-optic sampling, the external probe for high-speed integrated circuits. A brief comparison of internal- and external-probing schemes is presented. External probing is then explained with several applications.

Section VI provides two novel examples of nonsampling techniques that rely on unique ultrafast optical detection schemes. Both systems exploit electro-optic modulators as the initial interface to an unknown electrical signal.

II. Electro-Optic Modulation

Electro-optic sampling employs the oldest transducer technology of any current sampling system, the electro-optic effect. Pockels discovered that an electric field applied to some crystals changes the birefringence properties of the crystal and hence also changes the polarization of light that propagates through it. By placing the crystal between crossed polarizers, the transmitted light intensity changes as a function of the applied field. It is this simple principle coupled with the latest mode-locked laser technology that enables electro-optic sampling. With this technique, optical pulses are exploited directly as sampling gates in the electro-optic medium, and therefore the technique benefits naturally from the evolution of shorter laser pulses.

A. The Pockels, or Electro-Optic, Effect

1. The Electro-Optic Mechanism

The electro-optic effect causes a change in the optical dielectric properties of a medium (usually a crystal) in response to an applied electric field. The effect commonly changes the index of refraction. To more completely understand the limits of this mechanism, the behavior of the electronic charges within the medium must be investigated.

The application of an external electric field to a medium displaces both the ions in the lattice and the electron orbits from their unperturbed positions and orientations. These displacements create electric dipoles that are manifested as the electric polarization, **P**.

In any material, **P** is a function of the applied electric field $\mathcal{E}$. The polarization can be represented by the following power series:

$$\mathbf{P}_i(\omega_l) = \chi_{ij}^{(1)} \mathcal{E}_j(\omega_l) + \chi_{ijk}^{(2)} \mathcal{E}_j(\omega_m)\mathcal{E}_k(\omega_n) + \cdots \tag{1}$$

where $\chi^{(1)}$ and $\chi^{(2)}$ are tensors that relate the vectors $\mathbf{P}$ and $\mathscr{E}$. i, j, and k are the cartesian indices that run from 1 to 3, while l, m, and n represent different frequency components (ω). Each term is summed over all repeated indices according to the Einstein sum convention. Only terms of interest are included here. χ is a tensor, because in many materials, especially crystals, $\mathscr{E}$ and its induced $\mathbf{P}$ are not necessarily colinear.

The $\chi^{(1)}$ term is dominant and is an extremely good approximation for $\mathbf{P}$ with small applied electric fields. This term applies to all linear optics and yields the common index of refraction, n, and optical dielectric constant, ε, as follows:

$$n = \sqrt{\varepsilon} = (1 + 4\pi\chi^{(1)})^{1/2}. \tag{2}$$

The $\chi^{(2)}$ term (containing 27 elements) gives rise to optical mixing, second harmonic generation, and the Pockels effect. It exists only for crystals lacking inversion symmetry. Otherwise, all components of the tensor vanish. In the general case, $\mathbf{P}$ is written

$$\mathbf{P}_i \begin{bmatrix} \omega_m + \omega_n \\ \omega_m - \omega_n \end{bmatrix} = \chi^{(2)}_{ijk} \mathscr{E}_j(\omega_m)\, \mathscr{E}_k(\omega_n), \tag{3}$$

which relates fields of three different frequencies. If one of the fields is at DC, then $\chi^{(2)}$ gives rise to the Pockels effect, where the input and output frequencies are the same, and the effective $\chi^{(2)}$ becomes a function of the DC field, i.e.,

$$\mathbf{P}_i(\omega_l) = (\chi^{(2)}_{ijk}\mathscr{E}_k(DC))\mathscr{E}_j(\omega_l). \tag{4}$$

Thus, similarly to the $\chi^{(1)}$ term, $\chi^{(2)}$ produces an index of refraction change that depends on the magnitude of the applied field.

$\mathscr{E}_k$ does not have to be at DC, but can actually be extended well into the radio frequency (RF) regime without appreciably changing the frequency of the output field. As long as the input and output frequencies are essentially equal, the change of index is still considered to be the Pockels effect.

In most electro-optic materials, the linear dielectric constant is a strong function of frequency. Figure 1 depicts the general behavior of ε and displays several resonances where ε changes markedly. There are usually several acoustic resonances that depend upon physical parameters of the crystal, such as size and mounting configuration. High-frequency behavior is caused by lattice resonance in the THz regime. Between resonances, dielectric dispersion is negligible.

It is not surprising that the nonlinear coefficients, χ, exhibit similar resonance behavior. Above the lattice resonance, in the optical regime, the only contribution to the electro-optic effect is electronic. Below the lattice

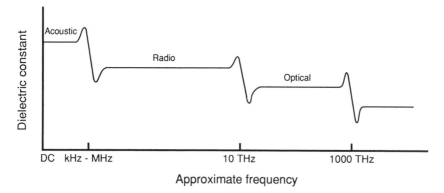

FIG. 1. General behavior of the dielectric constant as a function of frequency for electro-optic crystals. For modulator applications, the upper frequency limit is usually set by the first lattice resonance that occurs in the THz regime.

resonance, the lattice contributes the bulk, ~90%, of the electro-optic effect. Below acoustic resonances, the Pockels effect accrues two more contributions from the elasto-optic and piezo-optic effects. These are experienced only if the crystal is "free" (constant stress) and therefore can be avoided by either physically clamping the crystal or simply operating at a frequency above the acoustic region.

Thus, the mechanism responsible for the electro-optic effect is extremely fast, extending up to 1000 THz. The limiting physical phenomenon, however, is the anomalous behavior of $\chi^{(2)}$ near the lattice resonance. This resonance behavior indicates that the electro-optic mechanism limits temporal response only when electrical-signal characteristics are in the regime of hundreds of femtoseconds. This issue is addressed more specifically later, in Section III.B.

The electro-optic effect is represented as a change in the relative optical impermeability B_{ij}, proportional to the applied electric field $\mathscr{E}_k$. By definition,

$$B_{ij} = \left(\frac{1}{\varepsilon}\right)_{ij} = \left(\frac{1}{n^2}\right)_{ij}. \tag{5}$$

It is through this quantity that the susceptibility, $\chi^{(2)}$, relates to the commonly tabulated electro-optic coefficient r_{ijk}. The resultant tensor relationship is given by:

$$\Delta B_{ij} = r_{ijk} \mathscr{E}_k. \tag{6}$$

In this form, r_{ijk} has 27 components that correspond to the 27 components of the susceptibility tensor $\chi_{ijk}^{(2)}$, From the permutation relations, it can be shown that (ijk) is equivalent to (jik) and, hence, a reduced form of the electro-optic tensor, r_{pk}, can be employed. This tensor contains 18 elements of which many are usually equal or identically zero owing to crystal symmetry properties. p refers to the contracted notation for (ij), i.e., $1 = (11)$, $2 = (22)$, $3 = (33)$, $4 = (23)$ $5 = (13)$, $6 = (12)$. This results in a simplified form of equation (6):

$$\Delta B_p = r_{pk} \mathscr{E}_k , \tag{7}$$

which can be transformed to relate the change in index of refraction to the applied electric field.

$$\Delta B_p = r_{pk} \mathscr{E}_k = \Delta \left(\frac{1}{n_p^2} \right) = -\frac{2 \Delta n_p}{n_p^3} \tag{8}$$

for $\Delta n_p \ll n_p$ as is usually the case. Thus we obtain the common electro-optic relation:

$$\Delta n_p = -\frac{1}{2} n_p^3 r_{pk} \mathscr{E}_k . \tag{9}$$

Two values of r_{pk} are commonly specified. If the crystal is at constant stress, or "free," then the electro-optic coefficient is denoted with a superscript T, e.g., r_{pk}^T. The "free" condition is exemplified at low frequencies, below acoustic resonances, where the elasto-optic and piezo-optic effects can contribute. If r_{pk} is determined at constant strain, or "clamped," the electro-optic coefficient is written with a superscript S. r_{pk}^s is the value of the electro-optic coefficient obtained in the RF regime, i.e., between the acoustic and lattice resonances.

To understand how the tensor index changes, Δn_p, alter the birefringent characteristics of the medium, the index indicatrix or index ellipsoid must be introduced. The indicatrix, a mathematical construct, depicts the anisotropic index properties of birefringent crystals of which electro-optic media are a subset. In these crystals, for any particular direction, only two linearly and orthogonally polarized waves propagate. They usually experience different indices and hence travel at different velocities. In general, the index indicatrix is a triaxial ellipsoid. Any central plane cross section is an ellipse. The lengths of any pair of orthogonal axes within that ellipse are proportional to the indices experienced by waves polarized along those same axes and travelling normal to the plane of the ellipse.

The general form of the index ellipsoid in an arbitrary cartesian coordinate system is

$$\frac{x^2}{N_1^2} + \frac{y^2}{N_2^2} + \frac{z^2}{N_3^2} + \frac{2yz}{N_4^2} + \frac{2xz}{N_5^2} + \frac{2xy}{N_6^2} = 1 \tag{10}$$

where the subscripts of 1–6 again are the reduced form of the cartesian coordinates, $1 = xx$, $2 = yy$, $3 = zz$, $4 = yz$, $5 = xz$, and $6 = xy$. By reorienting the axes, Eq. (10) takes a simpler form:

$$\frac{x^2}{n_1^2} + \frac{y^2}{n_2^2} + \frac{z^2}{n_3^2} = 1. \tag{11}$$

In this case, the major and minor axes of the ellipsoid are oriented along the cartesian axes. These are called the principal axes of the crystal and have corresponding principal indices n_1, n_2, and n_3.

Many common electro-optic crystals are uniaxial. This implies that for one particular direction within the medium, the indices for both polarizations are equal. Thus, the plane cross section of the index ellipsoid normal to this direction is a circle. This direction is usually referred to as the optic axis or c-axis, and in a cartesian coordinate system is oriented along the z-axis. Figure 2 illustrates the index ellipsoid and its principal cross sections for a uniaxial crystal in its principal axis coordinate system. The indicatrix equation in the principal axis system becomes

$$\frac{x^2}{n_o^2} + \frac{y^2}{n_o^2} + \frac{z^2}{n_e^2} = 1. \tag{12}$$

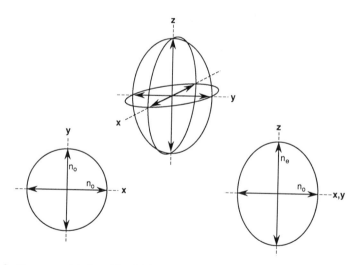

FIG. 2. The general index ellipsoid for a uniaxial crystal in its principal axis coordinate system. In this construct, the lengths of the major and minor axes of any elliptical cross section normal to the direction of propagation are proportional to the refractive index for light polarized along that same axis. Light polarized purely along the crystal's optic axis, or z-axis, experiences the extraordinary index, n_e, while light polarized purely along either the x- or y-axis experiences the ordinary index, n_o.

Light travelling in the x-y plane is resolved into two orthogonal polarizations along the principal axes of the crystal. The polarization component perpendicular to the optic axis is referred to as the ordinary ray and experiences the ordinary index, n_o. The other wave experiences the extraordinary index, n_e, hence is called the extraordinary ray. If light propagates along the z-axis, it experiences no birefringence, because in this direction $n_e = n_o$.

Any electro-optic crystal in its unperturbed state can be represented by Eq. (10). The application of an electric field can distort the indicatrix by changing not only its symmetry, but also its overall size and orientation due to the tensor nature of the electro-optic coefficient.

The previously derived Eq. (8) describes the change of the indices N_1–N_6 with respect to the applied field. The equation can be rewritten in a more convenient form applicable to use with the indicatrix equation as follows:

$$\Delta\left(\frac{1}{n_p^2}\right) = r_{pk}\mathscr{E}_k \tag{13}$$

where, as before, $p = 1, 2, \ldots 6$ and $k = 1, 2, 3$.

2. Intensity Modulation

Coherent light that impinges on an electro-optic crystal resolves into two coherent orthogonal components, each experiencing, in general, a different index. After propagating through a length L, of the medium, a phase difference of δ between the waves is accrued according to the equation

$$\delta = \frac{2\pi}{\lambda} L \Delta n \tag{14}$$

where λ is the free space wavelength and Δn is the birefringence. Thus, the state of polarization of the emergent beam is altered from the input unless δ is an integer multiple of 2π. This change in polarization can be converted to an intensity change by placing the crystal between crossed linear polarizers. By orienting the input polarization to the crystal at 45 degrees to its principal axes, the most efficient change of intensity for a given δ is attained. The resultant intensity modulation function is periodic in δ with period π:

$$I_o/I_i(\sin^2(\delta/2)) \tag{15}$$

where I_o and I_i are the output and input intensities. T is the transmission defined by $T = I_o/I_i$, as shown in Fig. 3. Thus by the application of an external electric field, the birefringence and, hence, δ can be modulated, thereby achieving intensity modulation.

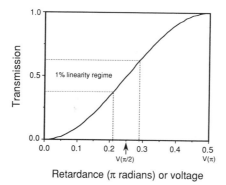

Retardance (π radians) or voltage

FIG. 3. The intensity-transmission function for an electro-optic modulator, $T = \sin^2(\delta/2)$, where δ is the net phase retardance between polarizers. For linear response of maximum sensitivity, the modulator is biased at its quarterwave point $V_{\pi/2}$.

As shown in Fig. 3, the intensity modulator operates best at its 50% transmission point. In this regime, the slope of the modulation function (equivalent to sensitivity) is at its maximum and the response is also linear for small values of induced retardance relative to π. To operate the modulator at this point, the static birefringence (i.e., with no applied field) between polarizers, δ, should be an odd multiple of $\pi/2$ radians. A retardance of $\pi/2$ radians is the same as a retardance of one quarter wavelength and hence 50% transmission is referred to as the quarterwave point. For 1% maximum linearity error, the total retardance including that from the applied field must be less than 0.08π radians.

For many modulator embodiments, the exact static retardance due to the electro-optic crystal alone may not be $\pi/2$ radians. It may not even be easily predetermined or varied. In this case an additional birefringent element, called a compensator, is introduced between the polarizers. The compensator can be adjusted to exactly offset the static birefringence of the electro-optic crystal except for a residual $\pm\pi/2$ radians that bias the modulation system at the 50% point.

Phase retardance (δ) relates to an applied voltage V. Figure 4 shows the general case of a crystal in a transverse geometry, having length L and electrode spacing (thickness) d. The applied electric field is given by

$$\mathcal{E} = \frac{V}{d} \tag{16}$$

where V is the voltage between the electrodes. Once the material and the configuration it will be used in are known, δ can be linearly related to V by

$$\delta = -\frac{\pi}{\lambda} n^3 r \frac{L}{d} V. \tag{17}$$

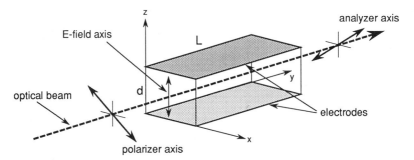

Fig. 4. The general transverse electro-optic modulator configuration. The electric field is applied along the z-axis, and the optical beam propagates normal to the z-axis, polarized at 45°.

Because of this linear relationship, the horizontal axis in Fig. 3 can be labeled by either δ or V.

The sensitivity of a particular intensity modulator to an applied voltage is specified in two related ways. One can specify either the modulator's halfwave voltage, V_π, or the slope of the transmission function at the quarterwave point. The halfwave voltage is the voltage necessary, in a specific geometry, to induce a retardation of $\delta = \pi$ radians and hence achieve 100% intensity modulation. A nongeometry-specific, or reduced, halfwave voltage V_π^* can be defined by

$$V_\pi^* = V_\pi L/d, \tag{18}$$

which is the halfwave voltage necessary for a crystal with $L = d$.

Given the halfwave voltage, the birefringence induced by a voltage, ΔV, can be calculated from the following relation:

$$\delta = \pi \frac{\Delta V}{V_\pi}. \tag{19}$$

The subsequent intensity modulation can then be determined by using Eq. (19) as follows:

$$I_o = I_i \sin^2\left(\frac{\delta}{2}\right). \tag{20}$$

The more direct method by which to specify the sensitivity is merely to determine the slope of the transmission function, $\Delta T/\Delta V$. The slope is easily determined experimentally, and immediately relates an applied voltage to a change in transmission. By evaluating the derivative of the transmission function at the quarterwave point, the slope to the halfwave voltage can be related by

$$\frac{\Delta T}{\Delta V} = \frac{\pi/2}{V_\pi}. \tag{21}$$

Thus, merely by measuring ΔT and knowing either the slope of the transmission function or the modulator's halfwave voltage, there is enough information to determine the amplitude of an unknown applied voltage. (Note: This holds only for $\Delta V \ll V_\pi$.) For this measurement, Eq. (21) is usually rewritten as

$$\Delta V = \frac{\Delta T}{1.57} V_\pi. \tag{22}$$

Additional discussion is found both in Chapter 5 and later in this chapter.

B. ELECTRO-OPTIC MATERIALS AND THEIR PROPERTIES

There are literally hundreds of crystals that exhibit an electro-optic effect, many of which are suitable for electro-optic modulation (Kaminow, 1974). The most common crystals fall into two different categories of modulation configuration: longitudinal and transverse. These designations represent the relative orientations between the electric field applied to the crystal and the direction of propagation of light through the crystal. If these two directions are parallel, then it is referred to as longitudinal modulation; whereas if they are orthogonal, it is called transverse modulation.

Table I compares several crystals suitable for electro-optic measurements. Although many parameters determine whether a particular crystal is useful in any application, the major considerations are sensitivity and linearity over a broad frequency range. In this table, sensitivity is gauged by comparing the reduced halfwave voltages, V_π^*. For those crystals that are best used in a transverse geometry (as marked by an $*$ in the table), one must remember that V_π is inversely proportional to the ratio of length to thickness of the modulator active area. Because the modulator is part of an electrical circuit, the dielectric constant of the material in the direction (i) of the applied field ε_i must also be considered. The table shows that the halfwave voltages of these crystals is roughly inversely proportional to ε_i. Thus, higher sensitivity would imply more electrical loading.

Lithium tantalate ($LiTaO_3$) is a common crystal often used in bulk modulator applications. It is closely related to lithium niobate ($LiNbO_3$), a very popular material for integrated optic waveguide applications. Both lithium niobate and lithium tantalate have rhombohedral crystal structure and are in the point group $3m$; they are most efficiently used as transverse modulators. The crystals grow at high temperatures and, once poled, can

TABLE I

ELECTRO-OPTIC MODULATOR CRYSTAL PROPERTIES.*

Crystal	Group	$\gamma(\mu m)$	ε_i	$V_\pi(kV)$
GaAs	43 m	1.06	13	10.8
KD*P	42 m	0.63	48	3.8
LiTaO$_3$	3 m	0.63	43	2.7*
Sr$_{0.75}$Ba$_{0.25}$Nb$_2$O$_6$	4 mm	0.63	3400	0.047*
ZnTe	43 m	0.63	10.1	2.7
ZnO	6 mm	0.63	8.2	19.4*

Source: Data compiled in R.J. Pressley, ed. (1971), *Handbook of Lasers*, 447–459.

* Comparison of properties for a variety of crystals used as electro-optic modulators. Note those with halfwave voltages marked with an asterisk are optimally used in the transverse geometry and, as such, have a half-wave voltage calculated for a modulator with unity [length ÷ thickness].

be readily fabricated using conventional techniques. Environmentally stable, they are cleaned using common solvents. Both crystals exhibit superior electro-optic coefficients relative to other materials. An advantage of lithium tantalate is that it has a static birefringence 18 times less than that of lithium niobate. The smaller birefringence relaxes fabrication tolerances, enables a larger entrance cone angle when focusing through the crystal, and also reduces multiple-order retardance effects for large-bandwidth optical pulses. Lithium tantalate is also much less susceptible to optical damage when exposed to intense laser beams than is niobate.

Most of the work discussed in this chapter used lithium tantalate as the electro-optic medium, because of its high sensitivity, ease of fabrication, transparency in the visible and IR, and high-frequency response. Lithium tantalate has been fabricated into a wide variety of modulator configurations ranging from hybrid travelling-wave cells to external noncontact probes. It has also proven itself at cryogenic temperatures. The details of electro-optic modulation in lithium tantalate are addressed in the next section.

KDP and its isomorphs are in the point group $\overline{4}2m$ and, as such, are used effectively as longitudinal modulators. They are widely used in large-aperture, bulk-modulator applications where nanosecond switching speeds are adequate. They are readily obtained and exhibit excellent optical qualities of transparency, freedom from strains, and high resistance to optical damage. However, KDP-type crystals grow from a water solution and hence are soft. They require special handling and must be environmentally isolated to prevent surface degradation. Even though ADP and deuterated KDP have good electro-optic coefficients, handling problems make their experimental use difficult. As of the late 1980s, these crystals have had only limited use in electro-optic sampling applications.

GaAs, a semiconductor, is not usually considered an electro-optic modulator material, but it is included for comparison in Table I because of its unique role in the technique of internal electro-optic sampling, as discussed in detail elsewhere in this book. In this case, GaAs integrated circuits have the fortuitous propety that the substrate they are fabricated on itself is electro-optic. Thus, the substrate essentially becomes a built-in modulator for probing internal points of these circuits.

GaAs, in point group $\bar{4}3m$, is most effectively used as a longitudinal modulator and thus exploits fields in the circuit substrate that are normal to the circuit faces. In this geometry, if a circuit line is probed from the back of the substrate, the intensity change of the probing light is proportional to the integrated electric field between the two surfaces of the circuit and that corresponds to the actual voltage on the line being measured. One drawback to this technique is that both surfaces of the circuit must be of optical quality.

The details of lithium tantalate as it is employed in most modulator configurations is discussed next. The electro-optic tensor, r_{pk}, and its component values are shown in Fig. 5.

As mentioned before, the largest electro-optic effect in lithium tantalate is realized in the transverse geometry when the electric field is applied along the z-axis and light is propagated in the x-y plane normal to the

$$r_{pk} = \begin{bmatrix} 0 & -r_{22} & r_{13} \\ 0 & r_{22} & r_{13} \\ 0 & 0 & r_{33} \\ 0 & r_{51} & 0 \\ r_{51} & 0 & 0 \\ -r_{22} & 0 & 0 \end{bmatrix}$$

$r_{12} = -r_{22} = r_{61}, \ r_{13} = r_{23}, \ r_{42} = r_{51}$

$r_{33} = 30.3 \times 10^{-12} \text{ m/V at } 632.8 \text{ nm}$
$r_{13} = 7.0 \times 10^{-12}$
$r_{51} = 20.0 \times 10^{-12}$
$r_{22} \approx 1.0 \times 10^{-12}$

FIG. 5. The electro-optic tensor, r_{pk}, and its component values for lithium tantalate, LiTaO$_3$, when used with light at 632.8 nm.

electric field. Applying Eq. (13) to the principal axis form of the index indicatrix for this electric-field direction yields

$$\left(\frac{1}{n_o^2} + r_{13}\mathscr{E}_z\right)(x^2 + y^2) + \left(\frac{1}{n_e^2} + r_{33}\mathscr{E}_z\right)z^2 = 1 \qquad (23)$$

where

$$\Delta\left(\frac{1}{n_o^2}\right) = r_{13}\mathscr{E}_z \quad \text{and} \quad \Delta\left(\frac{1}{n_e^2}\right) = r_{33}\mathscr{E}_z. \qquad (24)$$

For small perturbations, the direct index changes are

$$\Delta n_o = -\frac{1}{2}n_o^3 r_{13}\mathscr{E}_z \qquad (25)$$

where

$$(\mathscr{E}_Z = \mathscr{E}_3)$$

and

$$\Delta n_e = \frac{1}{2}n_e^3 r_{33}\mathscr{E}_Z. \qquad (26)$$

The change of the indicatrix is depicted in Fig. 6. For a positive $\mathscr{E}_Z$, both principal indices decrease, and no cross terms are introduced. Only the size and symmetry of the ellipsoid change. No rotation is induced, and it is still rotationally symmetric about the optic axis. This is an important property for velocity-matching considerations. The field-induced retardance is

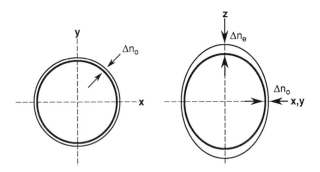

FIG. 6. The change in the principal cross sections of the index ellipsoid of lithium tantalate. Note the change in the x-y plane is isotropic; hence, no phase retardance is experienced for light travelling in the z direction. The largest effect is seen by light travelling normal to the z-axis where the phase retardance is proportional to $(\Delta n_e - \Delta n_o)$.

written as

$$\delta = \frac{2\pi}{\lambda}(\Delta n_e - \Delta n_o)L = \frac{\pi}{\lambda}(n_e^3 r_{33} - n_o^3 r_{13})\, V \frac{L}{d}. \tag{27}$$

Using this equation for δ and Eq. (18), V_π^* for the crystal can be calculated as follows:

$$V_\pi^* = \frac{\lambda}{(n_e^3 r_{33} - n_o^3 r_{13})}. \tag{28}$$

This yields a $V_\pi^* = 2580$ V for lithium tantalate, where $\lambda = 630$ nm.

The static birefringence introduces two effects that can affect proper operation of the intensity modulator. The temperature-dependence of the birefringence necessitates that for an intensity modulator to remain linear to within $\pm 0.05\%$, the temperature of a one-mm–long crystal must be maintained within one degree Celsius. Short sampling pulses that have a large optical bandwidth pose the second problem. In this case, all wavelengths do not experience the same amount of retardance and hence have different polarizations. If the wavelengths all fall within a 0.5 radian range, then this is of minimal consequence because they are all still on the linear portion of the transmission function (Eq. 15). If the retardance is too great, the wavelengths can emerge in different retardance orders and render the modulator ineffective. Both birefringent effects can be compensated for by introducing a compensating crystal as mentioned earlier, to "unwind" the accumulated static retardance.

III. Principles of Electro-Optic Sampling

This section describes the optical and electronic systems that are part of the generalized electro-optic sampling system (Valdmanis and Mourou, 1986) and discusses characteristics that limit the temporal resolution and sensitivity of the technique. The most common types of electro-optic modulator configurations are introduced, and methods for synchronizing the optical system to the electrical system under study are considered.

A. The Optical System

1. The Sampling Principle

Figure 7 depicts the general layout of an electro-optic sampling system. The active element in the system is a high-repetition-rate ($\sim$100-MHz) picosecond or subpicosecond pulse laser system. The optical pulse train is divided into two beams. One beam repetitively triggers the generation of the electrical signal to be measured, while the other beam synchronously

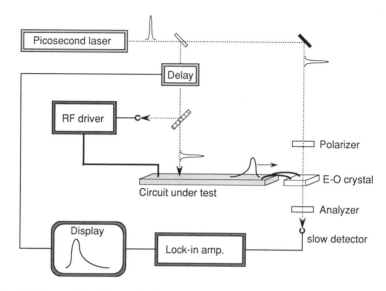

FIG. 7. A generalized schematic of the electro-optic sampling system. The figure shows a generic circuit under test connected in some non specific way to an electro-optic crystal. The general principle applies to any of the specific modulators described in the text. The circuit can be driven in either of two ways: (1) directly by an optical pulse or (2) by a standard RF source synchronized to the laser pulse train.

samples the field-induced birefringence in the electro-optic medium due to the electrical signal. The optical trigger beam passes through a modulator operated at a frequency below that of the laser repetition rate to facilitate lock-in detection of the sampling beam. The trigger beam generates the electrical signal to be measured either directly from a photosensitive device or indirectly by using the electrical pulse generated by a photodetector to drive the electrical device under study. A microwave or RF synthesizer can also be synchronized with the laser to drive devices and circuits.

The sampling optical beam passes through a variable delay line that is synchronized with a display device to provide scanning of the electrical signal. The sampling pulse train is then focused through the electro-optic modulator where the intensity of each pulse is changed in proportion to the amplitude of the segment of the electrical signal that it encountered as it passed through the electro-optic medium. A variable compensator between the polarizers offsets the the static birefringence of the lithium tantalate and optically biases the modulator for 50% transmission, as discussed earlier. The sampling beam intensity is measured by a conventional "slow" photodiode that does not resolve individual optical pulses, but whose speed of response is dictated only by the frequency of the trigger beam modulator. A lock-in or tuned amplifier measures the amplitude of the modulated intensity and yields a DC output proportional to the amplitude of the sampled electrical signal. The output signal is plotted as a function of the optical delay, thus yielding an equivalent time representation of the electrical signal under study. If necessary, further signal-to-noise enhancement can be achieved by averaging successive traces.

In the preceding description, the coupling between the circuit or device under study and the modulator is represented arbitrarily. The specific configuration used depends on the particular device and associated modulator chosen, as will be elaborated upon in Section III.C on modulators and in Section IV on applications. All electro-optic modulator configurations described in this chapter, including the external electro-optic prober, are based on this fundamental optical system.

The flexibility of this system comes not only from the variety of modulator arrangements, but also from the freedom to choose almost any type of high-repetition-rate picosecond laser source.

2. Picosecond Laser Sources

Today, there are many lasers that produce picosecond optical pulses at wavelengths spanning the entire visible and near-infrared spectrum. Their level of complexity is also large. However, the laser chosen is usually quickly identified by the requirements of the sampling system and the circuits under study. The most common parameters considered are the temporal resolution required, the method of synchronization between the laser and the circuit under study, the wavelength to be used, the system's ease of use, and the cost.

Most picosecond lasers used for electro-optic sampling are one of three types: dye, solid-state, or diode. They are vastly different in their operation, but each has distinct advantages. Only a cursory overview of these broad classes is given here. For a detailed description of these and other systems, refer to Kaiser (1988).

Visible-dye lasers offer the shortest pulses available to date, <100 fs, and thus enable unprecedented temporal resolution with some electro-optic configurations (<300 fs). However, dye lasers require another laser (typically an ion laser) as a pump source and are quite expensive and complex.

Mode-locked solid-state lasers (e.g., Nd:YAG) are simpler and more reliable, but they typically operate in the infrared range and require elaborate pulse compression and stabilization techniques (Rodwell et al., 1986) to attain picosecond resolution in electro-optic sampling systems.

Diode lasers are compact and power efficient. They can generate pulses ~15 ps long by gain-switching techniques (White et al., 1985) and as such

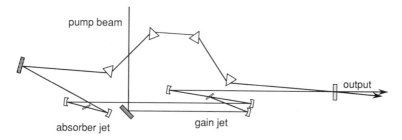

Fɪɢ. 8. The balanced colliding-pulse–mode-locked ring dye laser. Pumped by a CW argon-ion laser, this passively mode-locked laser produces pulses as short as 30 fs, at 630 nm, at a repetition frequency of ~100 MHz. The average output power is ~30 mW per beam. The prisms are used to balance out the dispersive effects of group velocity dispersion and self-phase modulation present in the cavity.

can have continuously variable repetition rates and low jitter up into the GHz regime (Taylor et al., 1986b). Their use is limited, however, by temporal resolution and low-power IR operation.

All the applications described in this chapter use the colliding-pulse–mode-locked (CPM) ring dye laser (Valdmanis et al., 1985). This laser, depicted in Fig. 8 is one of the most stable, reliable sources of sub-picosecond optical pulses available. It routinely generates 100-fs pulses at ~100-MHz repetition rates with 10–50-mW average power per output beam. Because it is a ring laser, it conveniently generates two precisely synchronized pulse trains, ideal for electro-optic sampling systems. The CPM laser uses dyes dissolved in a solvent for the gain medium. Best performance is achieved with Rhodamine 6G (Rhodamine 590), which lases in the red at 630 nm when pumped with an argon ion laser. The ion pump laser operates CW, but the pulses in the dye laser are formed by passive mode-locking with another dissolved dye solution, typically DODCI. Pulse duration and shaping in the cavity is also controlled by a prism sequence that allows variability of the intracavity dispersion, a critical parameter for subpicosecond pulses. Because the laser is passively mode-locked, it does not offer wavelength tunability, but red light is well suited to use with lithium tantalate as the electro-optic medium. The pulse repetition rate is not easily adjusted over large ranges, but operation at 100 MHz is ideal for synchronization to many devices, circuits, and waveform synthesizers as is described in Section III.E.

B. Temporal Resolution

An ideal intensity modulator would impress an electrical signal onto an optical beam with a one-to-one correspondence between temporal points.

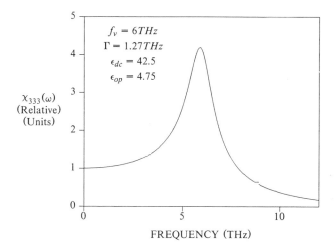

FIG. 9. The resonant behavior of χ_{333} in lithium tantalate. The lattice resonance at 6 THz also limits the frequency response of the linear electro-optic effect in the RF regime which , in turn, limits the ultimate temporal resolution of a lithium tantalate modulator to ~300 fs. (Figure courtesy of D. Auston, Columbia University)

Such a device would have infinite bandwidth and infinitely narrow temporal resolution. Experimentally, however, the temporal resolution of any electro-optic sampling configuration depends on several factors that are discussed in order of increasing limitation.

First, consider the intrinsic frequency response or, conversely, the temporal resolution of the electro-optic material employed. Usually this limit is dictated by the onset of absorption in the terahertz regime due to a lattice resonance, as discussed in Section II.A.1. Specifically, it was shown that in lithium tantalate, a vibrational resonance exists at 6.3 THz (Cheung and Auston, 1986). This results in an absorption coefficient of 10 cm^{-1} at 0.5 THz, which implies that for signals with temporal characteristics under one ps, propagation distances must be limited to 0.23 cm to retain even 10% of the original signal. The electro-optic effect is also enhanced due to the resonance, as shown in Fig. 9, but in this regime of strong absorption, it cannot be exploited. The resonance limits the response time of the electro-optic effect via the resonance damping rate (1.1 THz) at approximately 300 fs, which was experimentally corroborated with the external electro-optic probe. (See Section V.B.2.) However, in most applications, this intrinsic material frequency response well exceeds the limitations of the following other factors.

Secondly, consider the resolution of the modulator interaction geometry. Its temporal resolution is determined by the convolution time of the optical probe pulse and the travelling electrical signal as they copropagate through the electro-optic material. If these two signals travel orthogonally, as shown in Fig. 10, the temporal resolution is merely the time, τ_o, it takes for the optical sampling pulse to traverse the active region of the crystal, h_e, convolved with the transit time of the electrical signal, τ_e, across the optical sampling beam waist ω_o. Thus, temporal resolution can be improved by simply using shorter optical pulses and tighter focusing of the sampling beam. The convolution time can be further minimized by implementing a velocity-matched geometry. In the matched geometry, the optical beam travels through the electro-optic medium at an angle such that the optical velocity in the direction of the travelling electric field and the velocity of that travelling field are the same. Under these conditions, the interaction distance of the optical and electrical signals is irrelevant and the temporal resolution is determined entirely by the temporal and spatial extent of the optical sampling pulse in the crystal. For example, by using 100-fs optical pulses focused to a 10-μm spot in lithium tantalate under velocity-matched conditions, a temporal resolution of less than 400 fs can be expected.

The third and most serious limitation for temporal resolution, common

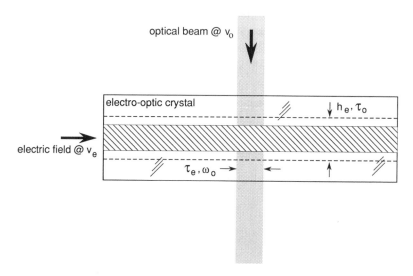

Fig. 10. Schematic diagram for visualizing optical and electrical transit times in a travelling-wave electro-optic modulator. The electrical transit time, τ_e, is the time it takes for the electrical signal, travelling at v_e, to cross the optical beam waist ω_o. The optical transit time, τ_o, is the time it takes for the optical beam, travelling at v_o, to cross the extent of the applied electric field, h_e. h_e is wider than just the electrode width because of fringing field contribution.

to all measurement techniques, is imposed by electrodes used to convey the signal under study from its generation point to the sampling point. Although travelling wave electrodes are employed because they neatly avoid the effects of slow, lumped capacitance element it must be noted that at the extremely high frequencies encountered in picosecond signals, dispersive effects due to the electrode geometry become significant and can render subpicosecond resolution useless. The applications section shows how these dispersive effects manifest themselves and how they can be avoided by limiting propagation distances or by the unique ability of electro-optic sampling to characterize the electrical signal *in situ* with electro-optic probes or electro-optic substrates.

C. MODULATOR GEOMETRIES

Most electrical signals that we are interested in measuring are generated by high-speed devices. As such, these devices, and integrated circuits operating in the GHz frequency regime, were designed with electrode structures that can support and propagate the bandwidth of frequencies required. Most commonly, these structures are either "microstrip," coplanar-strip, or balanced-strip transmission lines. Unless the circuit dimensions are extremely small, as in integrated circuits, the electrodes operate as travelling-wave structures and avoid lumped circuit element capacitances and inductances. Coaxial geometries are usually avoided because of the geometric difficulty in coupling to two-dimensional device structures. Because of this, electro-optic modulator geometries that adapt to two-dimensional devices and ciruits are discussed.

Three basic groups of electro-optic modulator configurations have evolved: hybrid, internal, and external. The hybrid modulators attach directly to the device under study and hence become an integral part of the electrical circuit. These modulators usually have an electrode structure that is matched to the device under study forming a hybrid arrangement. Internal and external configurations have been developed for application to integrated circuits where it is not possible to physically attach conventional measurement devices. Integrated circuits pose unique measurement problems because electrode structures are exceedingly small ($<10 \ \mu m$). This makes it very difficult to use conventional metal contact probes at arbitrary internal points. A complete discussion of these integrated-circuit probing techniques is given in Section V of this chapter.

The remainder of this section concentrates on describing modulator configurations commonly encountered. Figure 11 shows four types of electro-optic modulators commonly used in hybrid configurations. Figure 11c is also used for internal electro-optic sampling on GaAs circuits. Figure 12

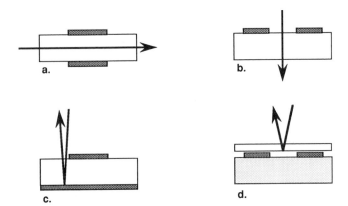

FIG. 11. Four types of travelling-wave hybrid electro-optic modulator used for ultrafast pulse measurement. (a) Transverse balanced-line modulator. (b) Transverse coplanar-stripline modulator. (c) Longitudinal microstrip modulator. (d) Transverse superstrate modulator. The configuration in (c) is also used for electro-optic probing in the substrate of GaAs circuits.

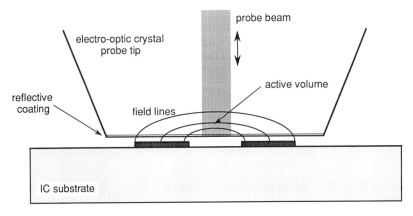

FIG. 12. The basic external electro-optic probe tip used for noncontact probing of integrated circuits on any type of substrate material. The figure shows a transverse-type electro-optic tip dipped into the fringing field between two coplanar-strip transmission lines. The probe beam is reflected at the tip face as it passes through the applied electric field. A typical tip is ~50 μm square at the apex.

shows the configuration for external probing. The performance capabilities of all the following modulators are addressed in Section IV.A. All modulator configurations described (except the longitudinal microstrip modulator in GaAs) have attained subpicosecond temporal resolution by virtue of simply minimizing the optical, τ_o, and electrical, τ_e, transit times. The following describes the pros and cons of each modulator design.

1. Transverse Balanced-Line Modulator

The transverse-field, balanced-line modulator (as shown in Fig. 11a) is a conventional travelling-wave modulator with equal size electrodes on either side of an electro-optic substrate. Initial sampling systems (Vald-manis et al., 1982) used this modulator design with lithium tantalate crystals having dimensions 0.5 mm thick by 0.7 mm wide by 25 mm long with the crystal optic axis, or z-axis, oriented normal to the electroded faces. Electrodes are slightly narrower than the width of the crystal and result in an impedance of ~20 Ω. The sampling beam is focused through the crystal and linearly polarized at 45 degrees to the z-axis. Its angle of incidence with the face of the crystal can be adjusted to realize velocity matching and hence minimize the electrical–optical convolution time. In many applications, this type of modulator is used as a separate sampling head whereby "slow" electrical signals, i.e., those having a bandwidth less than ~40 GHz, can be applied to the modulator through high-speed co-axial connectors. For higher-speed sampling, connectors are avoided, and the modulator crystal is attached directly to the device generating the signal to be measured. This hybrid design, although much faster, quickly reveals the dispersive nature of striplines when the wavelength of the electromagnetic signal is of comparable size to the separation of the electrodes. These dispersive effects are elaborated upon later. To minimize dispersion so as to measure subpicosecond waveforms with this geometry, it is necessary to reduce the crystal thickness and the electrode separation to approximately 0.1 mm.

As shown in the applications section, this modulator type is still used when performing measurements on discrete devices that have electrodes on opposite faces of the substrate.

2. Transverse Coplanar-Line Modulator

An alternate method of fabricating travelling-wave modulators that avoids the difficulty of a balanced line on a thin substrate is the transverse-field, coplanar line modulator (Mourou and Meyer, 1984) as depicted in Fig. 11b. Coplanar waveguide and coplanar strip transmission lines have been used frequently in microwave integrated circuit applications. They are generally more easily fabricated than balanced microstrip lines because both signal and ground lines can be made in the same process step on the same surface of a relatively thick substrate. An even greater advantage, however, is that if the electrodes are fabricated photolithographically, the electrode dimensions can be reduced to a few microns. In particular, the electrode separation, which is analogous to the substrate thickness in the balanced microstrip geometry, can be made more than an order of magnitude smaller than is possible for the balanced stripline. This implies that the dispersion for a given signal is correspondingly decreased. The coplanar stripline consists of two parallel lines (as in the figure) while coplanar waveguide consists of three. In the waveguide, the center line carries the signal, while the two outer lines are grounded and are often extended as semi-infinite planes for better performance.

In the coplanar geometry, the crystal optical axis of the lithium tantalate is in the plane of the electrodes parallel to the direction of the electric field between the electrodes. However, contrary to the standard configuration for integrated optical modulators, the probe beam is perpendicular to the electrode plane and again linearly polarized at 45 degrees to the z-axis. Nevertheless, this geometry implements the same electro-optic coefficients as for the balanced-line case.

3. Longitudinal Microstrip Modulator

Another configuration that uses the electro-optic effect in the substrate of a travelling wave transmission line is the longitudinal microstrip modulator shown in Fig. 11c. The microstrip electrode geometry is similar to the balanced stripline in that the electrodes are on opposite sides of a substrate, but the microstrip is simplified by use of a common, planar ground electrode covering the entire bottom surface. The longitudinal field sampler exploits the fact that this electrode structure has a fringing field adjacent to the electrodes. The sampling beam enters the substrate in the fringing field from above and is then reflected back up from the ground-plane electrode. In this way, the optical beam is parallel to the electric field in the crystal, and the longitudinal electro-optic effect is realized.

Longitudinal electro-optic sampling is conveniently used with GaAs as the electro-optic medium and 1.06-μm picosecond pulses as the sampling beam (Kolner and Bloom, 1986). (See also Chapter 5.) In this arrangement, the GaAs substrate is cut for the standard $\langle 100 \rangle$ orientation. That is, electrodes and devices are formed on the (100) face and the sampling beam travels parallel to the [100] axis, polarized either parallel or perpendicular to the [001] axis. Pulses of 1.06 μm are used because GaAs is absorbing in the visible wavelengths, thus necessitating the use of radiation with a photon energy below that of its band gap. This embodiment makes possible fabrication of GaAs electronic devices in the same material as is being sampled, thus avoiding a discontinuity in transmission-line struture. Longitudinal sampling in GaAs also eliminates the need for a variable compensator because there is no static birefringence for zero applied field. However, a fixed compensator (quarterwave plate) is still required to bias the modulator at its 50% transmission point. Sensitivity improves compared to a transverse balanced line of the same impedance in lithium tantalate using 1.06-μm sampling pulses.

4. Transverse Superstrate Modulator

In general, ultrahigh-speed circuits and devices are better characterized by not disturbing the electrode structures they were designed to operate in. However, unless the circuit has an optically accessible electro-optic substrate, an external electro-optic crystal must be employed. The transverse superstrate modulator (see Fig. 11d) is an electro-optic modulator that relies on the placement of a thin plate of electro-optic material near the surface of the circuit or device to be evaluated (Meyer and Mourou, 1985). The plate has a dielectric reflective coating on the bottom surface through which the electrical signal in the circuit generates a fringing field that penetrates slightly into the electro-optic material above. The birefringence that is consequently induced is sampled in reflection mode from above, similar to the arrangement used for microstrip sampling. The crystal and sampling-beam orientations are the same as for the transverse coplanar modulator. The absence of electrodes on the electro-optic crystal enables a complete two-dimensional investigation of the electrical signal in the circuit below.

Preliminary studies of this embodiment employed electronic devices fabricated with coplanar electrode structures. In this way, the electric field produced is parallel to the surface of the crystal and, hence, the transverse sampling geometry is realized. With lithium tantalate as the electro-optic medium, 100-fs optical pulses in the visible wavelengths can be used for sampling and, hence, the temporal resolution is limited only by the convolution time through the region of induced birefringence. Because the fringing field only penetrates a short distance, subpicosecond resolution can be attained.

5. External Probe Tip Modulator

For integrated circuit applications, a configuration known as external electro-optic probing has been developed (Valdmanis and Pei, 1987; Valdmanis; 1987; Nees and Mourou, 1986). External electro-optic probing uses an extremely small electro-optic crystal as a proximity electric-field sensor near the surface of an integrated circuit, as shown in Fig. 12. This technique exploits the open electrode structure of two-dimensional circuits where there is a fringing field above the surface of the circuit between metallization lines at different potentials. "Dipping" an electro-optic tip into a region of a fringing field induces a birefringence change in the tip that can be measured from above by an optical beam directed through the tip. In this way, the electro-optic tip is employed as the modulator in a conventional electro-optic sampling system. Because the electro-optic medium is separated from the integrated circuit, we call this technique *external* electro-optic sampling. By using an external electro-optic medium, the sampling system does not rely on any optical properties of the integrated circuit itself, hence making it generally applicable to a wide variety of circuit embodiments. This technique also allows the use of electro-optic crystals that are transparent in the visible portion of the spectrum, thus enabling use of visible laser sources. This technique is discussed in more detail with several applications later in this chapter.

D. DETECTION AND SENSITIVITY

Because of the ultrafast response of the electro-optic effect, Pockels cells are routinely used as high-speed optical modulators in many applications even though they generally possess large (typically kilovolts) halfwave voltages, i.e., the applied voltage necessary for 100% modulation. As such, they were also used to measure kilovolt picosecond pulses (Auston and Glass, 1972). The high voltage requirement would seem to preclude the use of electro-optic techniques for picosecond characterization of devices that typically generate peak signals of less than one volt and, in turn, would modulate a transmitted optical beam by roughly 0.1%. To detect minute intensity modulations, electro-optic measurement techniques rely upon the high sampling rates, ~100 MHz, afforded by picosecond laser systems. At these rates, it is possible to employ extremely powerful electronic signal enhancement techniques such as lock-in detection and signal averaging. This section discusses the effects of noise on the sensitivity of the modulator and detection system.

Ultimately, noise limits the detectability of signals. Figure 13 depicts the simplified detection scheme used in this sampling system. A reverse biased photodiode receives the DC optical signal intensity, I_s, from the modulator and converts it into a proportional current, i_s, that flows in the detection

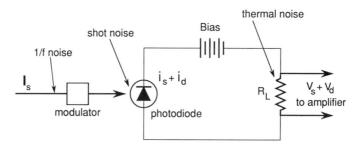

FIG. 13. Schematic diagram of a simple detector circuit showing I_s, the optical signal intensity; i_s and V_s, the signal current and voltage; i_d and V_d, the dark current and voltage; R_L, the load resistance; and the three basic sources of noise: $1/f$, shot, and thermal.

circuit. The current flows through the load resistor R_L, and generates a voltage, V_s. The photodiode also generates a dark current, i_d, that produces a voltage, V_d, at the load resistor. The total output voltage, V_L is given by:

$$V_L = V_d + V_s. \tag{29}$$

The fundamental detection limit is set by shot-noise and thermal noise. In this system, there are two components to shot-noise: one arising from the signal current, i_s, and the other from the constant dark current, i_d. Therefore, the shot-noise for this scheme is given by

$$i_{noise}(rms) = [2qB_f(i_s + i_d)]^{1/2} \tag{30}$$

where q is the electron charge (1.6×10^{-19} coulombs) and B_f is the bandwidth of the measurement system. The current noise is converted to a voltage fluctuation by applying Ohm's law as follows:

$$V_{shot}(rms) = R_L i_{noise}(rms). \tag{31}$$

The thermal noise voltage is calculated from:

$$V_{thermal}(rms) = (4kTR_L B_f)^{1/2} \tag{32}$$

where k is Boltzmann's constant (1.38×10^{-23} J/°K) and T is absolute temperature in degrees Kelvin.

The dominant source of signal noise is known as ($1/f$) or flicker noise. It has a spectrum whose amplitude increases toward DC so that there is equal noise power per decade of frequency. Noise with this spectral dependence is commonly referred to as pink noise. Pink noise arises from many sources and is usually noticeable, and disturbing, at frequencies below one kHz. Typical sources of this noise in the measured signal are temperature fluctuations, air currents, powerline voltage fluctuations, and variations in laser intensity. In the electro-optic sampling system, the dominant source of $1/f$ noise is laser intensity, I_s (DC–100 MHz). Although the amplitude of this noise cannot be mathematically predicted, it is the most significant experimental obstacle.

The noise amplitude measured depends on bandwidth and hence can be characterized by a noise-voltage spectral density, v_n. v_n is the noise voltage per unit bandwidth and is written

$$v_n = \frac{V_{noise}}{B_f^{1/2}} \; (\text{volts}/\sqrt{\text{Hz}}). \tag{33}$$

For white noise, v_n is constant with frequency whereas pink noise has a v_n that drops off at 3 dB/octave for increasing frequency. Noise power (voltage squared) per unit bandwidth is also commonly specified and is defined by

$$v_n^2 = \frac{V_{noise}^2}{B_f} \, (\text{volts}^2/\text{Hz}). \tag{34}$$

The usual method for specifying how much noise is present in a certain signal is the signal-to-noise ratio (S/N). It is the ratio, in decibels, of the rms voltage of the desired signal to the rms voltage of the noise that is also present. Mathematically, the S/N is defined as

$$S/N = 20 \log \left[\frac{V_s(\text{rms})}{V_n(\text{rms})} \right] d \, B_f \tag{35}$$

Because all the preceding noise sources are unavoidable, an electronic system to limit their effects must be used. As explained, the amplitude of the measured noise voltage depends upon the bandwidth of the detection system. The simplest means to reduce the effects of noise, then, is to reduce the detection bandwidth. For white noise, this technique works equally well for any frequency, but for pink noise, power per unit of bandwidth increases toward DC. Thus, the bandwidth must be narrow and the central frequency of the passband must be away from DC. For $1/f$ noise, as described before, a center frequency near one kHz should avoid the largest effects. The lock-in amplifier is specifically designed for this purpose.

The bandwidth of a lock-in amplifier is controlled by its time constant (TC) setting that controls the bandwidth of the final filtering stage. TC is simply related to the bandwidth by

$$B_f = \frac{1}{8(TC)} \tag{36}$$

for a filter with a 12-dB/octave roll-off. It can be shown that to increase V_s/V_n by n times for white noise, the detection bandwidth must decrease by a factor of n squared. This relationship implies that for a 20-dB S/N improvement, the bandwidth must decrease by a factor of 100. A typical operating TC is 300 ms, which corresponds to a bandwidth of 0.42 Hz. For a photodiode circuit with an upper frequency limit of 2 kHz, for example, a TC of 300 ms reduces the bandwidth by 2000/0.42 or 4762. V_s/V_n is enhanced by a factor of $\sqrt{4762} = 69$, which implies a S/N enhancement of 37 dB. If the entire 350-MHz bandwidth of a fast photodiode is the input to the lock-in amplifier, a bandwidth of 0.42 Hz will enhance the S/N by almost 90 dB.

The S/N can be further increased by repetitive signal averaging. A signal averager is basically a storage oscilloscope that averages successive traces together. If the desired signal is synchronized with the sweep of the averager, the signal will add coherently for integrated sweeps while random noise fluctuations will add incoherently. This process is equivalent to bandwidth narrowing. Thus, analogous to the lock-in amplifier, V_s/V_n increases with the square root of the number of sweeps.

The signal-recovery technique of bandwidth narrowing, however, is not free. Restricting the bandwidth produces a trade-off between S/N and the time taken to attain that particular S/N. To achieve a specific noise level, either a long TC (narrow bandwidth) must be set and a few sweeps taken very slowly, or many scans must be repeated and averaged when using a shorter TC.

Although $1/f$ noise from the laser is significantly reduced by the preceding techniques, in this system, an additional noise-reduction technique using differential detection is possible. Two matched detectors are used. Each senses one of the orthogonal polarization components that are separated by the analyzing polarizer. The outputs are subtracted by the differential input amplifier of the lock-in detection system. This way, common mode noise (most significantly laser-intensity fluctuations) is conveniently rejected. An added advantage is that the intensity change for an applied electro-optic signal modulates the intensity of each polarization component with opposite polarity. Thus, upon subtraction at the amplifier, the signal of interest is twice as large.

Initial sampling systems used standard mechanical choppers to modulate the trigger beam at kilohertz frequencies where only laser noise played a significant role. Typical sensitivity (signal-to-noise ratio, S/N, of unity) at a modulation frequency of 2 kHz for lithium tantalate modulators having a halfwave of ~2 kV is approximately one $mV\sqrt{Hz}$. This implies that a one-volt signal measured with a bandwidth of one Hz achieves a S/N of 1000. By increasing the modulation frequency into radio frequencies (Kolner and Bloom, 1986), the sensitivity can be enhanced further still. At about 10 MHz, typical laser systems are so quiet that detector shot-noise limited sensitivity of nearly 10 $\mu V\sqrt{Hz}$ can be attained. This implies that, for a 2-kV halfwave voltage, it is possible to detect an intensity modulation of 10^{-8}, which results in index changes in the electro-optic medium of only 10^{-11}. The ability to sample voltages in excess of one kV gives the electro-optic sampling system a dynamic range of eight orders of mangitude or 160 dB.

E. DEVICE SYNCHRONIZATION

Because electro-optic sampling uses picosecond optical pulses as "sampling gates," it is imperative that the signal under measurement be not only repetitive, but also precisely synchronized with the optical pulses. Depending upon the device under study, this can be implemented in one of two ways: optical triggering or radio-frequency synchronization.

The optical-triggering technique is schematically represented in Fig. 7, where a photo-sensitive device is triggered by a portion of the same pulse train used to sample the modulator. This jitter-free technique assures precise phase-stability between trigger and sample. Because it is an entirely optical technique, the relative delay between trigger and sample must also be introduced optically. Optical delays are commonly implemented by stepper-motor–driven delay lines incorporating multiple pass retroflecting optical elements. The driver is synchronized with the display unit to enable temporal mapping of the sampled signal.

Indirect optical triggering can also be used for devices or circuits that are not photo-sensitive. In this case, an external optical detector senses the triggering optical pulse train. Then, its output drives the device under test. Single-nanosecond pulses with amplitudes near one volt are easily generated with commercially available photodiodes.

A more conventional method of driving devices, particularly integrated circuits, is to use high-frequency continuous-wave signals, usually radio-frequency or microwave sinusoids generated by electronic waveform synthesizers that can control amplitude, phase, and frequency. For electro-

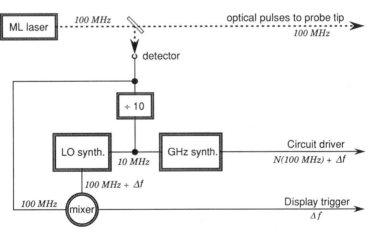

FIG. 14. One method of synchronizing CW RF and microwave synthesizers to a mode-locked laser. Part of the laser pulse train at 100 MHz is split off, detected, and divided by 10 to act as an external clock for a pair of synthesizers. The high-frequency synthesizer is operated at an integer multiple of the laser repetition plus an offset, Δf. This produces a natural scanning of the electrical signal by the optical pulses at the beat frequency, Δf. A trigger signal at Δf for the display device is generated by mixing the fundamental 100 MHz with another synthesizer operating at 100 MHz + Δf.

optic sampling, there is an added requirement that the synthesizer be precisely phase-locked to the repetition frequency of the optical pulses used. This requires a synthesizer that can accept an external reference frequency input and lock to it with extremely low phase jitter (i.e., low phase noise). For picosecond stability, this usually necessitates top-quality equipment.

Pulsed lasers can be phase-locked to synthesizers in two ways. If the laser itself is actively driven by a radio-frequency waveform to produce pulses (as in active mode-locking or gains-switching), then a master clock is used to drive both the laser and the circuit radio-frequency drivers. However, with actively mode-locked lasers, the pulse-to-pulse jitter within the laser itself (which, for example, in a Nd:YAG laser is approximately 5–10 ps) must be considered (Rodwell et al., 1986). In this case, active feedback techniques must be used to stabilize the laser cavity. This technique is discussed in detail in Chapter 5 of this book.

For lasers with acceptably low jitter, a simple scheme for radio-frequency synchronization is possible as depicted in Fig. 14. The basic repetition rate is detected optically and is then used, with suitable division, as the master clock for the microwave synthesizer. An added benefit of this type of system is that delay scanning can be implemented electronically at rates high enough to provide a "live" display of the signal. By offsetting the synthesizer frequency by a small amount, Δf, with respect to the laser frequency, the sampling pulses "walk" through the signal being measured at the beat frequency. This is depicted in Fig. 15. The output signal is, therefore, scanned at the beat frequency also. The beat-frequency trigger

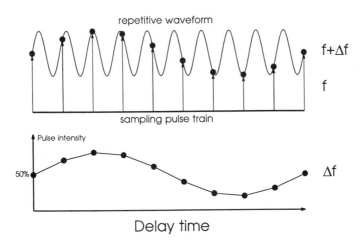

FIG. 15. Schematic concept of offset frequency sampling. The electrical signal has a repetition frequency of $f + \Delta f$, while the sampling pulses are at f. Thus, the electrical waveform is reconstucted at the beat frequency, Δf.

is generated by mixing the laser repetition rate frequency with the signal from another phase-locked synthesizer operating at the same offset frequency as the main driver. The mixer output is a sinusoid at the beat frequency that triggers the x-axis of the display unit. The electronic frequency offset technique is entirely analogous to optical delaying, except that it can scan at much higher rates and thus enable visually "live" displays. This electronic scanning technique also effectively translates the "measurement frequency" away from DC by Δf. A tuned amplifier is used as a receiver with its Q determining the bandwidth about Δf.

IV. Hybrid Sampling of Devices and Materials

This section describes a variety of measurements made with hybrid electro-optic techniques. Hybrid applications involve making measurements ón discrete devices where the electro-optic modulator becomes an integral part of the experimental arrangement. For several applications, this is still the preferred method for making ultrahigh-speed measurements. Hybrid measurements are differentiated from those that use the external electro-optic probing technique. External probing, described in detail in the next section, is best applied to integrated circuits and other two-dimensional devices usually comprised of many discrete components. The performance of several hybrid modulator geometries is assessed first. All the work described here is performed with the CPM laser described in Section III.A.

A. Performance of Hybrid Geometries

The three most popular types of hybrid modulator are the balanced-line, the coplanar-line, and the superstrate modulator. The particular advantages and disadvantages of each type were discussed earlier in Section III.C. The performance of each type is evaluated here. The temporal response time, or resolution, of each modulator was determined by using a chromium-doped GaAs photoconductive detector attached directly to the modulator structure. The detector generates a step function on the picosecond time scale when illuminated by 100-fs optical pulses. The sampled risetime is a direct measure of the temporal resolution of the modulator.

1. Balanced-Line Modulator

The balanced-line modulator is applied to devices that have their signal and ground-line electrodes on opposite sides of the substrate material.

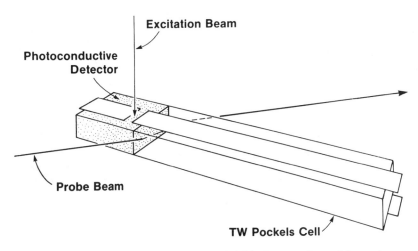

FIG. 16. The travelling-wave, balanced-line hybrid electro-optic modulator shown attached to a photoconductive detector, with the optical beam propagating in the velocity-matched geometry.

Figure 16 shows this style modulator connected to the GaAs photoconductive detector (Valdmanis et al., 1982). The detector is illuminated from above and the resulting electrical pulse propagates down the electrode over the lithium tantalate crystal. For best temporal resolution, the optical beam passes through the crystal at the velocity-matching angle determined by the ratio of optical to electrical velocities in the crystal. Figure 17 shows the resultant risetime of 760 fs for a balanced-line modulator with crystal thickness of 100 μm. This latter dimension determines the frequency at which non-TEM modes propagate in the electrode structure and, hence, also sets the upper limit for frequency response. For more detailed information, see Section IV.B on transmission lines.

2. Coplanar-Line Modulator

Figure 18 shows the coplanar-stripline–modulator configuration as it is integrated with the GaAs photoconductive detector (Mourou and Meyer, 1984). The detector and modulator crystal are fabricated together so that smooth, continuous transmission-line electrodes are evaporated over the material boundary. In this case, electrode widths and spacings are both 50 μm, resulting in a transmission-line impedance of ~56 ohms on the modulator. Fig. 19 shows that the temporal response of this geometry is 460 fs, which is significantly better than that of the balanced-line modulator. The peak signal amplitude was approximately 30 mV. Calculations

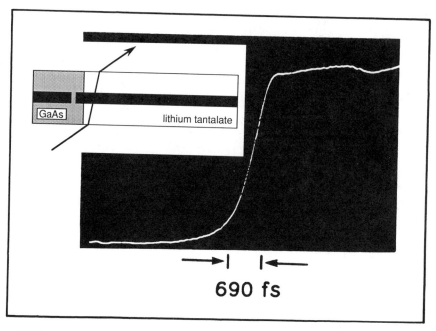

FIG. 17. A 760-fs–risetime waveform generated by a GaAs photoconductive detector and measured by a 100-μm–thick balanced-line electro-optic modulator as shown in the inset. Optical pulses for both excitation and sampling were 100 fs in duration. The signal level is ~1V in amplitude.

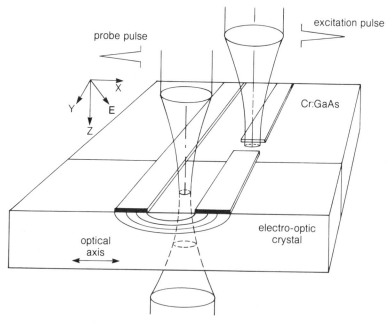

FIG. 18. The travelling-wave, coplanar-stripline hybrid electro-optic modulator shown attached to a GaAs photoconductive detector. Electrode width and spacing are ~10 μm. (Figure courtesy of K. Meyer, University of Rochester)

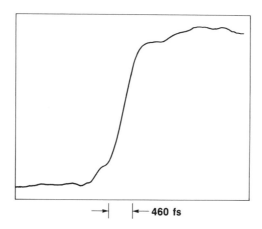

FIG. 19. A 460-fs–risetime waveform as measured by the coplanar hybrid electro-optic modulator shown in Fig. 18. (Figure courtesy of K. Meyer, University of Rochester)

from experimental results show that the penetration depth of the electric field into the crystal was only about 50 μm, i.e., approximately equal to the line spacing. Therefore, in this geometry, velocity-matching techniques would not offer any significant improvement in temporal resolution. This greatly relaxes the experimental complexity for obtaining subpicosecond temporal resolution.

3. Superstrate Modulator

The superstrate-type modulation technique, also known as reflection mode sampling, is shown in Fig. 20 (Meyer and Mourou, 1985). In this technique, the modulating crystal has no electrodes of its own, but is placed over the electrodes of the circuit itself, usually in contact. This is similar to the external probing technique except that the crystal is much larger and becomes an integral part of the testing geometry. In the figure, the lithium tantalate slab is placed over the GaAs photoconductive detector. Both excitation and sampling pulses come from above, with the sampling pulse reflecting off a high-reflection coating on the bottom of the slab. Both trigger and sampling beams focus to a diamater of ~15 μm. The electrode widths and spacings are 50 μm, which allow the electric field lines to extend about 50 μm into the lithium tantalate superstrate. The resultant risetime of 740 fs is shown in Fig. 21 with a signal amplitude of approximately 30 mV. Sensitivity is virtually the same as for the coplanar-stripline modulator with electrodes of the same dimensions. The degraded risetime as compared with the coplanar-stripline modulator is probably

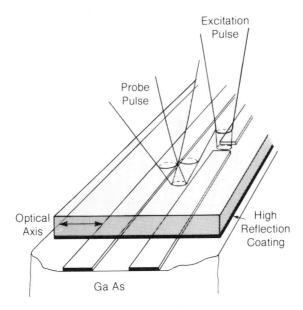

Fig. 20. A transverse superstrate-type modulator placed over coplanar-stripline electrodes on a GaAs photoconductor. The excitation pulse passes through a region of the electro-optic crystal with no HR coating, while the probing pulses are reflected back to the optical system. (Figure courtesy of K. Meyer, University of Rochester)

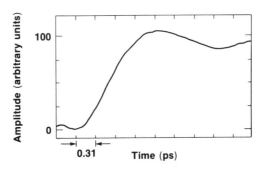

Fig. 21. A 740-fs risetime measured by the superstrate-modulator configuration shown in Fig. 20. The longer risetime, as compared with the coplanar modulator in Fig. 19 is likely due to the larger 50-μm electrodes used in this measurement. The optical pulses take longer to travel through the larger fringing field in this case. (Figure courtesy of K. Meyer, University of Rochester)

due to effects inherent in the superstrate principle. The first effect is caused by the sampling beam passing through the crystal twice, thus doubling the optical transit time. In this 50-μm geometry, the electric field extends into the crystal far enough to incur an optical transit time of several hundred femtoseconds. Smaller electrodes eliminate this problem. The second source of temporal degradation is due to the signal having to propagate some distance in both GaAs and lithium tantalate. The higher dielectric constant of lithium tantalate produces a more dispersive line. This effect can be avoided by making the electro-optic crystal as small as possible, as in the external probing technique, or sampling near the leading edge of the crystal.

A commercial instrument, the EG&G Princeton Applied Research Model 8300, employs the superstrate configuration. With this instrument, the user attaches the device to a glass substrate upon which is evaporated a known electrode structure. A lithium tantalate slab is placed over the electrodes and the sampling beam passes through the assembly. Because of the fixed sampling geometry, voltage calibration is assured. A temporal resolution of 4 ps is specified, making this the fastest commercially available instrument on the market at the time of this writing.

B. TRANSMISSION LINES

This section investigates the performance of several different types of transmission lines for high-speed electrical pulses. Such lines typically interconnect ultrahigh-speed devices over short distances. An ideal transmission line would transmit a signal without any distortion or loss. However, this is rarely the case in real systems where frequency-dependent losses and dispersive effects distort the signal as it propagates along the line. Electro-optic sampling offers a window into this high-speed electronic world through which to observe and investigate these complex effects.

Of the next two subsections, the first looks at three types of room-temperature transmission lines and compares their dispersive characteristics while the second looks at superconducting transmission lines. With superconductors as electrodes, effects due to resistivity are eliminated while other effects, such as modal dispersion, can be isolated and studied. Both indium lines and lines made from the new high T_c materials are described.

1. Room-Temperature Lines

Distortion of high-frequency signals in open–transmission-line electrode geometries can be considerable. There are several causes of deterioration, such as frequency-dependent loss in the substrate material, conductor loss, and modal dispersion. The most serious is modal dispersion due to a frequency-dependent effective dielectric constant, ε_{eff}, that is determined by the relative dielectric constant of the dielectric material, ε_{rel}, and the specific electrode geometry. In typical microstrip-electrode geometries, there exists a critical frequency, f_c, at which the onset of significant dispersive effects appears, due to the excitation of spurious modes in the form of surface-wave TM and TE modes (Yamashita and Atsuki, 1976). The critical frequency at which coupling occurs between the fundamental quasi-TEM mode and the lowest-order surface wave mode is given by

$$f_c = v_0/4h\sqrt{\varepsilon_{rel} - 1} \tag{37}$$

where v_0 is the velocity of light in a vacuum, and h is the electrode separation. The critical frequency occurs in the regime where the physical dimensions of the electrical wavelength in the dielectric compares to the transmission line cross-sectional dimension. Below the critical frequency, ε_{eff} assumes a constant nondispersive lower value, ε_{tem}. Above the critical frequency, ε_{eff} equals the higher relative dielectric constant, ε_{rel}, and is also nondispersive. Hence, the dispersion curve is basically two-valued, with a smooth transition regime between.

The electro-optic sampling system uniquely measures these dispersive effects in an electrical signal as it propagates along a transmission line. Figure 16 shows a balanced-stripline modulator. Its electrode geometry provides a Pockels cell with a long, open aperture that allows measurement of the electrical transient at various points as it travels along the stripline without disturbing the circuit (Valdmanis and Mourou, 1984). By translating the sampling beam by a known amount with respect to the stationary modulator, the modulator aperture can be interrogated at ever-increasing distances from the electrical signal source. Sampling at each successive point provides a progression of pulse shapes that displays the accumulating effects of the dispersive stripline on the electrical pulse. Figure 22a–c shows a pulse generated by a GaAs photoconductive detector as it propagates along a 250-μm–thick modulator. Figure 22d shows the theoretical waveform obtained by Hasnain et al. (1983) from computer modelling based upon the dispersive effects of the effective dielectric constant. Their findings show fair agreement with the experimental waveform although at slightly different propagation distances.

When a signal contains frequencies in the regime of the critical frequency for a particular geometry (as in this case), the dispersive charac-, teristics of the stripline conceptually divide the signal bandwidth into two "packets": one containing the lower frequencies and the other the higher frequencies. The lower frequencies experience a lower effective dielectric constant ε_{eff} and, hence, travel along the stripline faster. Higher frequencies, in turn, travel more slowly owing to their increased ε_{eff}. Thus, as the

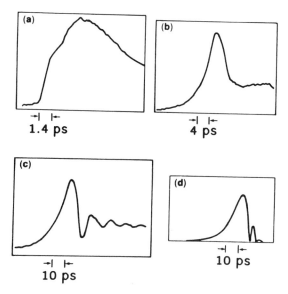

FIG. 22. Photoconductively-generated pulse propagation on a 250-μm–thick balanced-line travelling-wave electro-optic modulator. The configuration used is shown in Fig. 16. The pulse is sampled at (a) the beginning of the modulator, (b) after 1.9 mm, and (c) after 6.2 mm. Plot (d) shows a theoretical simulation of the pulse in (c).

pulse propagates, its higher-frequency components appear later in the waveform. In Figure 22b, high frequencies are near the peak and hence sharpen that feature. In later stages of propagation, the high frequencies contribute their effects only to the trailing side of the pulse as is evident in Fig. 22c. At this point, the trailing edge of the pulse is much faster than the low-frequency–dominated risetime.

Equation 37 shows that the dispersive characteristics of a microstrip transmission line are sensitive to the product of the electrode separation and the square root of the substrate dielectric constant. Figure 23 compares the magnitude of the dispersion for three different stripline electrode geometries by plotting the risetime of a test pulse as a function of propagation distance along the stripline. Curve 23a plots the risetime of a pulse travelling along a 500-μm–thick balanced transmission line on LiTaO$_3$ with $\varepsilon_{rel} = 43$ (Valdmanis et al., 1982). Significant dispersion is observed because of the relatively large electrode separation and high dielectric constant.

To decrease the electrode separation, Meyer and Mourou (1985) used a coplanar line on LiTaO$_3$ as represented in Fig. 18. Linewidths and spacings were 50 μm, the same as in the performance section. Representative propagating pulses are shown in Figure 24. Curve 23b plots the risetimes of

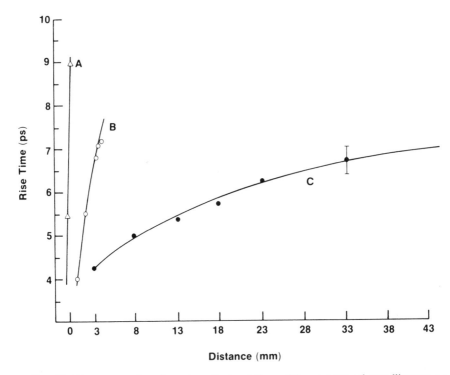

FIG. 23. The comparative dispersive effects of three different types of travelling-wave transmission line. The plot shows risetime as a function of propagation distance. (a) is for a 500-μm–thick balanced-line modulator; (b) is for a 50-μm–wide coplanar stripline; and (c) is for a 500-μm–thick air-spaced balanced line. (Figure courtesy of K. Meyer, University of Rochester)

this coplanar stripline. As predicted, there is a noticeable improvement over the balanced-line geometry, but dispersion effects are still significant. As shown in the superconducting portion of this section, modal dispersion is the most serious effect in distorting signals. This is because in the open-electrode geometry, a true TEM mode cannot propagate. This causes extraneous modes that travel at significantly different phase velocities.

Kryzak et al. (1985) implemented a similar arrangement of this electro-optic sampling technique to study and successfully develop new transmission lines that can propagate picosecond electrical signals with less dispersion than was previously achieved. Their design uses an air-spaced microstrip geometry where the upper electrode is supported by a 125-μm–thick glass superstrate 500 μm above the ground plane. The electrode width is 750 μm. In this design, the travelling field is confined as closely as

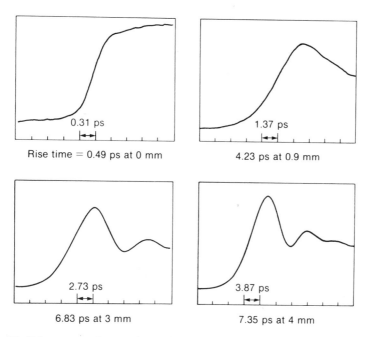

Rise time = 0.49 ps at 0 mm   4.23 ps at 0.9 mm

6.83 ps at 3 mm   7.35 ps at 4 mm

FIG. 24. Pulse propagation on the coplanar-stripline electrode configuration shown in Fig. 18 with linewidths and spacing of 50 μm. (Figure courtesy of K. Meyer, University of Rochester)

possible to a true TEM mode by arranging it so most of the fields are contained in a single dielectric medium (in this case, air). Measurements were done by sampling a small piece of $LiTaO_3$ inserted into the electrode air gap at ever-increasing distances from the source. Therefore, the signal propagates in air until it is measured. Due to the elimination of a high-ε dielectric between the electrodes, the electric field exists predominantly in the uniform air gap, thus significantly reducing dispersion. Curve 23c shows the improvement is nearly two orders of magnitude.

2. Superconducting Lines

i. Conventional Superconductors

By studying superconducting transmission lines, we can separate the distorting effects caused by conductor loss mechanisms from those caused by modal dispersion and other effects. Electro-optic sampling is the only technique to achieve subpicosecond resolution in a cryogenic environment (Dykaar et al., 1985, 1987). The electro-optic configuration used to study conventional superconductors (indium-based) is the coplanar stripline described in Section III.C. (Hsiang et al., 1987). Electrical pulses are generated and launched, as before, from an optically triggered GaAs photoconductive detector. However, in this case, the electrodes connecting the detector to the lithium tantalate crystal are made of indium, which when immersed in superfluid helium (<2 Kelvin) become superconducting. The electrodes were 500 nm thick with the standard 50-μm width and spacing used for other coplanar samplers. Superfluid helium allows the laser beams to propagate through the cryogenic environment without disturbance due to bubbling. A risetime of 360 fs was achieved with superconducting electrodes, compared to 460 fs at room temperature with aluminum electrodes.

For lithium tantalate, as discussed earlier, the dielectric loss is minimal for frequencies up to about one THz. Thus, only conductor loss (Mattis-Bardeen effect) and modal dispersion could significantly contribute to signal distortion. Qualitatively, both effects have a characteristic frequency at which deleterious effects set in. In addition, both effects manifest themselves in the same manner, by reducing the phase velocity for frequency components approaching the critical frequency, which results in signal dispersion. For the geometry described, significant dispersive effects occur at 5 GHz for modal dispersion and at 40 GHz for the Mattis-Bardeen effects. Figure 25a shows a sequence of electrical transients measured in

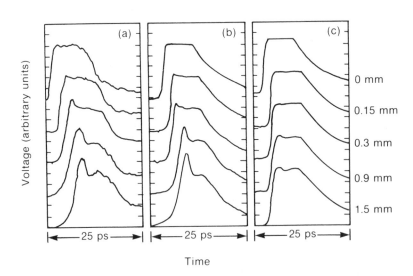

FIG. 25. Pulse-propagation waveforms for signals on superconducting coplanar striplines. Electrodes were indium, with 50-μm linewidth/spacing, cooled to ~2°K in superfluid helium. (a) shows the experimental results, (b) shows the theoretical predictions for the same propagation, and (c) shows the theory without modal dispersion (which does not agree with the experiment). (Figure courtesy of J. Whitaker, University of Rochester)

the preceding geometry. Significant risetime lengthening and ringing is evident as the pulse propagates. Figure 25b shows the theoretical model predictions for the same propagation. In this model, which includes both dispersive effects, the agreement is excellent. Figure 25c shows the predictions from a theoretical model that does not include model dispersion and does not fit the data. This study concludes that modal dispersion is a critical source of signal distortion and is much more important than conductor losses for signals in the GHz to THz range.

ii. High-T_c Superconductors

With the advent of thin films of the new ceramic superconductor ($YBa_2Cu_3O_{7-x}$), it is now reasonable to consider high-speed applications of this material (Dykaar, 1988). If a BCS-type behavior is assumed for this material, then the superconducting energy gap is about 20 THz! Of course, a BCS phase delay is expected as well, but this would be a small effect for distances normally encountered in electronic circuits. To measure the propagation characteristics of this material, a thin film 360 nm thick was fabricated (Lathrop et al., 1987) and patterned on a cubic zirconia substrate as shown in Fig. 26. As shown, high-speed measurements were made using the electro-optic sampling technique in a reflection-mode geometry (Dykaar et al., 1988). This approach allowed the actual signal which was launched onto the ceramic lines to be measured as the input signal. In this way, the effect of the wire bonds and the metal–ceramic interface at the

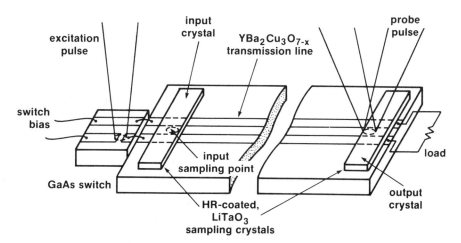

FIG. 26. The experimental configuration for electro-optically measuring signals on high-T_c superconducting transmission lines. A photoconductive detector wire-bonded to the lines provides the input signal, which is then measured by two separate crystals of lithium tantalate at the beginning and end of the lines. (Figure courtesy of D. Dykaar, University of Rochester)

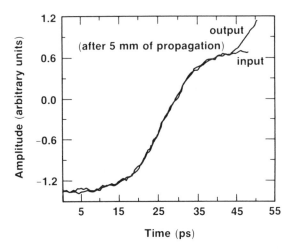

FIG. 27. Experimental data showing input and output waveforms from the high-T_c super-conducting transmission lines described in Fig. 26. (Figure courtesy of D. Dykaar, University of Rochester)

contact pads did not influence the measurement process. The wire bonds did, however, limit the input signal risetime to about 15 ps.

In Fig. 27, the input and output waveforms are shown superimposed. They display no distortion over the 4.5-mm propagation distance, and the only deviation occurs at the end of the output pulse and corresponds to the arrival of a reflection at the output sampling point.

In addition to risetime, a critical current measurement was obtained. For this measurement, the maximum current was applied while the temperature was increased until the risetime began to decay, indicating the maximum critical current for that temperature. For this sample, this effective critical temperature was about 17 K for a current density of 10^6 A/cm^2.

As mentioned previously, the material properties of the superconductor should introduce an increasing phase delay with increasing temperature, assuming the BCS model. Figure 28 measures this delay, as well as the pulse shape. The waveforms were simulated using a previously published modelling program, which yielded excellent results for ordinary superconducting transmission lines (Hsiang et al., 1987; Whitaker et al., 1988). The percent change in phase delay (both measured and simulated), calculated relative to the initial delay at a temperature of 1.8 K, was very weakly dependent on the substrate parameters and very strongly dependent on the YBCO superconducting parameters. The simulation slightly overestimates the value of current density for temperatures near T_c because the simulations were performed assuming zero current.

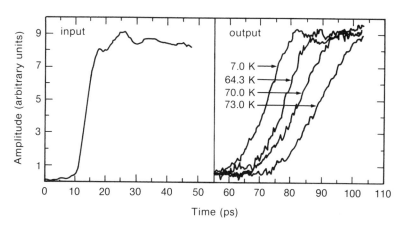

FIG. 28. Input and output waveforms for the high-T_c superconducting transmission lines described in Fig. 26 as a function of temperature. This shows, as expected, an increasing phase delay with increasing temperature. (Figure courtesy of D. Dykaar, University of Rochester)

These results indicate that this material may be applied where the need for extreme speed or reduced conductor size for the equivalent current-carrying capability takes precedence over other considerations.

C. HIGH-SPEED–DEVICE CHARACTERIZATION

Advances in material technology, especially GaAs molecular beam epitaxy, have spawned new classes of electronic materials now advancing the performance levels of solid-state devices. Many of these new devices have switching speeds in the picosecond regime. Unfortunately, conventional electronic-measurement techniques do not have the temporal resolution needed to test in the picosecond domain. Also, it is very difficult to interrogate individual devices with conventional techniques based on co-axial instrumentation techniques. Several device and material measurements based on hybrid electro-optic modulators are described next and are also described in Chapter 5.

1. MESFET

Typically, transistor-switching speeds are evaluated by ring oscillators that average the response of many individual devices. Electro-optic sampling enables single-device characterization with single-picosecond temporal resolution.

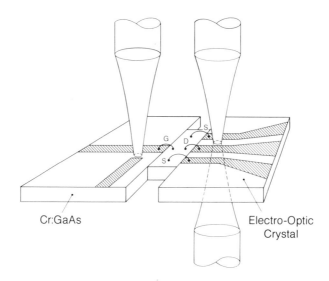

FIG. 29. Sampling geometry used for characterizing a GaAs MESFET. The transistor is sandwiched between a coplanar photoconductive detector, which provides the input signal, and a coplanar-line modulator for sampling. Wire bonds are made as short as possible to avoid parasitics. (Figure courtesy of K. Meyer, University of Rochester)

Figure 29 shows the experimental configuration for characterizing the response of a GaAs MESFET (Meyer et al., 1985). The device is sandwiched between a pulse generator and a coplanar electro-optic modulator. The gate of the MESFET is wire-bonded to the output of an optically triggered GaAs photoconductive detector and a DC bias line in a microstrip geometry. The drain connects to the center electrode of a coplanar waveguide electrode geometry on the electro-optic crystal. The source is grounded to both outside lines of the waveguide. The sampling beam passes through one of the gaps next to the center electrode. The modulator is calibrated by applying a known voltage directly to its electrodes. The input signal is measured by directly attaching the detector to the modulator. The input signal, as shown in Fig. 30a, has an amplitude of 53 mV and a risetime of 5.4 ps. With drain and gate biases off, a portion of the input signal is coupled directly through the device to yield the trace in Fig. 30b. With both biases on, the output appears as in Fig. 30c. The signal is inverted, has slight gain, and has a risetime of ~25 ps, which corresponds to S-parameter measurements predicting a 20-GHz current-gain cut-off frequency. The output pulse corresponds very well with theoretical models of MESFET performance, even reproducing the initial negative portion of the pulse accurately.

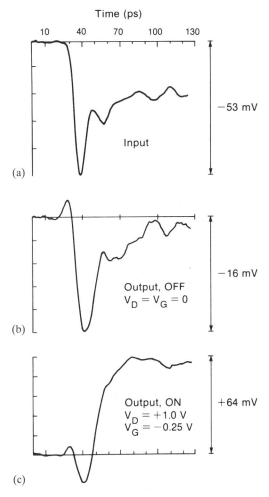

FIG. 30. MESFET response curves as measured in the geometry of Fig. 29. (a) shows the 5.4-ps–risetime input signal; (b) shows the output with gate and drain biases off; and (c) shows the output with biases on. (Figure courtesy of K. Meyer, University of Rochester)

2. Permeable Base Transistor

Another product of state-of-the-art fabrication techniques in MBE and submicron lithography is the permeable base transistor (PBT) (Bozler et al., 1979). In concept, it is similar to the vacuum-tube triode, except that a solid-state "grid" with sub-micron dimensions is fabricated between the MBE-deposited layers. As yields for this device were very low, ring

oscillators were not constructed to test switching speeds that were anticipated to yield oscillation frequencies in excess of 200 GHz. Once again,
electro-optic sampling techniques proved to be uniquely suited to characterization of the PBT (Dykaar et al., 1986).

Figure 31 shows the configuration for investigating PBT switching speed.
Because the PBT itself was fabricated in a microstrip geometry, a matching
electro-optic sampler was assembled by sandwiching the PBT between a
microstrip GaAs photoconductive detector and a balanced-line modulator.
The detector, when triggered by subpicosecond optical pulses, presents a
near step function pulse to the base of the PBT. Suitable bias levels are also
supplied to the base and the collector, with the emitter grounded. The
output signal is shown in Fig. 32. The risetime of 5.3 ps indicates that this is
one of the fastest three-terminal room-temperature devices. The initial dip
in the signal may be some direct coupling of the input signal through the
device, and it agrees with theoretical models of the device.

3. Resonant Tunneling Diode

Use of heterojunction double-barrier structures in fast devices has
attracted a great deal of attention (Whitaker, 1988). Quantum mechanical
tunneling is the fastest known charge-transport mechanism in semiconductor devices such as the resonant tunneling diode (RTD). These single-
quantum well devices exhibit a large negative differential resistance (NDR)

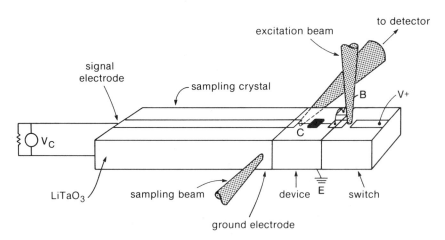

FIG. 31. Microstrip sampling geometry for characterization of the permeable base transistor. The PBT is sandwiched between, and wire-bonded to, a GaAs photoconductive detector
which supplies the input signal, and a balanced-line electro-optic modulator for measurement.
(Figure courtesy of D. Dykaar, University of Rochester)

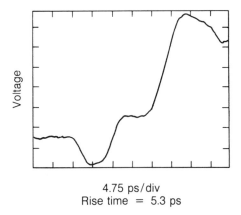

4.75 ps/div
Rise time = 5.3 ps

FIG. 32. The output signal of the PBT as measured in the geometry of Fig. 31. The 5.3-ps risetime is preceded by a feature that may be some direct coupling of the input signal through the device. (Figure courtesy of D. Dykaar, University of Rochester)

region, and since they have an extremely fast current response (Solner et al., 1983), they display a high-frequency output when applied as oscillators (Solner et al., 1987). In addition, experiments on the double-barrier diodes indicate the potential use of these devices in high-speed, digital logic circuits (Liu and Coon, 1987).

The electro-optic sampling technique was applied to a measurement of the switching time for a resonant tunneling diode. Figure 33 shows the electro-optic sampling configuration used. Both the photoconductive pulse generator and the lithium tantalate modulator are in coplanar stripline form. The device is mounted on an edge adjacent to the output electrode of the detector. One side of the RTD is connected to the detector; the other is connected to one electrode of the coplanar stripline of the modulator. The entire other electrode of the stripline is grounded. A DC bias is also applied to the input of the RTD to control its operating point on the I-V characteristic curve (Fig. 34).

For this demonstration, the bias is adjusted to operate the RTD in its bistable regime. This mode of operation is conceptually shown by the use of AC load lines superimposed on the current-voltage characteristic in Fig. 34. When the device experiences the fast step output from a photoconductive switch, the load line rapidly increases from its DC-bias position, causing a sudden drop in current from point B to C in Fig. 34. When this waveform is compared to the output when no DC-bias is applied (so current only increases with the step input), the switching time from point B to C is found. The rapid decrease in current is shown in Fig. 35, where the

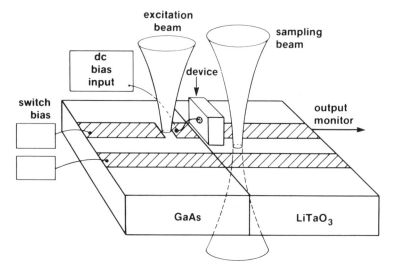

FIG. 33. Sampling geometry for characterization of a resonant tunneling diode. Both the input photodetector and the electro-optic crystal are of coplanar-stripline geometry. The RTD chip is mounted on edge at the input to the modulator and the RTD mesa itself is connected by a whisker to the output of the detector. (Figure courtesy of J. Whitaker, University of Rochester).

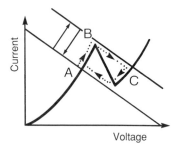

FIG. 34. Bistable operational characteristic of a typical resonant tunneling diode (with AC load lines). (Figure courtesy of J. Whitaker, University of Rochester)

10–90% characteristic time is 2 ps. This is the fastest known operation of an electronic device measured by electro-optic techniques.

D. ELECTRO-OPTIC CHERENKOV PULSES

As described earlier, the temporal resolution of any electro-optic sampling system is most often limited by geometric factors associated with the

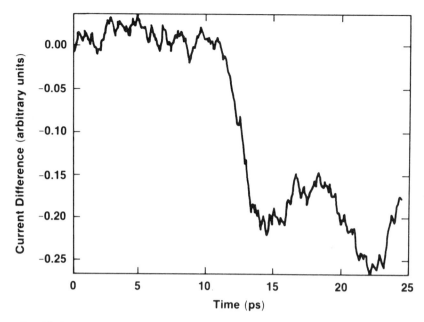

FIG. 35. Experimental result from the resonant tunneling diode measurement configuration shown in Fig. 33. This shows a switching time (from point B to C of Fig. 34) of ~2 ps and represents the fastest electronic device operation to date. (Figure courtesy of J. Whitaker, University of Rochester)

sampling convolution time. The electrical signal can also be distorted by dispersive effects inherent in the propagation of the signal along a transmission line. The following experiment demonstrates the temporal resolution that can be achieved when a precise velocity-matched geometry is employed to measure freely radiating electro-magnetic transients in the terahertz regime.

In this experiment by Auston et al. (1984), an extremely fast electrical transient is generated by the propagation of a subpicosecond optical pulse in electro-optic lithium tantalate. Known as the electro-optic Cherenkov effect, the optical pulse generates electromagnetic radiation through the inverse electro-optic effect. The radiation travels slower than the optical generating pulse and, hence, creates a Cherenkov-like cone that is almost a single cycle of terahertz radiation.

Because the host material is lithium tantalate, electro-optic sampling can be employed to characterize the radiated pulse. The insert of Fig. 36 shows the experimental configuration employed. Pulses of 100 fs from a CPM laser operating at 150 MHz are used for both generating and sampling

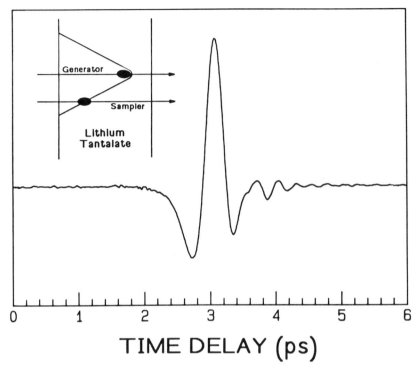

TIME DELAY (ps)

FIG. 36. The waveform generated by electro-optic Cherenkov radiation in lithium tanta-late. The wave has a half-cycle width of 225 fs and a peak electric field strength of ~10 V/cm. The spectral bandwidth is greater than 4 THz. The inset shows the experimental configuration where two 100-fs optical pulses propagate on parallel paths in the electro-optic medium. The generating pulse is of much larger amplitude, and both pulses are focused to ≤10 μm. (Figure courtesy of D. Auston, Columbia University)

functions. As before, the crystal is placed between crossed polarizers. The sampling pulse travels parallel to the generator, separated by a small distance, typically 100 μm. Since both pulses travel at the same velocity, the sampling pulse essentially "surfs" on the same point of the Cherenkov wavefront throughout the crystal, precisely matching the velocity of the generator. Hence, the temporal resolution is now solely determined by the extent of the optical sampling pulse in the crystal.

Figure 36 shows the measured waveform. It demonstrates the ability to resolve a waveform having a half-cycle width of only 225 fs and peak amplitude of 10 V/cm. Fourier analysis of the waveform reveals that its spectral bandwidth extends to 4 THz, corresponding to an electrical wavelength of 75 μm in air or 20 μm in $LiTaO_3$. This is the highest temporal-resolution, coherent measurement of an electromagnetic signal that has been performed to date. Using the preceding generation and detection technique, Cheung and Auston (1986) measured the absorption and dispersion of lithium tantalate in the terahertz spectral regime and found a resonant frequency of 6.3 THz for the lowest transverse optic lattice mode. This resonance sets an upper limit for the useful frequency range of electro-optic sampling in lithium tantalate.

E. GaAs Carrier Mobility Measurements

The ultrahigh temporal resolution of the electro-optic sampling technique when used with optical pulses of less than 100-fs duration allows investigation of carrier phenomena in semiconducting materials on the subpicosecond time scale. Studying carrier behavior on these very short time scales can yield important information about the high-speed dynamics of new devices. In the following experiment, electro-optic sampling is used to measure hot photo-injected carriers in GaAs (Nuss et al., 1987).

The experimental setup is depicted in Fig. 37. It is based on the principles of Cherenkov radiation discussed in the previous section. A generating and probing beam are provided as in the previous experiment for generating and measuring the optically produced conical radiation wave.

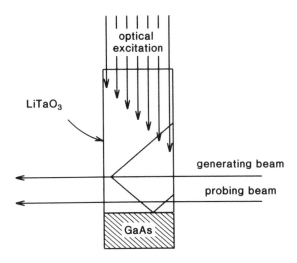

FIG. 37. Experimental configuration for using electro-optic Cherenkov radiation to characterize THz properties of electronic materials. The generating beam produces the THz wave that reflects off the material interface. The probing beam measures the wave before and after reflection. The material properties under investigation can also be actively changed by illuminating the surface with another short optical pulse that has a tilted wavefront to match the Cherenkov wave. (Figure courtesy of M. Nuss, AT&T Bell Laboratories)

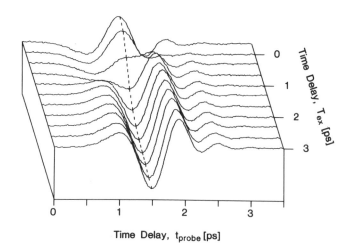

Fig. 38. Experimental results from the Cherenkov measurement geometry in Fig. 37. This plots the reflected waveform from photoexcited GaAs as a function of time delay between excitation and sampling. The 180° phase change of the waveform indicates a change in the GaAs surface from insulating to conducting in about 2 ps. (Figure courtesy of M. Nuss, AT&T Bell Laboratories)

The beams are focused to 8-μm diameter at separation of 20 μm. One end of the lithium tantalate is contacted to the sample of GaAs to be characterized. The Cherenkov wave reflects off the GaAs interface and passes through the probe beam for a second time. Thus, the probe beam measures the Cherenkov pulse before and after reflection. The difference in the two waveforms indicates carrier phenomena at the GaAs surface. Because the technique is coherent, both amplitude and phase information about the waveforms are obtained. To obtain information about induced carrier behavior, a third, intense optical pulse illuminates the GaAs surface to inject hot carriers. It is synchronized with the generating and probing pulses. By varying the arrival time of the intense pulse, carrier behavior in time relative to the injection instant is sampled. The intense pulse is brought in with a tilted wavefront to keep spatial synchronization with the sweeping conical Cherenkov wave.

Figure 38 shows reflected Cherenkov waveforms as a function of carrier-injection delay. The rear-most waveform, taken before any carrier injection, indicates the insulating behavior of the GaAs. As the time delay increases, both phase and amplitude change. In particular, the phase changes by 180 degrees, indicating a transition from insulator to conductor. From these data, carrier mobility as a function of time and of injected carrier densities is determined. This study shows that the electron mobilities are very low immediately after injection, due to fast transfer into low-mobility satellite valleys. They only approach equilibrium mobilities with a time constant of ~2 ps. This implies that delays of a few picoseconds could be experienced in detectors illuminated by light having a photon energy above the intervalley separation.

V. Electro-Optic Probing of Integrated Circuits

Interrogating integrated circuits (ICs) adds yet another degree of complexity to the problem of making high-speed measurements—i.e., how to access points of interest that are *internal* to circuits. Integrated circuits can consists of hundreds of thousands of devices on a single chip. For the sake of this discussion, an integrated circuit can contain a single device or thousands, but it is either fabricated *for* integration or it consists of any number of integrated elements. That is, any individual element is not designed to be connected to the outside world by itself, and it is often buried between many other elements. The problem is compounded because the elements and their connections are of micrometer size. Not only are conventional coaxial high-speed instruments out of the question due to resolution limits, but also are all the hybrid electro-optic techniques described in the previous sections. To probe ICs with picosecond resolution, two electro-optic techniques that solve the preceding problems are used.

A. EXTERNAL VERSUS INTERNAL ELECTRO-OPTIC PROBING

To probe internal points of ICs electro-optically, an electro-optic crystal must be adjacent to the conductor of interest so as to interact with the local electric field and invoke the electro-optic effect in either the longitudinal or transverse geometries.

External electro-optic probing (Valdmanis, 1987) uses an extremely small crystal as a proximity electric-field sensor that is brought close to the circuit from above as depicted in Fig. 39. Internal electro-optic probing (Kolner and Bloom, 1986) uses the circuit substrate itself as the electro-optic medium as depicted in Fig. 40. Internal probing has the obvious drawback that it can only be used on circuits whose substrates are electro-optic including GaAs and InP, but excluding silicon, ceramic hybrids, and others. External probing works on any IC material. The advantage to internal probing is that it does not require any hard probe to be near the circuit; it is truly an optical technique. However, because the sampling light must penetrate the circuit, the surfaces must be of optical quality, and the light must not be absorbed within the circuit. With external probing, light never reaches the circuit. These and other differences are summarized

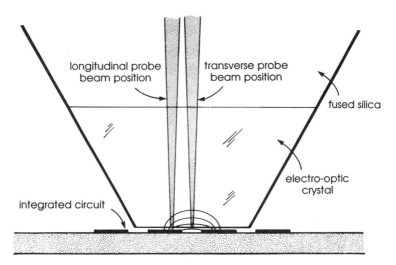

FIG. 39. The external electro-optic probe tip, showing two different modulator configurations: transverse and longitudinal. With a transverse electro-optic crystal, the optical beam direction is perpendicular to the electric field lines. With a longitudinal crystal, the optical beam and electric field lines are parallel.

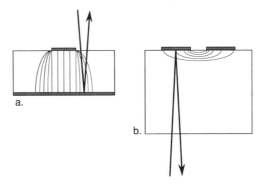

FIG. 40. Internal, or substrate, electro-optic probing geometry for use with circuits whose substrate is electro-optic and optically accessible (e.g., GaAs and InP). (a) shows the frontside probing arrangement typically used for analog microwave circuits, while (b) shows the backside probing geometry used for digital circuits. Both techniques rely on the reflection of the optical beam from the circuit metallization.

TABLE II

COMPARISON OF EXTERNAL AND INTERNAL PROBING.*

Parameter	External	Internal
Temporal resolution	<300 fs	≈2 ps
Bandwidth	>1 THz	≈175 GHz
Sensitivity	1–10 mV/$\sqrt{\text{Hz}}$	<100 μV/$\sqrt{\text{Hz}}$
Laser wavelength	Visible-IR	IR only
Circuit materials	Any	GaAs, InP
Circuit access	Front only	Front and/or back
Circuit preparation	None	Backside polish (substrate thinned)
Voltage calibration	Possible	Easy from back, possible from front

*Table compares several parameters for external and internal electro-optic probing techniques based on demonstrated performance.

in Table II. However, it should be noted that both techniques are currently being developed for use in industry.

Internal electro-optic probing is an elegant technique and has been widely applied to GaAs technology ICs. For a detailed discussion of the principles of this technique and its applications, refer to the following chapter in this book. The remainder of this section discusses the principles of external electro-optic probing and its applications.

B. EXTERNAL ELECTRO-OPTIC PROBING

Ideally, a nonperturbative means of probing integrated circuits is needed. It must have high temporal and spatial resolution that is *generally* applicable to circuits fabricated on any substrate. This section describes external electro-optic probing—a noncontact, picosecond, electro-optic technique for probing internal nodes of high-speed integrated circuits fabricated on any substrate (Valdmanis, 1987). This technique has achieved a temporal resolution of less than 300 fs without relying on optical interaction within the substrate, and it offers potentially single-micron spatial resolution. This temporal resolution corresponds to a measurement bandwidth in excess of one THz (1000 GHz). The system operates at the wafer level with conventional wafer probing equipment and no special circuit preparation. Optical pulses from any picosecond, high-repetition-rate laser throughout the visible and near-infrared can be used. The system has been applied to internal nodes of GaAs digital circuits (Valdmanis and Pei, 1987), silicon circuits, as well as discrete ultrafast devices fabricated on non–electro-optic substrates (Valdmanis et al., 1987). A similar system, with 30-ps resolution, that uses a diode laser, was used to study waveforms around integrated-circuit packaging pins and in coaxial cable (Ness and Mourou, 1986).

1. Principles of Operation

External electro-optic probing exploits the fact that two-dimensional circuits have an open electrode structure that gives rise to fringing fields above the surface of the circuit. Dipping a minute electro-optic crystal into these fields changes the crystal's birefringence, which can then be measured optically. Figure 39 depicts two arrangements for external probing: transverse and longitudinal. Transverse probing relies on local field components that are parallel to the surface of the circuit and hence transverse to the direction of the optical sampling beam. Longitudinal probing interacts with field components that emanate normal to the metallization lines (Valdmanis, 1988).

Most external probes to date have relied on the transverse e-o effect in lithium tantalate. A transverse field probe is sensitive only to fields parallel to the surface of the circuit that exist between conductors at different potentials. Because two-dimensional circuits have all their conductors in the same plane, overlapping transverse fields can exist, leading to crosstalk at the probing point. Sensitivity of the probes also depends on conductor spacings and widths, as both parameters affect transverse field strength.

Ideally, one would like to probe fringing fields unique to the conductor of interest and proportional to the voltage on the line. Longitudinal e-o probing offers these advantages and was first demonstrated in GaAs circuits, where the substrate itself is longitudinally electro-optic.

Initial longitudinal external probes are made of the e-o material KD*P (deuterated potassium dihydrogen phosphate, KD_2PO_4), although there are many materials that exhibit the longitudinal effect, as shown in Table III. In these materials, when light is propagated along the optic axis (z-axis), an e-o effect is experienced only for fields applied along the same axis, parallel to the optical beam. When used as a probe tip, the crystal's optic axis is perpendicular to the circuit and hence is sensitive only to fields normal to the surface of the circuit. These fields are the strongest directly above a conductor, so probing is also done directly over the conductor of interest.

The measured signal for this type of longitudinal e-o probe is proportional to the integral of the electric field in the e-o medium along the line of the optical beam. If the upper face of the e-o material is at zero potential, the integral of the electric field is exactly the potential at the conductor. Thus, the signal is proportional to the conductor's voltage. A similar situa-

TABLE III

EXTERNAL LONGITUDINAL TIP CRYSTALS*

Crystal	$r_{ij} \cdot 10^{-12}\ m/V$	$n(\Delta n)$	$\lambda(\mu m) \geq$	ε_i	$V_\pi(kV$ dbl. pass)	
KDP	8.8	1.5(0.03)	Visible	21	5.2	0.63 μm
KD*P	24	1.5(0.03)	Visible	48	1.9	0.63
GaAs	1.2	3.5	1.0	12	5.2	1.06
ZnTe	4.3	3.1	0.6	10	1.3	0.63
$Bi_{12}SiO_{20}$	5.0	2.5	0.5	56	1.9	0.63
CdTe	6.8	2.8	0.9	9.4	2.2	1.3
CuCl	2.4	2.0	Visible	7.5	8.2	0.63

Source: Data compiled in (R.J. Pressley, ed.) (1971), *Handbook of Lasers*, Chemical Rubber Co., Cleveland, 447–459; and A. Yariv and P. Yeh (1984), *Optical Waves in Crystals*, Wiley, New York, 230–234.

*Comparison of properties for selected longitudinal electro-optical crystals. Halfwave voltages are calculated for a double pass modulator geometry as usually encountered in both internal and external probing applications.

tion is achieved in e-o GaAs substrate probing when performed in the backside probe configuration.

Figure 41 compares the measured signal dependence on line separation for a lithium tantalate probe and a KD*P probe. The test circuit consisted of two 10-micron–wide lines separated by varying amounts, with one line grounded and the other driven with a one-volt p-p sine wave at one kHz. The figure clearly shows that the lithium tantalate probe signal is strongly affected by line spacing, while the KD*P longitudinal probe exhibits a signal independent of line spacing, indicative of a constant longitudinal field and constant potential, or voltage.

The advantage of these external configurations is that they can be applied to almost any two-dimensional circuit: GaAs, silicon, ceramic hybrid, and even printed circuit boards. Because the interaction is based on a field effect, no charge is removed from the circuit under study, and, therefore, no contact with the circuit is needed. Only a small capacitive coupling with the circuit is induced, which has been shown experimentally to have minimal effect (subpicosecond) in low-impedance circuits. Transverse probes are suited to simple electrode geometries, while longitudinal probes offer advantages in complex circuits.

As shown in Fig. 42, the basic principles for external electro-optic probing are the same as for conventional electro-optic sampling (Valdmanis and Mourou, 1986) except that the modulator is the electro-optic probe tip. Mechanically, the system was adapted for application to integrated circuits, both packaged and in wafer form. The entire optical system (minus

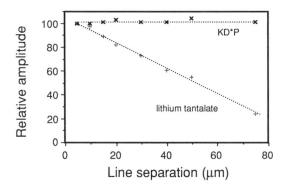

FIG. 41. Comparative voltage-sensitivity for transverse and longitudinal external probes for the electrode configuration shown. It can be seen that the signal level measured by the longitudinal KD*P probe is independent of line spacing, whereas the transverse lithium tantalate probe yields a signal level inversely proportional with line spacing. The plots are normalized to 100% at 5-μm line spacing.

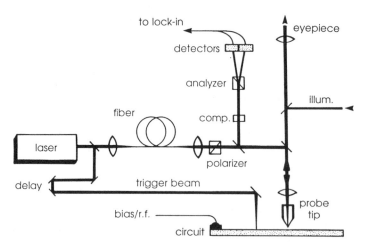

FIG. 42. The general system layout for external electro-optic probing. The circuit can be driven either optically or by a synthesizer. The optical fiber provides a flexible coupling between the laser and the prober optics. All the prober optical elements and detectors are assembled on a single, positionable plate that allows the probe tip to be located at any test point of the circuit without realignment. Illumination and an eyepiece are included to enable simultaneous viewing of the tip, the circuit features, and the laser spot.

FIG. 43. Photograph of the external probing system described in Fig. 42 mounted on a conventional integrated-circuit probing station, complete with probe card. The laser input fiber can be seen on the left. The eyepiece and detectors are at the top of the plate, and an illumination fiber optic, for viewing of the circuit through the tip, is brought in from the right. The view through the eyepiece simultaneously shows the tip endface, the circuit features beneath, and the sampling laser spot. The electro-optic tip is held (out of view) by a small brass stalk below the microscope objective in the center of the probe card. The entire optical assembly is on a precision x-y-z translation stage for positioning the tip..

laser) is an integral unit supported by a precision x-y-z translation stage to position the tip at any point on the circuit under test. Optical pulses are brought into the optical unit via a single-mode, polarization-preserving optical fiber. A microscope enables simultaneous viewing of the circuit and the laser spot. Figure 43 is a photograph of the external electro-optic probing system installed on a conventional integrated circuit probing station, complete with probe card for driving unpackaged circuits in wafer form.

The probe tip used in the following experiments is of lithium tantalate, which has the transverse electro-optic effect. Photographs of the tip are shown in Figs. 44 and 45. It is a 7-mm–long by 3-mm–diameter fused silica cylinder with a 100-μm–thick lithium tantalate plate bonded to one end. It

FIG. 44. Photograph of a typical external electro-optic probe tip polished as a four-sided pyramid. The electro-optic crystal is bonded to a cylindrical fused silica substrate 2–3 mm in diameter before polishing. For this particular tip, the side faces are formed at a 30° angle with respect to the cylinder axis. For transverse-type crystals, a high-reflectivity coating is evaporated over the tip.

is oriented with its optical axis in the plane of the plate. The tip is polished to form a four-side pyramid with the remaining flat on the end about 40 μm square. Finally, the tip is coated with a dielectric reflector.

KD*P tips, for longitudinal probing, are fabricated similarly, except they are oriented with the optical axis normal to the plate. For circuits not requiring high spatial resolution, sampling pulses are focused into the tip via a 10 × microscope objective creating a spot diameter of ~5 μm (at 630 nm). Typically, the proximity of the probe tip to the surface of the circuit must be less than the particular line spacing involved. Sensitivity is a few millivolts in a one-Hz measurement bandwidth. Figure 46 shows a lithium tantalate probe tip *in situ* over a packaged silicon multiplexing circuit. This measurement is described in section V.B.3.b.

2. Temporal Resolution

The temporal resolution of the lithium tantalate probe is determined by applying a step-function waveform generated by a photoconductive detector, much like the technique used for the hybrid geometries discussed

FIG. 45. Close-up view of Fig. 44's electro-optic tip showing the 100-μm–thick piece of lithium tantalate bonded to the fused silica. The tip face dimensions are ~40 μm square.

FIG. 46. Photograph of the external electro-optic probe tip in position over a packaged silicon multiplexer chip. Virtually any point within the circle of wire bonds is accessible.

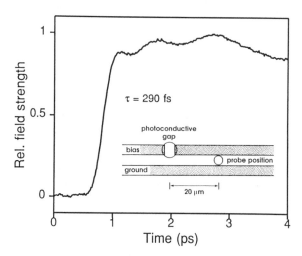

F<small>IG</small>. 47. The electrode geometry (inset) and the resultant measurement risetime of 290 fs from a GaAs photoconductive detector when illuminated by 100-fs optical pulses and measured by the external electro-optic probe technique. This represents a measurement bandwidth of 1.2 THz.

earlier. Optical pulses for both triggering the detector and sampling the probe tip are produced by a balanced colliding pulse mode-locked laser that generates 100-fs pulses at 630 nm at a repetition rate of 100 MHz with an average power of 30 mW in each of two beams. Synchronization and scanning delay between the two beams are implemented by standard optical delay line techniques. The circuit consists of two one-mm–long by 5-μm–wide Ti/Au electrodes evaporated 5 μm apart on a semi-insulating GaAs substrate to form a 100-ohm coplanar strip transmission line. (See Fig. 47 insert.) A single 5-μm gap in one of the lines forms the photoconductive detector. One side of the gap is biased to 10 V, and the other side is terminated off of the chip. The gap is illuminated with 100-fs pulses having a pulse energy of ~30 pJ. The resulting electrical pulse has a risetime on the order of the optical pulse duration and a $1/e$ falltime of ~90 ps.

The electrical pulse is measured by simply positioning the probe tip over the transmission line so the laser spot is between the striplines in the region of the strongest transverse fringing field. Figure 47 shows the resultant risetime of 290 fs as measured by the external electro-optic prober 20 μm from the gap. This risetime is a convolution of four effects: the sampling pulse width, the optical (τ_o) and electrical (τ_e), transit times, the probe tip material response time, and the actual risetime of the electrical pulse. The first two effects are ≤100 fs, while the material response is limited to ~300 fs by the first lattice resonance of lithium tantalate, which occurs at

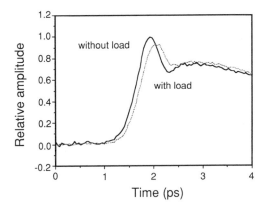

FIG. 48. Experimental result from investigating the effect of circuit loading due to the presence of the external probe tip. ($-$) is without load; (. . . .) is with load. The presence of a dummy probe crystal placed between the signal generator and the probing point lengthens a 480-fs risetime signal only to 520 fs with slight attenuation and appropriate phase delay.

6.3 THz (Cheung and Auston, 1986). Thus, the measured risetime, and, in this case, the probe resolution are dominated by the material limit, indicating that the actual risetime of the test pulse is significantly less than 300 fs.

To test the loading effect of the probe tip, the probe is moved to a distance of 500 μm from the detector gap and the risetime of the step function at that point is measured to be 480 fs. Then, a similar, 200-μm–wide, piece of lithium tantalate is placed on top of the stripline, between the gap and the probe tip. The resultant risetime, as shown in Fig. 48, was lengthened only to 520 fs, indicating that the invasiveness of a probe tip with these dimensions is negligible for picosecond measurements in low-impedance circuits.

3. Applications

This section presents three applications of the external electro-optic probing system. The first two demonstrate measurements made at internal nodes of small-scale and medium-scale integrated circuits. Neither of the circuits was specially designed to accommodate electro-optic testing, but suitable test points naturally occurred in the circuit layouts. This is true because every active device must be connected to a power supply and to ground lines. Hence, these reference lines for transverse probing are always nearby. A GaAs selectively doped heterostructure transistor (SDHT) prescaler IC probed at six points and preliminary measurements on a silicon NMOS circuit are presented.

The third application demonstrates the characterization of the shortest guided electrical pulses generated as of the late 1980s. Although not really an integrated circuit, the application demonstrates the ability to probe on any substrate with subpicosecond resolution. Once again, the following applications use a CPM laser, as described in Section III.A.

a. GaAs SDHT Prescaler

Initial measurements investigated the operation of a GaAs SDHT prescaler (Valdmanis and Pei, 1987). Measurements were made on the circuit input buffer consisting of two banks of three inverter stages, which generate clock and complementary clock-output signals. A scale drawing of the inverter stages and output lines is shown in Fig. 49. Linewidths are 3.5 μm with a closest spacing of 3.5 μm. A ground bus separates the two banks of inverters with the upper bank generating the clock signal and the lower bank the complementary clock signal. Each inverter has two SDHTs, one operating as the switching stage, and the other as an active load. The reference input enables limited adjustment of the relative phase between the clock outputs. V_{dd} is the supply bus at 2 volts, and the switching threshold of each inverter is 0.4–0.5 volts.

The external probe was used to monitor the propagation and shaping of an input pulse as it progressed through the input buffer inverter stages. The input signal was a single pulse generated by a fast photodiode brought to

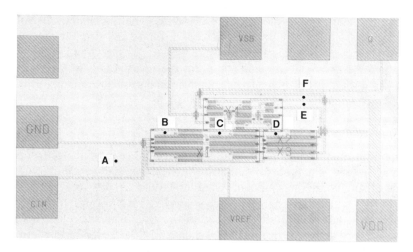

FIG. 49. Schematic diagram of GaAs SDHT prescaler circuit consisting of a three-stage input buffer and subsequent divider. Measurements were made at the input (A), at the output of each inverter stage (B, C, and D), and at the clock and complementary clock inputs to the divider stage (E and F, respectively).

the circuit by a miniature coaxial cable and terminated at the chip. The input pulse risetime was about one ns with a peak amplitude of ~0.8 V. Waveforms were measured at points A–F in Fig. 49 by bringing the probe tip to within a few microns of the surface of the circuit at each of the these points. To make a measurement, the probe tip was positioned between the line that carries the signal of interest and a line that behaves as a DC reference, i.e., typically either V_{dd} or ground. Point A measures the field between the input signal line and the ground bus. Points B, C, and D measure the output signal of inverter stages 1, 2, and 3, respectively, by monitoring the field between the source metallization of the switching transistor and V_{dd}. The clock and complementary clock outputs were measured with respect to V_{dd} at points E and F.

Figure 50 shows the waveforms obtained at points A through D labeled

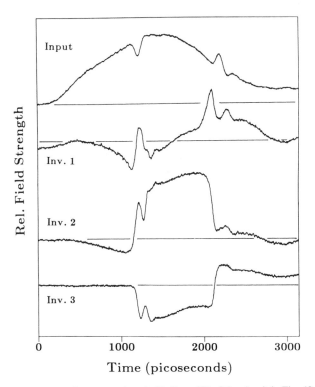

FIG. 50. Measured waveforms at points A, B, C, and D of the circuit in Fig. 49. The one-ns input pulse shows extra structure that occurs at the switching threshold points. The outputs of the three inverter stages show progressive shaping of the input pulse into a square pulse with ~60-ps rise-and falltimes and a 0.9-ns duration.

in the figure as Input, Inv. 1, Inv. 2, and Inv. 3, respectively. The input waveform was measured with the circuit powered up. It exhibits extra structure near the first-stage switching points. This structure arises from the switching transistor's input capacitance charging and discharging at the threshold levels. The next three traces show that for each successive stage, the switching transitions become more clearly defined, the pulse polarity alternates (as it should), and a clear clock pulse evolves with a 56-ps risetime, a 66-ps falltime, and a 0.93-ns duration.

Two other effects are observed in Fig. 50. The first is an apparent skew of the traces from low to high, especially for stages 1 and 3. This is attributed, in part, to crosstalk at the measurement point and in part to loading of the V_{dd} line (which has a nonnegligible resistance) when the inverter stages are drawing current. The latter effect is most evident for Inv. 1 and 3, because they draw current simultaneously. The second effect is the propagation of the structure on the leading edge through the entire input buffer, which does not occur for the trailing edge. This is attributed to the switching characteristic of the first inverter which, in this case, sees an input-pulse amplitude slightly less than what is specified for proper operation. This results in an extra oscillation at the switching point.

Figure 51 shows the propagation delay between stages 2 and 3 to be

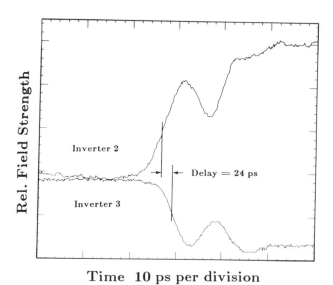

Time 10 ps per division

FIG. 51. Direct measurement of the single-stage propagation delay between inverter stages 2 and 3 (points C and D) of the circuit in Fig. 49. The delay of 24 ps agrees very well with ring-oscillator measurements made on the same types of devices.

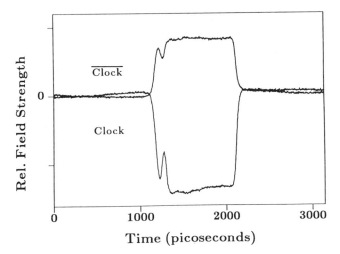

Time (picoseconds)

Fɪɢ. 52. The final clock and complementary clock waveforms from the input buffer section of the circuit in Fig. 49. The internal node probing technique allows precise monitoring of the critical phase relationship of these two signals before they enter the divider section of the circuit.

24 ps, which agrees well with ring oscillator measurements on the same type of device. Figure 52 shows the clock and complementary clock waveforms measured at points E and F. Here, almost no skew in the traces is seen and there is very good symmetry. The clock-line signal is larger because its spacing from the V_{dd} line is larger, which allows more field to enter the probing crystal, even though the field should be less due to the increased spacing.

Figure 53 shows the signal at point E when the circuit is driven by a CW RF signal at one GHz for various bias levels. We see symmetric switching at the proper bias level of 500 mV and, as before, an extra oscillation if the bias is reduced so the peak input signal is too low. Note also that switching time is phase-delayed with respect to the input waveform for smaller bias levels.

b. Silicon NMOS Multiplexer

External electro-optic probing has also been applied to silicon circuits. In this simple test, a 1.7-Gbit NMOS, 12-bit multiplexer circuit was studied. It was driven in synchronism with the mode-locked ring dye laser via the RF driving scheme presented in Section III.E. The tenth harmonic of the laser (at one GHz) was the test frequency. A 300-mV p-p signal at 1.000001 GHz was applied between the clock and V_{dd} supply pins of the

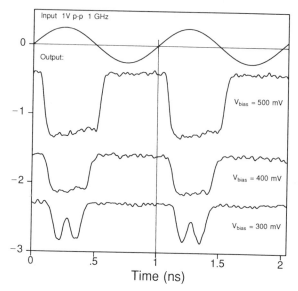

FIG. 53. Waveforms obtained by driving the circuit in Fig. 49 by a CW RF waveform at one GHz. The circuit was probed at point E as a function of DC bias voltage on the input signal. For a proper bias level of 500 mV, a nearly symmetric clock waveform is obtained, while for decreasing bias levels, the waveform becomes asymmetric until anomalous switching occurs.

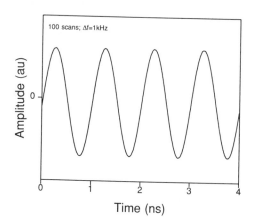

FIG. 54. A one-GHz clock input signal to a 12-bit, 1.7-GHz silicon, NMOS multiplexer, measured on chip by an external electro-optic probe.

packaged chip. The driving frequency was offset from the true tenth harmonic by one kHz to produce the electronic scanning of the measured waveform at a one-kHz rate. This produced a sample signal waveform that visually appears "live," allowing easy adjustment of the probing system. The clock signal is measured internally to the circuit at a 4-μm gap between clock and V_{dd} lines. Sensitivity was adequate to produce the "real-time" trace (at one kHz), with a signal-to-noise ratio of ~20. Figure 54 shows the resultant signal after averaging 100 scans.

c. Subpicosecond Electrical-Pulse Characterization

The temporal resolution of <300 fs attainable by the external electro-optic probe is excessive for virtually all integrated circuits. However, the materials that compose these circuits (i.e., gallium arsenide and silicon) have fundamental physical effects that occur well below one picosecond. One such effect is the photogeneration of carriers or the "photoconductive effect." This effect in GaAs generated a subpicosecond step function for testing the temporal resolution of the modulator structures presented earlier (Valdmanis, 1987). The temporal behavior of photoconductive effects in silicon can be altered by damaging the material on an atomic scale. Typically, this is done by bombarding the material with ions that displace atoms from their normal crystalline sites. It was shown (Ketchen et al., 1986) that in silicon, the carrier lifetime can be reduced to less than one picosecond. Thus, when subpicosecond optical pulses generate carriers in the ion-bombarded material, subpicosecond electrical pulses can be generated. Such short pulses are usually measured by using electronic autocorrelation techniques, but this only estimates the pulse duration and gives no information about the shape of the electrical pulse. A more accurate method of measuring these pulses is with the external electro-optic probe.

A layer of silicon was grown epitaxially on a sapphire substrate to a thickness of 0.6 μm. A coplanar stripline of 5-μm dimensions was evaporated on top of the silicon, and then the entire sample was bombarded by 2-MeV argon ions at a fluency of 3×10^{15} cm^{-2}. Figure 55 shows the experimental arrangement. One electrode was biased at ~10 volts, and the trigger beam was focused between the electrodes in a "sliding-contact" geometry. The resulting electrical pulse was switched to the other electrode and propagated down the line. By placing the electro-optic probe between the electrodes, the electrical pulse was characterized. Figure 56 shows the measured pulse. It had a 375-fs risetime, a 615-fs $1/e$ falltime, and a 750-fs width. This represents the first nonautocorrelation measurement of subpicosecond electrical pulses that enables the true pulse shape to be observed (Valdmanis et al., 1987).

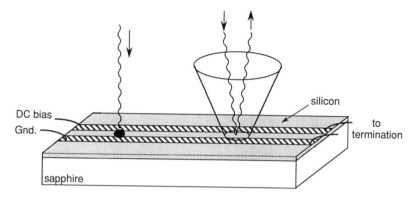

FIG. 55. Experimental geometry for generating and measuring subpicosecond electrical pulses using the "sliding-contact" arrangement on a silicon-on-sapphire substrate. The 100-fs excitation pulse is focused between two 5-μm coplanar-strip transmission lines. The resultant pulse travels down the line and is measured by the external electro-optic probe 30 μm away.

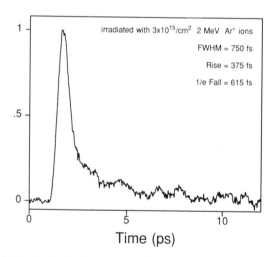

FIG. 56. A 750-fs electrical pulse generated by an ion-bombarded silicon-on-sapphire photodetector in the geometry shown in Fig. 55 and measured with an electro-optic tip.

VI. Nonsampling Techniques

As was discussed in the introduction, there are two converse approaches to making high-speed measurements optically. One uses a "slow" (long-pulse or CW) laser with a fast detector; the other uses a "fast" (short-pulse) laser with a slow detector. The responsibility of attaining resolution lies with the "fast" element. The previous sections showed how a fast laser is used in sampling systems to attain temporal resolution. The resolution was relatively straightforward because of the ready availability of picosecond optical pulses. This section discusses two techniques that take the converse approach and derive their temporal resolution by using novel detectors to achieve picosecond performance.

A. THE PICOSECOND OPTICAL OSCILLOSCOPE

The picosecond optical oscilloscope measurement technique (Valdmanis, 1986) is generally applicable to the characterization of picosecond and subpicosecond optical modulation phenomena. The technique, as it is applied to the investigation of picosecond electrical transients, is described here. The system is based on the real time modulation of a relatively long, frequency-swept (chirped) optical probe pulse. This method encodes temporal information as a function of frequency or wavelength. Then, a spectrograph converts the wavelength-encoded temporal information to the spatial domain for readout. This method could be used to analyze picosecond signals on a single-shot basis.

Figure 57 depicts the experimental arrangement for characterizing electrical signals, which is similar to that for conventional electro-optic sampling. A femtosecond pulse-laser output is split into two beams: a trigger beam and a probe beam. The trigger beam is directed, via a fixed delay line, to the test device where it triggers the generation of the electrical transient to be characterized. The resulting electrical signal then propagates across a travelling-wave Pockels cell and modulates the optical probe pulse.

The probe beam is directed through a dispersive medium, in this case a length of single-mode polarization preserving fiber. Because of the large optical-pulse bandwidth, dispersion readily lengthens the pulse and introduces a frequency sweep along the pulse in time, i.e., chirp. As the probe beam propagates through the electro-optic medium, its temporal profile is modulated in proportion to the profile of the synchronous electrical transient as depicted in Fig. 58. The fixed delay line is adjusted to ensure that the electrical signal arrives within the optical measurement "window." Thus, the entire electrical signal is measured with every optical pulse.

Optical phase compensation is included between the polarizers to ensure the modulator operates in the zeroth order of net phase retardance. The intensity-modulated output pulse is directed into a spectrometer coupled to a detector array and spectrum analyzer for signal retrieval. Spectra with and without modulation, as switched by the chopper, are subtracted to extract the waveform of interest.

Optical pulses of 70 fs with a nontransform limited bandwidth of ~13 nm at 630 nm are generated at 100 MHz by a balanced, colliding-pulse dye

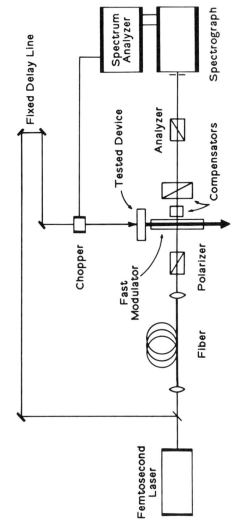

FIG. 57. Experimental arrangement for the picosecond optical oscilloscope. Femtosecond optical pulses are stretched to (as much as) hundreds of picoseconds in an optical fiber and then intensity-modulated by an unknown signal via an electro-optic modulator. As the optical pulses impinge on the grating in the spectrograph, their natural frequency sweep (due to the dispersive stretching process) spreads the temporal modulation information spatially. This spatial modulation, as read by an array of detectors, corresponds to the waveform of the unknown electrical signal.

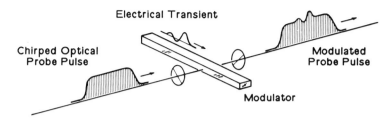

FIG. 58. Visualization of the modulated, stretched optical pulse used in the picosecond optical oscilloscope of Fig. 57.

laser. These probe pulses are subsequently stretched to ~300 ps in 50 m of optical fiber to provide the measurement window. Because the dispersion of glass is very nearly linear over the pulse bandwidth, the chirp of the probe pulse is also linear. No change of the pulse spectrum occurs between the fiber input and output spectrum occurs.

Temporal resolution of this system is ultimately determined by convolving the original pulse duration with a time given by the stretched pulse duration divided by the ratio of pulse bandwidth to spectrometer resolution. In this experiment, a 0.3-m spectrometer coupled to a 1024-element Reticon array and EG&G PARC OMA-III spectrum analyzer was used. The system resolution is 0.04 nm and can resolve 325 points within the spectrum of the probe pulse, yielding a resolution of less than one picosecond. By stretching the pulse less, this limiting temporal resolution can be as short as the original pulse duration. Another temporal resolution limit arises from the time–wavelength duality of the chirped pulse. Additional frequencies or wavelengths (i.e., sidebands) introduced by the impressed modulation at a particular location in time (and hence wavelength) will appear to have come from "neighboring" times, thus broadening the modulation. This temporal limit can be approximated by the square root of the product of the stretched pulse duration and the original pulse duration. In this example, this resolution is on the order of 3 ps. It is also interesting to note that as the chirped optical pulse impinges on the spectrometer grating (at 100 MHz), the diffracted light effectively reconstructs a wavefront that sweeps across the detector array, analogous to the electron beam in a real-time oscilloscope or streak camera. However, in the optical "oscilloscope," the effective sweep speed can well exceed the speed of light!

The modulator arrangement was optically biased close to zero transmission to maximize the modulation depth of the transmitted light, even though the overall intensity level at this point is significantly lower (Williamson and Mourou, 1985). This is necessary because even near zero transmission, the modulation depth resulting from picosecond signals can

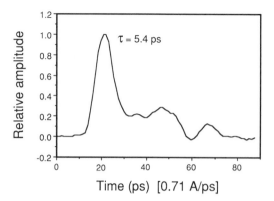

FIG. 59. Electrical signal measured by the picosecond optical oscilloscope shown in Fig. 57.

often be less than 10^{-3} and the dynamic range of our multichannel analyzer is limited to ~10^4. This trick to increase the modulation depth sacrifices overall light level, hence decreasing signal-to-noise ratio, but does not adversely affect linearity for small signals compared to conventional e-o modulation. By synchronizing the trigger beam chopper with alternating scans of the array detector system, background spectrum subtractions were accumulated "on the fly" to enhance the signal-to-noise ratio for small signals. Larger signals were detected essentially "live" on a single scan basis. Typical scan times were 100 ms per scan.

An electro-optic application of the optical oscilloscope is demonstrated by measuring the impulse response of an ion-bombarded GaAs photoconductive detector. The standard hybrid balanced-line–type modulator was used, as depicted in Fig. 16. In this case, the electrical signal propagates across a 250-μm–thick balanced-line travelling-wave LiTaO$_3$ modulator with the resulting electric field along the c-axis. The chirped optical pulse is polarized at 45 degrees and propagates normal to the c-axis. The detector is attached directly to the end of the modulator crystal. Figure 59 shows the resulting electrical waveform averaged over 1000 scans. The temporal scale was 1.15 ps per channel or 1.4 ps per Angstrom. The risetime of 5.4 ps is limited by the modulator configuration. The displayed waveform in the figure was adjusted to compensate for the spectral shape of the measurement pulse even though, in this case, the effect is small.

B. The Electro-Electron Optic Oscilloscope

Another nonsampling technique for measuring ultrafast electrical signals that exploits the high speed of electro-optic modulators is the picosecond electro-electron optic oscilloscope (Williamson and Mourou, 1985). It is

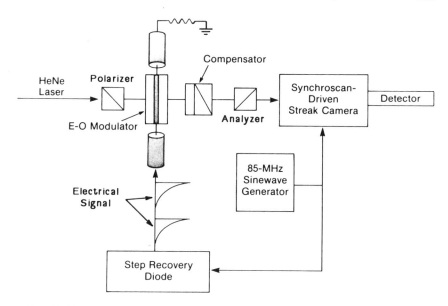

FIG. 60. Experimental configuration of the electro-electron optic oscilloscope. Unlike sampling systems, this scheme uses a CW optical source and obtains its temporal resolution by employing an ultrafast optical detector (a streak camera) to display the electro-optic signal modulation in real time. The signal-to-noise ratio is enhanced by averaging at MHz rates enabled by synchronous operation of the camera and the signal source. The resultant signal is read out by a linear detector array. (Figure courtesy of S. Williamson, University of Rochester)

schematically depicted in Fig. 60. This technique also reverses the temporal roles of the optical source and its detector. The picosecond pulse source is replaced by a continuous source, such as a conventional helium-neon laser, and the "slow" detector is replaced by an ultrafast synchroscan streak camera. A streak camera is comprised of an electron image converter tube that converts an optical pulse into an electron pulse replica via the photoelectric effect. The system is like a standard oscilloscope, but the role of the vertical amplifier and deflection plates is replaced by the ultrafast electro-optic modulator, and instead of modulating the vertical position of the beam, its intensity is modulated.

The Pockels cell can be of any design, but instead of conventional repetitive, delayed sampling as described earlier, the continuous-wave laser beam is repetitively modulated by the electrical signal to be measured. The modulated optical beam is then incident on the photocathode of the streak camera, which replicates the optical signal as an intensity-modulated electron beam. The electron beam is synchronously swept, or

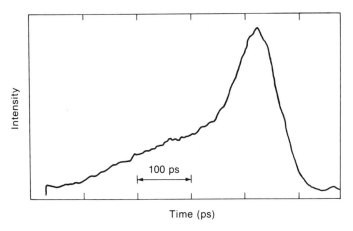

FIG. 61. Electro-electron optic measurement of the signal from a step-recovery diode. The risetime of 40 ps is limited by the Pockels cell used in this experiment. Also note that time in this figure is reversed and runs from right to left. (Figure courtesy of S. Williamson, University of Rochester)

streaked, on a phosphor screen, thus reproducing the original electrical signal as an intensity-versus-displacement display. A linear detector array reads out the resultant display into a multichannel signal averager that averages successive traces at MHz repetition rates to increase the signal-to-noise ratio. The key element in this system is the averaging of traces that is possible because the streak camera and the signal source are driven in precise synchronism by a common radio-frequency source.

In this experiment, the temporal response of a step-recovery diode is characterized. The Pockels cell is a commercially available unit with 40-ps resolution. The laser is a one-mW helium-neon CW laser. Both the step-recovery diode and the plates of the streak camera are driven by a master oscillator at 85 MHz. A phase shifter sets the temporal synchronization between the diode and the camera. Figure 61 shows the resultant trace after 10 seconds of accumulation, which shows the risetime of the Pockels cell to be the limiting element in this case at 40 ps.

The temporal resolution of this technique is limited only by the resolution of the streak camera. Current models as of 1989 achieve resolutions of approximately 2 ps and Hamamatsu Photonics Co. has recently introduced a camera that achieves subpicosecond performance. Sensitivity is presently on the order of 100 mV. This system can be operated with lasers at any wavelength between 250 and 1000 nm. Elimination of the need for a picosecond laser source makes the system small enough to be portable.

VII. Conclusion

This chapter presented a high-speed measurement technique based on use of electro-optic modulators in a wide variety of configurations. As of the late 1980s, electro-optic sampling is the fastest technique available for the measurement of electrical signals either in discrete devices or in integrated circuits. The inherent high-speed response of electro-optic materials coupled with the availability of picosecond and subpicosecond layer systems enables electrical measurements to be made with unprecedented bandwidths exceeding one THz.

References

Auston, D.H., and Glass, A.M. (1972). *Appl. Phys. Lett.* **20** (10), 398–399.

Auston, D.H., Cheung, K.P., Valdmanis, J.A., and Kleinman D.A. (1984). *Phys. Rev. Lett.* **53** (16), 1555–1558.

Berenz, J.J. (1987). In *Picosecond Electronics and Optoelectronics* **II**, F.J. Leonberger, C.H. Lee, F. Capasso, and H. Morkoc, eds. Springer–Verlag, New York, 165–168.

Bloom, D.M., and Sollner, G., eds. (1989). *Picosecond Electronics and Optoelectronics* **III.** Springer-Verlag, New York.

Bozler, C.O., and Alley, G.D., (1980). *IEEE Trans. Electron Devices* **ED-27,** 1128.

Chen, Y.K., Nottenburg, R., Panish, M., Hamm, R.A., and Humphrey, D.A. (1989). *IEEE Elect. Dev. Lett.* **10,** 267.

Cheung, K.P., and Auston, D.H. (1986). *Infrared Phys* **26** (1), 23–27.

Dykaar, D.R. (1988). By permission. Unpublished.

Dykaar, D.R., Hsiang, T.Y., and Mourou, G.A. (1985). *IEEE Trans. Magn.* **MAG-21,** 230.

Dykaar, D.R., Sobolewski, R., Whitaker, J.F., Hsiang, T.Y., Mourou, G.A., Hollis, M.A., Clifton, B.J., Nichols, K.B., Bozler, C.O., and Murphy, R.A. (1986). In *Ultrafast Phenomena* V, G.R. Fleming and A.E. Siegman, eds. Springer-Verlag, New York, 103–106.

Dykaar, D.R., Sobolewski, R., Hsiang, T.Y., and Mourou, G.A. (1987). *IEEE Trans. Magn.* **MAG-23,** 767.

Dykaar, D.R., Sobolewski, R., Chwalek, J., Hsiang, T.Y., and Mourou, G.A. (1988). In *Advances in Cryogenic Engineering* 33, R.W. Fast, ed. Plenum, New York, 1097–1104.

Fleming, G.R., and Siegman, A.E., eds. (1986). *Ultrafast Phenomena* V. Springer-Verlag, New York.

Freeman, J.L., Diamond, S.K., Fong, H., and Bloom, D.M. (1985). *Appl. Phys. Lett.* **47,** 1083–1084.

Hasnain, G., Arjavalingam, G., Dienes, A., and Whinnery J.R. (1983). *Proc. SPIE Conf. on Picosecond Opto-electronics*, **439** SPIE, Bellingham, WA.

Hsiang, T.Y., Whitaker, J.F., Sobolewski, R., Dykaar, D.R., and Mourou, G.A. (1987). *Appl. Phys. Lett.* **51,** 1551.

Kaiser, W. (1988). *Ultrashort Laser Pulses and Applications.* Springer-Verlag, New York.

Kaminow, I.P. (1974). *An Introduction to Electro-optic Devices.* Academic Press, New York.

Ketchen, M.B., Grischkowsky, D., Chen, T.C., Chi, C.C., Duling I.N., Halas, N.J., Halbout, J.M., Kash, J.A., and Li, G.P. (1986). *Appl. Phys. Lett* **48** (12), 751–753.

Koechner, W. (1976). In *Solid State Laser Engineering.* Springer-Verlag, New York, 489.

Kolner, B.H., and Bloom, D.M. (1984). *Electron. Lett.* **20,** 818–819.

Kolner, B.H., and Bloom, D.M, (1986). *IEEE J. Quan. Elect.* **QE-22** (1), 79–93.

Lathrop, D.K., Russek, S.E., and Buhrman, R.A. (1987). *Appl. Phys. Lett.* **51** (19), 1554.

Kryzak, C.J., Meyer, K.E., and Mourou, G.A. (1985). In *Picosecond Electronics and Optoelectronics*, G.A. Mourou, D.M. Bloom, and C.H. Lee, eds. Springer-Verlag, New York, 244–248.

Leonberger, F.J., Lee, C.H., Capasso, F., and Morkoc, H., eds. (1987). *Picosecond Electronics and Optoelectronics* **II**. Springler-Verlag, New York.

Liu, H.C., and Coon, D.D. (1987). *Appl. Phys. Lett.* **50**, 1246.

Meyer, K.E., and Mourou, G.A. (1985). In *Picosecond Electronics and Optoelectronics*, G.A. Mourou, D.M. Bloom, and C.H. Lee, eds. Springer-Verlag, New York, 46–49.

Meyer, K.E., Dykaar, D.R., and Mourou, G.A. (1985). In *Picosecond Electronics and Optoelectronics*, G.A. Mourou, D.M. Bloom, and C.H. Lee, eds. Springer-Verlag, New York, 54–57.

Mourou, G.A., and Meyer, K.E. (1984). *Appl. Phys. Lett.* **45** (5), 492–494.

Mourou, G.A., Bloom, D.M., and Lee, C.H., eds. (1985). *Picosecond Electronics and Optoelectronics*. Springer-Verlag, New York.

Nees, J., and Mourou, G.A. (1986). *Electron. Lett.* **22**, 918–919.

Nuss, M.C., Auston, D.H., and Capasso, F. (1987). *Phys. Rev. Lett.* **58** (22), 2355–2358.

Pockels, F. (1906). In *Lehrbuch der Kristalloptic*. Teubñer, Leipzig. ·

Rodwell, M.J.W., Weingarten, K.J., Bloom, D.M., Baer, T., and Kolner B.H. (1986). *Optic Letters* **11** (10), 638–640.

Shah, N.J., Pei, S.-S., Tu, C.W., and Tiberio, R.C. (1986). *IEEE Trans. Electron. Dev.* **ED-33** (5), 543–547.

Sollner, T.C.L.G., Goodhue, W.D., Tannerwald, P.E., Parker, C.D., and Peck, D.D. (1983). *Appl. Phys. Lett.* **43**, 588.

Sollner, T.C.L.G., Brown, E.R., Goodhue, W.D., and Le, H.Q. (1987). *Appl. Phys. Lett.* **50**, 332.

Taylor, A.J., Tucker, R.S., Wiesenfeld, J.M., Burrus, C.A., Eisenstein, G., Talman, J.R., and Pei, S.S. (1986a). *Elect. Lett.* **22** (20), 1068–1069.

Taylor, A.J., Wiesenfeld, J.M., Eisenstein, G., Tucker, R.S. (1986b). *Appl. Phys. Lett.* **49**, 681.

Valdmanis, J.A. (1986). In *Ultrafast Phenomena* V. G.R. Fleming and A.E. Siegman, eds. Springer-Verlag, New York, 82–85.

Valdmanis, J.A. (1987). *Electron. Lett.* **23**, 1308–1310.

Valdmanis, J.A. (1988). *Proc. Conf. on Lasers and Electro-Optics*, Anaheim, California, Optical Society of America, Washington, D.C.

Valdmanis, J.A., and Mourou, G. (1984). In *Picosecond Optoelectronic Devices*, C.H. Lee ed. Academic Press, Orlando, Florida, 249–270.

Valdmanis, J.A., and Mourou, G. (1986). *IEEE J. Quan. Elect.* **QE-22** (1), 69–78.

Valdmanis, J.A., and Pei, S.S. (1987). In *Picosecond Electronics and Optoelectronics* **II**, F.J. Leonberger, C.H. Lee, F. Capasso, and H. Morkoc, eds. Springer-Verlag, New York, 4–10.

Valdmanis, J.A., Mourou, G., and Gabel, C.W. (1982). *Appl. Phys. Lett.* **41** (3), 211–212.

Valdmanis, J.A., Mourou, G., and Gabel, C.W. (1983). *IEEE J. Quan Elect.* **QE-19**(4), 664–667.

Valdmanis, J.A., Fork, R.L., and Gordon, J.P. (1985). *Optics Letters* **10**, 131–133.

Valdmanis, J.A., Nuss, M.C., Smith, P.R., and Li, K.D. (1987). *Proc. Int. Electron. Dev. Meeting*, Washington D.C. IEEE, New York.

Weingarten, K.J., Rodwell, M.J.W., Heinrich, H.K., Kolner, B.H., and Bloom D.M. (1985). *Electron. Lett.* **21**, 765–766.

Whitaker, J.F. (1988). By permission. Unpublished.

Whitaker, J.F., Sobolewski, R., Dykaar, D.R., Hsiang, T.Y., and Mourou, G.A. (1988). *IEEE* **MTT 36** (February), 277–285.

White, I.H., Gallagher, D.F.G., Osinski, M., and Bowley D. (1985). *Electronics Lett.* **21**, 197.

Wiesenfeld, J.M., Tucker, R.S., Antreasyan, A., Burrus, C.A., Taylor, A.J., Mattera V.D. Jr., and Garbinski, P.A. (1987). *Appl. Phys. Lett.* **50**, 1310.

Williamson, S. and Mourou, G.A. (1985). In *Picosecond Electronics and Optoelectronics*, G.A. Mourou, D.M. Bloom, and C.H. Lee, eds. Springer-Verlag, New York, 58–61.

Wolf, P. (1985). In *Picosecond Electronics and Optoelectronics*, G.A. Mourou, D.M. Bloom, and C.H. Lee, eds. Springer-Verlag, New York, 236–243.

Yamashita, E., and Atsuki, K. (1976). *IEEE Trans. Microwave Theory Tech.* **MTT-24**, 195–200.

Yariv, A., and Yeh, P. (1984). *Optical Waves in Crystals*. Wiley, New York, 220.

CHAPTER 5

Direct Optical Probing of Integrated Circuits and High-Speed Devices

J.M. Wiesenfeld

CRAWFORD HILL LABORATORY
AT&T BELL LABORATORIES
HOLMDEL, NEW JERSEY

R.K. Jain

AMOCO RESEARCH CENTER
NAPERVILLE, ILLINOIS

I. Introduction

A. General Motivation

The continuing increase in the operating speed of ultrafast electronic devices and integrated circuits (ICs) has outpaced the development of conventional electronic testing equipment (Gheewala, 1987). For example,

221

in the domain of digital devices, selectively doped heterostructure transistors (SDHT) which have 10-ps propagation delay at room temperature, as measured by a ring oscillator circuit, have been fabricated (Shah et al., 1986). A digital small-scale integrated (SSI) circuit with an 18-GHz clock frequency has been reported (Jensen et al., 1986), as has a 26.7-GHz frequency divider circuit (Mishra et al., 1988). In the domain of analog devices, a GaAs metal-semiconductor field-effect transistor (MESFET) with a 110-GHz oscillation frequency has been demonstrated (Tserng and Kim, 1985) and a pseudomorphic InGaAs/AlGaAs modulation-doped field-effect transistor (MODFET) with a maximum frequency of oscillation, f_{MAX}, that extrapolates to 200 GHz has been reported (Henderson et al., 1986). To evaluate such devices, measurement techniques that provide both high temporal resolution and, for ICs, the ability to probe noninvasively at nodes inside ICs are required. (See, for instance, Gheewala, 1987.) The need for ultrahigh temporal resolution is obvious from the operating speed of the devices just cited. The ability to probe internal nodes noninvasively will be critical to the development of very-high–speed GaAs large-scale integrated (LSI) circuits, where design models and knowledge of the layout dependence of the circuit parasitics are not well developed (Weingarten et al., 1988).

A measurement problem exists for conventional electronic instrumentation. Fundamental constraints imposed by circuit parasitics (for example, stray capacitances and inductances associated with leads) make the use of electronic instrumentation for the measurement of ultrahigh-speed electronic devices and circuits exceedingly difficult. Indeed, as elaborated by Weingarten et al. (1988), high-impedance electronic probes that could be useful for probing internal nodes of ICs necessarily have large parasitics at high frequencies, rendering them useless. Low-impedance probes, which can accommodate higher frequencies (Strid et al., 1985), can only be used at IC input and output pads and cannot probe internal nodes. The problem with electronic-measurement instrumentation may become more severe in the future, with a widening of the gap between the ultrahigh speed of electronic devices and the measurement capabilities of high-speed electronic instrumentation. Thus, new technologies for possible solutions to this measurement problem are required.

Optical sampling techniques using ultrashort pulse lasers are a promising solution, at least from the perspective of the desired bandwidth capabilities. For instance, laser pulses as short as 6 femtoseconds have been generated (Fork et al., 1987). Using such pulses, ultrashort optical phenomena that occur at time scales of less than 100 femtoseconds have been resolved. (See, for example, Becker et al., 1988). Furthermore, the spatial localization of a laser beam enables spatially specific probing at individual nodes inside ICs. In this chapter, we review *direct* optical probing techniques, mainly using ultrashort pulse lasers, for high-speed ICs and devices. By direct optical probing, we mean techniques that apply an optical beam to the device under test (DUT) and use a physical property of the IC or device itself as the basis for the sampling measurement. For ICs, the DUT is considered to be that section of the IC being tested and not necessarily the entire IC. In Chapter 4, Valdmanis describes *external* optical probing of devices and ICs using electro-optic sampling. In the external probing geometry, the optical probe beam is applied to an electro-optic crystal which interacts with the electric fields of the DUT, but which is part of the measurement system and not part of the DUT.

B. SUMMARY OF PROBING EFFECTS

The effects used to probe ICs and devices can be placed into two broad categories: those that are *charge-related* and those that are *field-related*. Charge-related effects encompass a variety of techniques, including creation of carriers by photoexcitation, or sensing of the presence of carriers by their effect on optical probe beams via changes in index of refraction or absorption. The first picosecond optical technique for ICs involved creation of carriers by photoexcitation of GaAs MESFETs plus observation of the effect of the injected carriers on the electronic operating characteristics of the circuit (Jain, 1982; Jain and Snyder, 1983a,b). More recently, the presence of carriers inside active semiconductor devices, such as diodes and field-effect transistors (FETs), has been interrogated by using the carrier-induced perturbations on the refractive index (Heinrich et al., 1986a) or absorption (Chemla et al., 1987) of the substrate at the position of an optical probe beam.

The field-related effect used for optical probing is the electro-optic effect, which was discovered by Pockels (1906) at the beginning of this century. In the electro-optic effect, the presence of an electric field across the device substrate, which is caused by a voltage applied to a contact on the device, causes a birefringence in the substrate. The birefringence affects an optical probe beam by altering its state of polarization, thus enabling the sensing of the voltage applied to the device. For direct probing, it is necessary that the substrate be electro-optic, which GaAs is and silicon is not. (Silicon DUTs could be probed with external samplers, however, as discussed in Chapter 4.) The use of the electro-optic effect to measure an ultrashort electrical waveform was first demonstrated by Auston and coworkers (Auston and Glass, 1972; LeFur and Auston, 1976). Subsequently, the technique has been extended by other workers (Alferness et al., 1980; Valdmanis et al., 1982) and was first used for direct optical sampling in a GaAs IC by Weingarten et al. (1985).

Reviews on optical sampling of high-speed electrical signals include Valdmanis and Mourou (1986), Kolner and Bloom (1986), Kolner (1987), Zhang and Jain (1987), Wiesenfeld et al. (1987a), Weingarten et al. (1988), and Auston (1988). In this chapter, we shall emphasize testing of devices fabricated from III-V semiconductors (GaAs, InP, etc.) and their alloys, although the discussion includes some testing results for silicon devices.

C. OUTLINE OF CHAPTER

In Section II, we discuss general considerations concerning sampling of DUTs. We cover common features of excitation of the DUT and probing systems, such as spatial and temporal resolution, voltage-sensitivity, accuracy and calibration, and invasiveness. Section III presents details of a class of techniques useful for digital circuits, in which circuit operation is controlled by injection of carriers into discrete points of the DUT by photoexcitation. Various properties of the DUT are studied by using either a second optical pulse and slow electronic instrumentation or fast electronic instrumentation to monitor the output of the DUT. Section IV describes electro-optic sampling and comprises the major portion of this chapter. The electro-optic sampling technique is capable of measuring voltage waveforms internal to ICs with picosecond temporal resolution and sub-mV/$\sqrt{\text{Hz}}$ sensitivity. It is applicable to a large number of general testing problems, but suffers the major limitation of requiring that the DUT be fabricated on an electro-optic substrate (for example, GaAs or InP). In particular, silicon devices cannot be studied by direct electro-optic sampling. We also include in Section IV, for completeness, discussion of some measurements of discrete devices, such as photodiodes, in which the voltage produced by the DUT is sent to a hybrid external sampler (which is fabricated on an electro-optic substrate) and measured there. Optical probing by free carrier refraction is discussed in Section V. In this method, the index of refraction of the DUT is perturbed by the presence of free carriers, which can be sensed optically. The technique has been applied to silicon and GaAs devices. A new technique, phase-space absorption quenching, which is applicable to devices based on quantum well structures, is discussed briefly in Section VI. In this technique, the absorption of a probe beam is modified by changes of carrier density in the quantum well of the active region of the device. Finally, in Section VII we compare and contrast the probing techniques, and we consider some anticipated needs for future optical probing systems.

In addition to discussion of the basic principles, we emphasize measurement-system considerations (such as resolution and sensitivity) throughout the chapter. The literature on optical probing is sufficiently recent (as of 1989) that we attempt to discuss all published measurements for each of the techniques.

II. General Considerations

A. PROBING SCHEMES

In most general terms, a probing system consists of a means of excitation of the device under test, the DUT, and a probe. The DUT may be a discrete device, an IC, or a section of an IC. In the lattermost case, other parts of the IC are considered external to the DUT, even though they are fabricated on the same chip. In this chapter, the emphasis is on techniques in which a laser beam is involved in the probing and/or excitation of the DUT. The various excitation and probing methods are briefly discussed next. We distinguish between sampling and real-time probing. In a sampling scheme, the waveform to be measured is reconstructed by multiple averaging of the signal at each time position of the waveform. In real-time probing, an entire signal waveform is captured as it occurs.

1. Excitation of the DUT

Conventionally, the DUT is excited by electrical input. If the DUT is an integrated circuit, it is necessary to provide proper input signals and timing as well as bias voltages. For wafer-level testing of high-speed ICs, it may be necessary to use special high-frequency input probes (Strid et al., 1985) to excite the circuit. Electrical excitation corresponds to normal operation of the DUT, and probing under such conditions is appropriate.

For photodiode DUTs, optical excitation is required. Also, in certain cases, an optical pulse can be used to provide on-chip excitation of an IC. In principle, high-speed photodetectors could be incorporated into future ICs as optical address points for generation of test electrical pulses. However, some standard circuit components on existing ICs are already photosensitive. For example, a GaAs MESFET acts as a photoconductive detector when a short light pulse illuminates the region between the source and drain, and such MESFETs have been used to generate electrical pulses optically on an IC for high-speed analysis of the IC (Jain and Snyder, 1983a,b). For excitation of the DUT by optical means, it is not necessary to provide all the electrical inputs to the DUT, although it is necessary to provide appropriate bias voltages.

2. Probing

Optical probes are spatially localized, enabling examination of individual nodes inside the DUT. As described previously, the optical-probing techniques are classified as field-sensitive or charge-sensitive. The field-sensitive technique is electro-optic sampling, which, for direct probing, requires that the substrate be an electro-optic material. (Electro-optically inactive substrates may be probed with external samplers, as discussed in

EXCITATION   PROBING

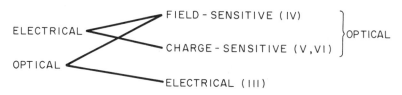

FIG. 1. Combinations of excitation and probing techniques described in this chapter. Roman numerals in parentheses refer to section in which the probing technique is discussed.

Chapter 4). The charge-sensitive techniques, on the other hand, will work for devices fabricated on any substrate material, because they rely on the interaction of the optical probe beam with carriers in the active region of devices. In certain schemes, the conventional electronic output of the device is measured as the device is excited optically at internal nodes.

The combinations of excitation and probing techniques discussed in this chapter are illustrated in Fig. 1. Subsequent sections in this chapter are arranged by probing method. In Section III, experiments using optical excitation and conventional electronic probing are discussed. Section IV presents the technique of electro-optic sampling, with excitation by both optical and electronic sources. The charge-sensitive probing techniques (charge-sensing by refractive index modulation and phase space absorption quenching (PAQ)) are discussed in Sections V and VI, respectively.

B. MEASUREMENT-SYSTEMS CONSIDERATIONS

Specifications for the optical-probing technique involve the spatial resolution, the temporal resolution or bandwidth, sensitivity or signal-to-noise ratio, accuracy, linearity, absolute calibration, and invasiveness. Each of these issues is discussed in general terms next. Not all factors will be appliable to all probing techniques.

Spatial Resolution. The spatial resolution is ultimately limited by the diameter of the focused laser beam used to probe and/or excite the DUT, which is comparable to the operating wavelength of the laser (about one μm for most experiments discussed in this chapter). The dimensions of active regions and contact lines in state-of-the-art high-speed ICs can be comparable to, or smaller than, one μm. Because of the small dimension, uncertainty in positioning of the probe (excitation) beam may lead to reduction of sensitivity of the probe (excitation) compared to theory. Diffraction effects, depending in detail on the precise probing (excitation) geometry, may further reduce spatial resolution. Crosstalk due to optical sampling of signals from nearby active lines or devices will also depend on spatial resolution.

Temporal Resolution. The temporal resolution (bandwidth) of the measurement is determined primarily by the pulse duration of the laser probe (excitation) beam or by the bandwidth of the photoreceiver that detects the probe beam. Other factors affecting the temporal resolution are the response time of the physical effect used for probing (excitation), jitter between the electronic signal exciting the DUT and the probe pulse train, and transit-time effects. The relevant transit-time effects are (1) the transit time of the optical probe beam through the region of the DUT in which the time-varying fields or charges are present and (2) for electro-optic sampling, the transit time of the electrical signal across the diameter of the focused probe beam.

Sensitivity. The sensitivity of the measurement (in $V/\sqrt{Hz}$) is determined by the magnitude of the physical effect used for probing or excitation, by the average optical power, and by the noise level of the electronics used to process the probe signal. For example, to achieve the quantum limit in signal detection, which is the shot-noise limit, it is advantageous to have high optical power.

Accuracy, Linearity, and Calibration. The accuracy of the measurement depends on the physical effect and somewhat on the spatial resolution. To convert the measured signal to a voltage or carrier density, it is necessary to measure the fractional modulation of the probe beam. For the techniques that measure carrier density (that is, plasma refraction and PAQ), it is necessary also to know the capacitance at the probe position in order to convert the measured signal to a voltage. Calibration of a signal observed at one position on the DUT with those observed at other positions will depend somewhat on geometric factors, such as the particular layout of the IC. The linearity of the measurement depends on the physical effect used for probing or excitation, but, in the case of electro-optic sampling, can extend over a range of 120 dB (Kolner and Bloom, 1986; Kolner, 1987). Crosstalk may be a problem in absolute calibration of observed signals.

Invasiveness. The major possibility for disturbing the device characteristics by optical probing is the generation of extraneous carriers by absorption of the probe beam in the DUT. Therefore, the probe beam is chosen to have a wavelength well below the bandgap of the device (except for PAQ), where direct band-to-band absorption is negligible, and only absorption due to defects is possible. For GaAs and InP, the absorption due to defects decreases as the wavelength of the probe increases; therefore, a probe wavelength as long as possible is desirable.

III. Direct Probing by Photocarrier Generation

In this section we discuss techniques in which carriers are introduced into an IC by optical excitation of a photoactive element in the IC, such that the photoactive element is external to the specific device being tested. Two such optoelectronic techniques have been employed for the measurement of on-chip response times and logic-gate propagation delay:

- Differential observation of output waveforms and
- Optically induced logic-level sampling.

Both techniques are applicable only to digital ICs and rely on the presence of particular circuit elements, such as photoactive FETs. In related work, optical addressing of transistors in very–large scale integrated (VLSI) circuits with weak probe beams has been used to extract timing diagrams for devices operating at rates up to tens of MHz (Henley, 1984). This method relies on the coupling of induced photocurrents to the power bus of the circuit, followed by extensive computer analysis of the results. We do not discuss this technique further.

Both of the picosecond optoelectronic techniques require the selective illumination of specific on-chip devices with picosecond optical pulses, often in pairs at controlled, jitter-free interpulse optical delays, t_d. As seen in Fig. 2, the optical pulse pairs are generated by (1) first passing a beam containing a continuous train of picosecond optical pulses from a mode-locked dye laser through a 50% beam-splitter, which splits each picosecond optical pulse into two equal parts, and then (2) redirecting the two pulse trains toward the desired devices on the circuit with reflecting mirrors and focusing lenses. The wavelength of the optical radiation is shorter than the band edge of the device material, so that the optical pulses are absorbed, creating free carriers. The delay between the pulse pairs can be varied by changing the path length of one optical beam with respect to the other. Micron-size control in the displacement (x) of the prism as well as submicron jitter in the vibration of the optical elements lead to the generation of optical pulse-pairs with subpicosecond control of the interpulse separations t_d ($=2x/c$) and negligible interpulse jitter.

A key step in the implementation of the optoelectronic-measurement techniques is the optical generation of ultrashort electrical pulses directly on-chip, which, in turn, behave as low-jitter sources and samplers. For

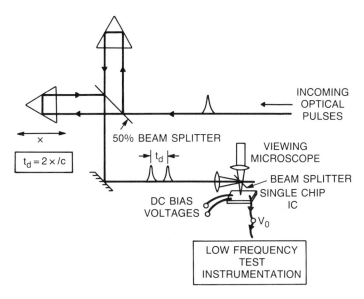

FIG. 2. Experimental arrangement for probing by photocarrier generation using two short pulses prepared by splitting one pulse into two and using an adjustable delay. (Jain, 1984. Reprinted from *Test & Measurement World.* © 1984 by Cahners Publishing, a division of Reed Publishing, USA.)

digital circuits, such ultrashort voltage pulses must have sufficient amplitude to correspond to logic-level swings. On-chip logic-level pulses (LLPs) may be generated by fabricating special on-chip detectors, such as photoconductors, photosensitive FETs, or Schottky-barrier photodiodes. However, these can be conveniently realized by using photosensitive elements (such as FETs) that may already exist within the IC. Such direct addressing of ICs has been demonstrated (Jain and Snyder, 1983a,b) in GaAs logic circuits of the buffered FET design, using above-bandgap pulses of ~10 pJ energy. The technique may also be applicable to various other logic families, circuit designs, and semiconductor materials.

In GaAs depletion-mode FET logic circuits, each of the FETs represent natural address points, since efficient carrier generation (and rapid carrier collection) can readily occur in the high-field region between the gate and drain electrodes. On-chip logic-level pulses (about 1 V amplitude) have been generated with less than 10 pJ of optical energy (<1 mW of average optical power in a 100-MHz pulse train) by focusing 5-ps–duration pulses from a mode-locked visible dye laser tightly on the gate-drain region of the addressed FETs. Typical laser spot sizes used are of the order of 5 μm. Figure 3 shows the circuit diagram of a GaAs depletion mode FET NOR gate. This gate is a basic logic unit of the IC studied by Jain and Snyder

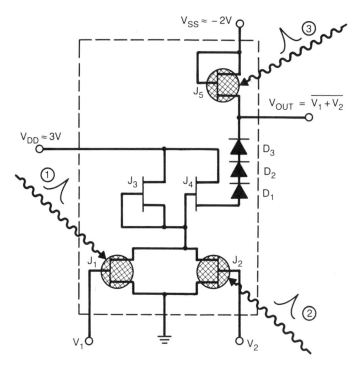

FIG. 3. Circuit diagram of GaAs depletion mode FET NOR gate studied by Jain and Snyder (1983b). FETs J_1, J_2, and J_5 are optical address points. (© 1983 IEEE)

(1983a,b),. If inputs 1 and 2, shown in Fig. 3, are maintained at logic 0 so that the outputs are normally 1, appropriate illumination of any one of J_1, J_2, or J_5 will result in a short 1–0 logic-level pulse (that is, the pulse that goes from a baseline value of logic 1 to a peak value of logic 0). For all the measurement techniques described here, it will be required that when more than one logic gate is illuminated by temporally distinct pulses, the same FET will be illuminated within each gate. Thus, the delay between the optical initiation of a logic-level pulse and its formation at the output of the logic gate may be assumed nearly identical for each gate that is optically addressed. This precaution circumvents errors due to mismatch of delays that are internal to the logic gates.

A. DIFFERENTIAL MEASUREMENT OF OUTPUT WAVEFORMS

In this technique, propagation delays are measured by observing differences in the timing or the width of optically triggered logic-level waveforms at the output of the circuit. Two examples are discussed next.

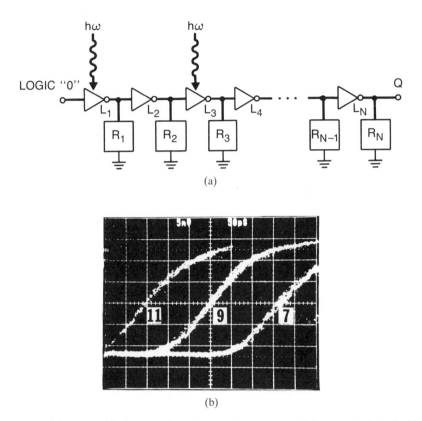

(a)

(b)

FIG. 4. (a) String of logic gates. In this example, gates L_1 and L_3 are illuminated with ultrashort optical pulses. (b) Experimental results. Horizontal scale is 50 ps/division. (Zhang and Jain, 1986)

In a simple implementation of this technique (Jain and Zhang, 1986), a string of logic gates is fabricated on a single chip as shown in Fig. 4a, and the propagation delay in inverting gates may be measured by (1) first optically switching one of the inverters (say, L_1), (2) recording the output waveform on a sampling scope, (3) optically switching another inverter (say L_3), and (4) observing the shift of this waveform relative to the first one, on the oscilloscope display. The sampling scope trigger signal in both cases is obtained from a fast-detector illuminated by a portion of the optical pulse. The difference in the observed transition is due to the propagation delay between the optically illuminated gates. Note that the resolution of the technique is not limited by the risetime of the oscilloscope but by its jitter, which can be as small as 10 ps. Thus, for the preceding measurement (two gate delays) the uncertainty can be as low as 5 ps per gate delay. Data obtained by this measurement are shown in Fig. 4b. The

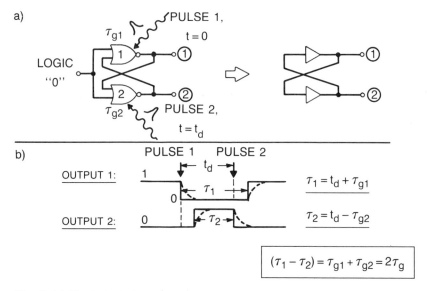

FIG. 5. (a) Simple latch circuit with two NOR gates. (b) Expected waveforms at the two outputs after optical excitation. Gate delays τ_g are inferred from measured electrical pulse widths, τ_1 and τ_2, as shown. (Jain, 1984. Reprinted from *Test & Measurement World*. © 1984 by Cahners Publishing, a division of Reed Publishing, USA.)

delay between gates 7 and 9 is 170 ps. Between gates 9 and 11, the delay is 150 ps. From these results, the average propagation delay is 80 ps per gate.

In another implementation of the differential technique (Stenersen and Jain, 1984), the two NOR gates of a simple latch (Fig. 5a) are switched with a pair of optical pulses separated by an interpulse delay t_d such that $t_d \gg t_{g1}, t_{g2}$, where t_{g1} and t_{g2} are delays of the gates comprising the latch. Gate propagation delays are then given by the difference in the widths of a complementary pair of pulses seen at the outputs, as depicted in Fig. 5b. We clarify this technique by considering a specific experiment that was performed by optically switching the gates in the output latch of the divide-by-two D-flip-flop circuit (Stenersen and Jain, 1984), as shown in Fig. 6. For this experiment, the external clock pulse input was set at a DC level of 0, so that the two NOR gates 7 and 8 forming the output latch of the circuit behaved essentially as the pair of cross-coupled inverters shown in Fig. 5a. Each NOR gate has the layout shown in Fig. 3. In both NOR gates (gates 7 and 8), FETs J_1 were illuminated, so that the optically induced switching always occurred in the same direction (from 1 to 0), while the internal delays of the switched outputs of each gate were matched. Thus, if the output Q is initially at logic 1, then the first optical

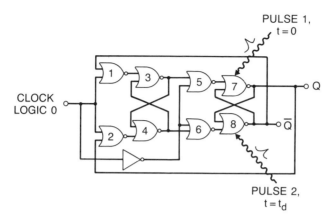

FIG. 6. Schematic of divide-by-two D-flip-flop circuit. NOR gates 7 and 8 are illuminated optically. (Stenersen and Jain, 1984; © 1984 IEEE.)

pulse (at $t = 0$) causes Q to go to 0, thereby causing $\bar{Q}$ to go to 1 after one NOR-gate propagation delay τ_{g8} (corresponding to gate 8). Note that the optical pulses initiate the direction of the logic switch and the circuit completes the latching action. Likewise, after the second optical pulse has illuminated NOR gate 8 (after a delay t_d), the output $\bar{Q}$ switches back to 0, while output Q returns to 1 after a gate delay τ_{g7} (corresponding to gate 7). Thus, optical pulse 1 acts as a "set" pulse, setting the output of Q to 0, while pulse 2 resets this output to the initial state 1. This produces a logic output pulse at Q of duratiion $\tau_1 = t_d + \tau_{g7}$. Similarly, a logic pulse of opposite polarity and duration $\tau_2 = t_d - \tau_{g8}$ is produced at $\bar{Q}$. For logic pulses with identical rise and falltime, precise measurement of the external pulsewidths yields accurate values for the propagation delays of gates 7 and 8; for pulses with differing rise and fall times, the difference in the pulse-widths measured at Q and $\bar{Q}$ gives an accurate value of the average propagation delay for these two gates. Data obtained from such measurements are shown in Fig. 7, which shows a propagation delay of 110 ± 15 ps for a circuit which exhibits complete logic swings. The uncertainty in the propagation delay is caused by the jitter in the sampling oscilloscope, as just discussed.

The differential measurement technique just described has the disadvantage of requiring high-speed coupling of the output waveforms for observation on a high-speed sampling oscilloscope. Nevertheless, neither a multigigahertz electronic-signal generator nor coupling of a high-speed electronic input is necessary. Also, as previously pointed out, the temporal uncertainty is not limited by the risetime but by the jitter in the sampling

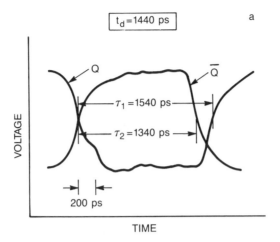

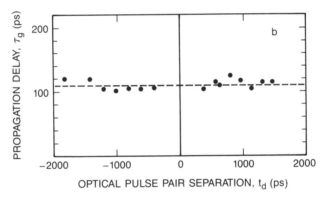

FIG. 7. (a) Measured waveforms for the circuits of Fig. 7. (b) Inferred gate propagation delay as a function of optical pulse separation, t_d. Propagation delay is 110 ps, independent of t_d. (Stenersen and Jain, 1984; © 1984, IEEE.)

instrumentation, so that with a long string of gates (say, 10) in the linear arrangement of Fig. 4, an uncertainty of about one ps/gate may be achievable with sampling scopes available as of 1989 (such as a Tektronix 7904, with an S-4 sampling head). Because the output signal is logic-level in amplitude (that is, volts), it is readily observed on a sampling oscilloscope with an excellent signal-to-noise ratio. Implementation of the technique does require sufficient energy in the optical pulse (10 pJ) to create enough carriers to cause a logic-level transition. The optoelectronic differential measurement technique allows direct measurement of specific individual gates within a relatively complex circuit (Stenersen and Jain, 1984); this flexibility of interrogating individual logic gates internal to a complex circuit stems from the ready accessibility of every gate in the IC to optical inputs. Also, τ_g versus fan-out studies can be performed easily to accurately assess the precise effects of loading on such gates L_1, L_2, and L_3 in the circuit of Fig. 4a and repeating the two-gate delay measurement by illuminating identical FETs in gates L_2 and L_3. The technique is not general, however, because it requires optically sensitive elements, such as GaAs MESFETs, as well as particular types of circuit logic elements.

B. OPTICALLY INDUCED LOGIC-LEVEL SAMPLING

This technique is related to the differential measurement technique because it depends strongly on the generation of logic-level pulses for successful operation. However, no high-speed output connection is required, because the desired measurement is effected by plotting a low-frequency electrical signal as a function of the optical pulse-pair separation, t_d. Instead of measuring a complete waveform, only changes in the logic state are sampled as a function of the pulse-pair separation. This requires nonlinearity in the response of the optically addressed logic gates.

We discuss two examples of this technique. In the first example (Jain, 1984), where the propagation delay through a nonlatching gate is to be measured, we may use a circuit similar to that shown in Fig. 8a, where the unknown propagation delay τ_X (corresponding to the circuit in the box marked X) may be inferred simply by plotting the average power of the output of the AND gate as a function of the pulse-pair separation, as shown in Fig. 8b. It is clear that the AND gate acts simply as a coincidence detector and that a maximum output will result when the pulse-pair separation is equal to the unknown propagation delay (τ_X), so that the two logic-level pulses (LLPs) arriving at the input of the AND gate are in exact temporal coincidence. Although the actual plot of P_0 versus t_d may consist of a relatively broad curve corresponding to an approximate convolution of the width of the optically generated LLPs with the response of the AND gate, the location of the maximum should provide a relatively accurate indication of the desired propagation delay τ_X. Because fast AND gates are not easily implemented in a large variety of logic families, an equivalent function can be performed by substituting the combination of a NOR gate followed by a inverter for the AND gate or by simply using a NOR gate.

A second variation of the logic-level sampling technique involves measurement of the latch-up time τ_L of a latching gate or a flip-flop. In one implementation of this technique for the measurement of latching times

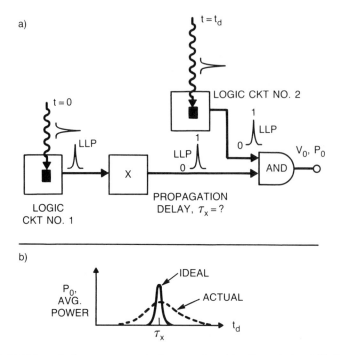

FIG. 8 (a) Schematic for measurement of propagation delay through circuit X. (b) Antici-pated output of AND gate as a function of time delay of the two optical pulses. (Jain, 1984. Reprinted from *Test & Measurement World*. © 1984 by Cahners Publishing, a division of Reed Publishing, USA.)

(Jain et al., 1984), a repetitive train (period T) of pairs of optical pulses, separated by t_d, is used to address an inverter. Logic-level pulses from its output are then fed to the clock input of an on-chip flip-flop. The output of the flip-flop switches states for each input logic-level pulse, provided that the pulse arrival rate is slower than the flip-flop maximum latching rate $(1/\tau_L)$. The external measurement (as shown in Fig. 2) is simply that of the average low-frequency $(\sim 1/T)$ power P_0 at the output of the flip-flop, as a function of t_d. For $t_d > \tau_L$, the flip-flop output is simply a rectangular pulse of width t_d, as depicted in Fig. 9a, and P_0 exhibits a linear dependence on t_d, as sketched in Fig. 9b. However, for $t_d < \tau_L$, the flip-flop only switches states once in response to each pair of pulses, and P_0 has a constant value (as a function of t_d) corresponding to a square wave of period $2T$, as depicted in Fig. 9a. The response time, τ_L, of the latching circuit is thus manifested by the strong nonlinearity in this P_0-versus-t_d plot, corre-sponding to a transition from a duty cycle of one-half during single switch-ing to a duty cycle of nearly 0 (or ~ 1) at the onset of dual switching (at $t_d \sim \tau_L$). An experimental plot obtained with a D-flip-flop (whose circuit

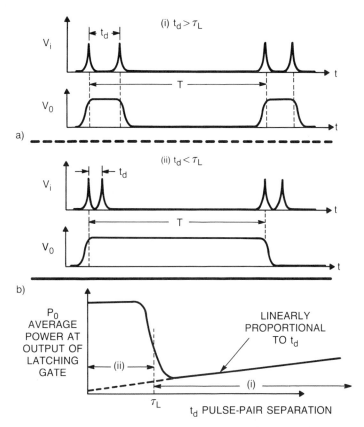

FIG. 9. (a) Input (V_i) and output (V_0) waveforms for a latching gate when the separation of optical pulses is (i) greater than and (ii) less than the latching time, τ_L. (b) Anticipated output of the latching gate as a function of relative delay between two optical pulses. (Jain, 1984. Reprinted from *Test & Measurement World*. © 1984 by Cahners Publishing, a division of Reed Publishing, USA.)

diagram is given in Fig. 6) is shown in Fig. 10, which depicts a latching time of ~475 ± 50 ps. A separate measurement on this circuit using a standard electronic divide-by-two technique resulted in a value of $f_{MAX} = 2.2 \pm 0.2$ GHz, indicating excellent agreement between the electronic and optical techniques (from both of which an averaged single-gate propagation delay of $\tau_g \sim 95 \pm 10$ ps may be inferred). An important point to remember about the optical technique, however, is that it does not require any coupling of high-speed electrical signals and thus may easily be implemented on undiced IC wafers in a low-frequency probe station. Moreover, it may be easily scaled to much faster gates and may remain a practical measurement technique when flip-flop latching rates exceed 10 GHz.

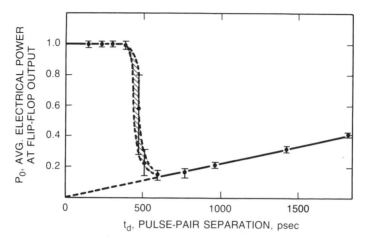

Fɪɢ. 10. Measurement of output power of a D-flip-flop versus optical delay between two pulses. The curves are guides to the eye corresponding to those shown in Fig. 9b. (Jain et al., 1984; © 1984 IEEE.)

IV. Direct Probing by Electro-optic Sampling

In this section, we discuss direct (or internal) electro-optic sampling. Of the techniques discussed in this chapter, electro-optic sampling has been the most widely applied. It has been used to measure waveforms and frequency response at internal nodes of integrated circuits, to measure the response times for high-speed FETs, and to measure ultrawide bandwidth responses for photodiodes. Its major limitation, for IC applications, is that it requires the IC to be fabricated on an electro-optic substrate (for example, GaAs or InP). Thus, it is not applicable to the direct optical probing of silicon ICs. However, external electro-optic probes may be useful for silicon devices, as discussed in Chapter 4.

This section has three subdivisions. In the first, the physical principles of electro-optic sampling are developed. In Section B, technical aspects (such as experimental arrangements, resolution, and sensitivity) are discussed. A summary of reported results using electro-optic sampling is presented in Section C.

A. Physical Principles

In a birefringent crystal, the index of refraction depends on the polarization of the optical beam and the direction of propagation. The refractive index of the crystal is described by the index ellipsoid (or indicatrix):

$$\frac{x^2}{n_1^2} + \frac{y^2}{n_2^2} + \frac{z^2}{n_3^2} = 1. \tag{1}$$

In writing Eq. (1), it is assumed that x, y, and z are the principal axes of the crystal. For propagation of an electromagnetic wave along the direction $\hat{\mathbf{k}}$, the index of refraction depends on the polarization of the wave and is described by the ellipse formed by the intersection of the index ellipsoid with the plane normal to $\hat{\mathbf{k}}$. The major and minor axes of this ellipse give the orientations and magnitude of the ordinary and extraordinary indices of refraction for the wave. The index of refraction for a particular polarization, displacement vector $\mathbf{D}$, of the wave is given by the magnitude of the ellipse in the direction $\mathbf{D}$.

The impermeability tensor, B, for a crystal is defined by

$$B_{ij} = \frac{\varepsilon_0}{\varepsilon_{ij}} = \frac{1}{n_{ij}^2},$$ (2)

where ε_{ij} is the dielectric tensor for the crystal and ε_0 is the permittivity of free space. The index ellipsoid is in general given by

$$B_{ij}x_i x_j = 1.$$ (3)

In Eq. (3) and henceforth, the convention of summation over repeated indices is used. x_i, $i = 1, 2$, or 3, is the cartesian coordinate. In the principal coordinate system, the dielectric tensor is diagonal and n_{ij} $n_i \delta_{ij}$. With this condition, Eqs. (2) and (3) reduce to Eq. (1).

The electro-optic effect is described in terms of the electric-field dependence of the impermeability tensor (Yariv and Yeh, 1984):

$$B_{ij}(\mathscr{E}) = B_{ij}(0) + r_{ijk}\mathscr{E}_k + s_{ijkl}\mathscr{E}_k\mathscr{E}_l.$$ (4)

$\mathscr{E}$ is a low-frequency (below-optical-frequency) electric field imposed on the crystal. The coefficients r_{ijk} are the linear or Pockels electro-optic coefficients and s_{ijkl} are the quadratic or Kerr electro-optic coefficients. For crystals with inversion symmetry, the transformation $\mathscr{E} \rightarrow -\mathscr{E}$ must cause no observable change in any tensor property of the crystal. Thus, all terms in the expansion of $B_{ij}(\mathscr{E})$ that are odd in $\mathscr{E}$ must vanish. Therefore, crystals with inversion symmetry (such as silicon) have vanishing linear electro-optic coefficients, and they manifest only the weaker quadratic electro-optic effect. In the remainder of this section, we consider only the linear electro-optic effect. Details of the much weaker quadratic electro-optic effect may be found elsewhere (Yariv and Yeh, 1984).

It can be shown from symmetry considerations (Kaminow, 1974; Yariv and Yeh, 1984) that i and j or k and l can be permuted. This reduces the number of independent elements of r_{ijk} from 27 to 18, and of s_{ijkl} from 81 to 36. To take advantage of the preceding permutation symmetries, we introduce contracted indices as follows: $(ij) \rightarrow k$; $(11) \rightarrow 1$, $(22) \rightarrow 2$, $(33) \rightarrow (3)$, $(23) = (32) \rightarrow 4$, $(13) = (31) \rightarrow 5$, $(12) = (21) \rightarrow 6$. Using contracted indices, the impermeability tensor in the presence of the linear electro-optic effect can be written as

$$B_i(\mathscr{E}) = B_i(0) + r_{ik}\mathscr{E}_k, \tag{5}$$

where $i = 1\text{--}6$ and $k = 1\text{--}3$. In the principal coordinate system in which ε_{ij} is diagonalized, Eqs. (2) and (5) may be combined to give the index ellipsoid in the presence of an applied electric field (Yariv and Yeh, 1984):

$$\left[\frac{1}{n_1^2} + r_{1k}\mathscr{E}_k\right]x^2 + \left[\frac{1}{n_2^2} + r_{2k}\mathscr{E}_k\right]y^2 + \left[\frac{1}{n_3^2} + r_{3k}\mathscr{E}_k\right]z^2$$

$$+ 2yzr_{4k}\mathscr{E}_k + 2xzr_{5k}\mathscr{E}_k + 2xyr_{6k}\mathscr{E}_k = 1. \tag{6}$$

Equation (6) is the basis for subsequent discussion.

The tensor r_{ik} has 18 elements. Crystal symmetry can reduce the number of independent elements. Table I gives the form of the linear electro-optic tensor for cubic $\bar{4}3m$ crystals (for example, GaAs and InP) and trigonal $3m$ crystals (for example, $LiNbO_3$ or $LiTaO_3$). Tables of the form of the linear electro-optic tensor for all noncentrosymmetric crystal classes are presented by Kaminow (1986). The electro-optic coefficients r_{ik} are also related to the second-order nonlinear optical coefficients (Kaminow, 1974).

The electro-optic coefficients just used assume that there is constant strain present in the crystal. This will be true when $\mathscr{E}$ is modulated above the acoustic resonances of the crystal (typically 10 MHz), and such will be the case for the high-speed operation of electrical devices discussed in this chapter. At lower-modulation frequencies, additional distortion of the crystal will occur due to piezoelectric and electrostrictive effects (Kaminow 1974), which will modify the numerical of r_{ik}.

Consider as our main example the electro-optic effect in a cubic crystal of the zincblende structure, symmetry $\bar{4}3m$ (for example, GaAs). In the absence of an applied electric field, the crystal is isotropic, and $n_i = n$ for all i. The principal axes correspond to the crystalline (100) axes. Let

TABLE I

ELECTRO-OPTIC TENSORS

$\bar{4}3m$			$3m$		
0	0	0	0	$-r_{22}$	r_{13}
0	0	0	0	r_{22}	r_{13}
0	0	0	0	0	r_{33}
r_{41}	0	0	0	r_{51}	0
0	r_{41}	0	r_{51}	0	0
0	0	r_{11}	$-r_{22}$	0	0

the external electric field be applied along the z axis, so $\mathcal{E} = \mathcal{E}_z \hat{z}$. The index ellipsoid is, from Eq. (6) and Table I,

$$\frac{x^2}{n^2} + \frac{y^2}{n^2} + \frac{z^2}{n^2} + 2r_{41}xy\mathcal{E}_z = 1. \tag{7}$$

The last term on the left-hand side of Eq. (7) mixes the x and y axes. New principal axes for the index ellipsoid are $\hat{x}'$, $\hat{y}'$, and $\hat{z}$, where (Yariv and Yeh, 1984)

$$\hat{x}' = \frac{1}{\sqrt{2}}(\hat{x} + \hat{y})$$

$$\hat{y}' = \frac{1}{\sqrt{2}}(-\hat{x} + \hat{y}).$$

Equation (7) becomes:

$$\left[\frac{1}{n^2} + r_{41}\mathcal{E}_z\right]x'^2 + \left[\frac{1}{n^2} - r_{41}\mathcal{E}_z\right]y'^2 + \frac{z^2}{n^2} = 1. \tag{8}$$

The indices of refraction along the new principal axes of the index ellipsoid, Eq. (8), are

$$n_{x'} = n - \frac{1}{2}n^3 r_{41}\mathcal{E}_z,$$

$$n_{y'} = n + \frac{1}{2}n^3 r_{41}\mathcal{E}_z, \tag{9}$$

$$n_z = n,$$

where, in arriving at Eq. (9), we have used the fact that the perturbation of the refractive indices is small compared to n. As can be seen, the application of an external electric field has caused the isotropic $\overline{4}3m$ crystal to become birefringent. The effects on the index ellipsoid in a zincblende-type crystal caused by application of an external electric field along directions other than $\hat{z}$ has been described by Namba (1961).

Consider an optical wave propagating along the z axis. The electro-optic effect manifested when the probe beam propagates parallel to the applied field is the *longitudinal* electro-optic effect. The difference in refractive indices of the components of this wave polarized along the x' and y' axes is

$$n_{x'} - n_{y'} = n^3 r_{41}\mathcal{E}_z. \tag{10}$$

For propagation of a distance L in the crystal, the difference in accumulated phase between the components polarized along x' and y' is the retardation Γ:

$$\Gamma = \frac{2\pi(n_{x'} - n_{y'})L}{\lambda} = \frac{2\pi n^3 r_{41} V}{\lambda}, \tag{11}$$

where $V = \mathscr{E}_z L$ is the voltage applied across the crystal, and λ is the vacuum wavelength of the optical wave. The retardation Γ causes the crystal to behave as a wave retardation plate.

It is possible to make a modulator from the crystal, as shown in Fig. 11. Let the input optical wave be polarized along the x axis, as required by the polarizer in Fig. 11. Then

$$\mathbf{D}(0) = D_0 \hat{\mathbf{x}} e^{-j\omega t} + c.c. = \frac{1}{\sqrt{2}} D_0 (\hat{\mathbf{x}}' - \hat{\mathbf{y}}') e^{-j\omega t} + c.c. \tag{12}$$

is the electric displacement vector for the input wave at the front surface of the crystal, $z = 0$. At $z = L$, the phase of the y' polarized component will advance by Γ relative to the x' polarized component. Thus, apart from an arbitrary phase factor,

$$\begin{aligned}
\mathbf{D}(L) &= \frac{1}{\sqrt{2}} D_0 (\hat{\mathbf{x}}' - \hat{\mathbf{y}}' e^{j\Gamma}) e^{-j\omega t} + c.c. \\
&= \frac{1}{2} D_0 \left[\hat{\mathbf{x}}(1 + e^{j\Gamma}) + \hat{\mathbf{y}}(1 - e^{j\Gamma}) \right] e^{-j\omega t} + c.c.
\end{aligned} \tag{13}$$

Neglect the static retardation plate in Fig. 11 for the moment. The field transmitted by an analyzer oriented along $\hat{\mathbf{y}}$ and, therefore, crossed to the input polarizer is

$$D_y(L) = \frac{1}{2} D_0 (1 - e^{j\Gamma}) e^{-j\omega t} + c.c. \tag{14}$$

The energy density transmitted through the analyzer is

$$\frac{1}{2\varepsilon_0} D_y D_y^* = \frac{1}{2\varepsilon_0} D_0^2 (1 - \cos \Gamma). \tag{15}$$

The optical intensity I is $cDD^*/2\varepsilon_0$. Using the identity $\sin^2(\theta/2) = \frac{1}{2}(1 - \cos \theta)$, we can write the intensity transmitted through the modulator:

$$I_T = I_0 \sin^2 \left(\frac{\Gamma}{2} \right), \tag{16}$$

where I_0 is the intensity incident on the initial polarizer in Fig. 11. A graph of Eq. (16), which is the transfer function for the electro-optic modulator of Fig. 11, is shown in Fig. 12. When $\Gamma = \pi$, the transmission of the

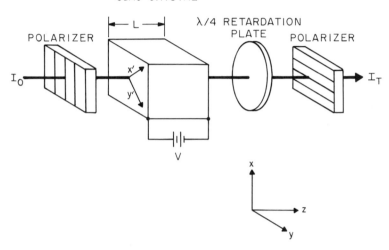

FIG. 11. Amplitude modulator based on the electro-optic effect. A voltage V is applied along the z axis of a GaAs crystal of length L. Principal axes of the induced birefringence x' and y', as well as principal axes (100) of the unperturbed crystal (x, y, z), are shown.

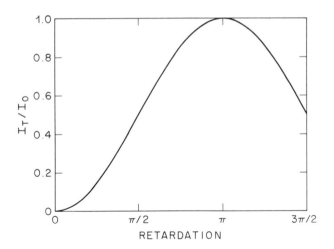

FIG. 12. Transfer function for the modulator of Fig. 11.

modulator is unity. The voltage applied to the modulator which produces this halfwave retardation is the halfwave voltage V_π, which is, from Eq. (11),

$$V_\pi = \frac{\lambda}{2n^3 r_{41}}. \tag{17}$$

In terms of V_π, Eq. (16) can be written

$$I_T = I_0 \sin^2(\pi V/2V_\pi). \tag{18}$$

For GaAs, $r_{41} = 1.2 \times 10^{-12}$ m/V and $n = 3.60$ at $\lambda = 1.0\ \mu$m (Kaminow, 1986), so $V_\pi = 8.9$ kV. For circuit applications, voltages applied across the GaAs substrate are approximately one V, so the retardation and fractional transmission changes caused by this voltage are $\sim 10^{-4}$.

To maximize the sensitivity of the modulator in Fig. 11 for small applied voltages, it is necessary to maximize $dI/d\Gamma$ (equivalently dI/dV). For the transfer function of Eq. (16), this occurs at $\Gamma_0 = \pi/2$. The transfer function linearity also is maximized at this value of Γ_0 (Kolner and Bloom, 1986). The value of Γ to be used in Eq. (16) is the total retardation caused by all components between the crossed polarizers in Fig. 11. The modulator can be biased with $\Gamma_0 = \pi/2$ by inserting a quarterwave plate between the polarizers, as shown in Fig. 11. The quarterwave plate causes a $\pi/2$ shift in phases between the $\hat{\mathbf{x}}'$, and $\hat{\mathbf{y}}'$ components of the optical field, and it transforms the initially linearly polarized light into circularly polarized light. The retardation between polarizers is then $\Gamma = \pi/2 + \pi V/V_\pi$. From Eq. (16),

$$I_T = I_0 \sin^2\left(\frac{\pi}{4} + \frac{\pi V}{2V_\pi}\right). \tag{19}$$

Eq. (19) can be rewritten as

$$I_T = \frac{I_0}{2}\left(1 + \sin\frac{\pi V}{V_\pi}\right). \tag{20}$$

The fractional modulation is $\Delta I/I = [I_T(V) - I_T(0)]/I_T(0)$ where $I_T(0)$ is the transmission through the analyzer when no voltage is applied to the modulator. Because $\pi V/V_\pi \ll 1$, we perform a series expansion of Eq. (20) to obtain

$$\frac{\Delta I}{I} = \sin\frac{\pi V}{V_\pi} = \frac{\pi V}{V_\pi} - \frac{1}{6}\left(\frac{\pi V}{V_\pi}\right)^3 + \cdots \tag{21}$$

The cubic nonlinearity in the fractional modulation at the quarterwave bias point is 10^{-8} times the linear term for a ~ 1V signal, and can be neglected.

A more significant nonlinearity in $\Delta I/I$ may arise when the bias Γ_0 is not exactly $\pi/2$. Suppose $\Gamma_0 = \pi/2 + \delta$. Eq. (19) becomes

$$I_T = I_0 \sin^2\left[\frac{\pi}{4} + \left(\frac{\pi V}{2V_\pi} + \frac{\delta}{2}\right)\right].$$

Following in parallel the development leading to Eqs. (20) and (21), the fractional modulation is

$$\frac{\Delta I}{I} = \frac{\pi V}{V_\pi} - \frac{\delta}{2}\left(\frac{\pi V}{V_\pi}\right)^2 - \frac{1}{6}\left(\frac{\pi V}{V_\pi}\right)^3 + \cdots, \tag{22}$$

which reduces to Eq. (21) when $\delta \to 0$. For $\delta \sim 1^0$, the second term in Eq. (22) is approximately 10^{-6} times smaller than the first term and 100 times larger than the third term.

Thus far, we have considered the electro-optic effect for $\bar{4}3m$ crystals, such as GaAs and InP. Eqs. (7)–(11) refer to such crystals when the field is applied along a (100) axis. This is the situation encountered for direct optical probing of integrated circuits on GaAs and InP, since the circuit is fabricated on a (100) crystal plane of the substrate. For measurement of the high-speed electrical waveforms of discrete devices, hybrid external electro-optic samplers have been fabricated on $LiNbO_3$ (Valdmanis et al., 1982), $LiTaO_3$ (Valdmanis et al., 1983a; Kolner et al., 1983a), and GaAs (Kolner and Bloom, 1984; Taylor et al., 1986a; Tucker et al., 1986).

The hybrid electro-optic samplers are fabricated in a microstrip geometry and are shown in cross section in Fig. 13. The electrical waveform to be measured is applied to the top circuit line in both hybrid samplers, and the bottom of the substrate is a ground plane. Hence, the electrical field is generated along the z-axis, as shown. For the GaAs sampler, shown in Fig. 13a, the optical probe beam propagates along the z axis, and the analysis previously presented applies in full. For the $LiTaO_3$ (or $LiNbO_3$) sampler shown in Fig. 13b, however, application of an electric field along the z axis causes a birefringence for a beam propagating along the y axis, because of the symmetry of the electro-optic tensor for $3m$ materials (Table I). This is the *transverse* electro-optic effect.

For completeness, we now analyze the transverse electro-optic effect in $3m$ crystals. $LiTaO_3$ and $LiNbO_3$ have a natural (i.e., field-independent) birefringence, with the c axis parallel to the z axis. Thus, the index ellipsoid for a $3m$ crystal, with a field applied along the $z(c)$ axis is, from Eq. (6) and Table I,

$$\left(\frac{1}{n_0^2} + r_{13}\mathscr{E}_z\right)x^2 + \left(\frac{1}{n_0^2} + r_{13}\mathscr{E}_z\right)y^2 + \left(\frac{1}{n_e^2} + r_{33}\mathscr{E}_z\right)z^2 = 1, \tag{23}$$

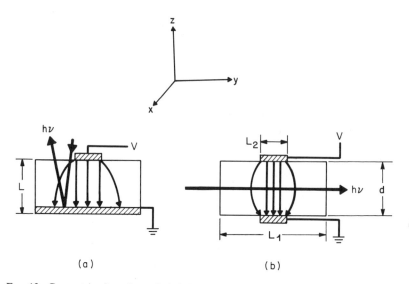

FIG. 13. Geometries for microstrip hybrid samplers. (a) Longitudinal sampling: $\bar{4}3m$ crystal symmetry. (b) Transverse sampling: $3m$ crystal symmetry.

where n_0 and n_e are the ordinary and extraordinary indices of refraction, and x and y are the principal axes orthogonal to the c axis. The principal axes are unchanged by the field, but the indices of refraction are modified as follows:

$$n_x = n_0 - \frac{1}{2} n_0^3 r_{13} \mathscr{E}_z$$

$$n_y = n_0 - \frac{1}{2} n_0^3 r_{13} \mathscr{E}_z$$

$$n_z = n_e - \frac{1}{2} n_e^3 r_{33} \mathscr{E}_z .$$

For an optical beam propagating along the y axis, the retardation is, analogous to Eq. (11),

$$\Gamma = \frac{2\pi}{\lambda} (n_e - n_0) L_1 - \frac{\pi}{\lambda} (n_e^3 r_{33} - n_0^3 r_{13}) \frac{L_2}{d} V, \tag{24}$$

where L_1 and L_2 are the widths of the crystal and the region of the electric field (approximately the width of the electrode), d is the thickness of the crystal, and $V = \mathscr{E}_z d$ is the potential between the signal and ground electrodes. The first term is the retardation caused by the static birefringence and the second term is the retardation caused by the electro-optic effect. The static retardation can be removed or converted into a bias $\Gamma_0 = \pi/2$ by a compensator (Valdmanis et al., 1982). The electro-optic halfwave voltage, V_π, derives from the second term in Eq. (24), and is

$$V_\pi = \frac{\lambda d}{L_2(n_e^3 r_{33} - n_0^3 r_{13})}. \tag{25}$$

For the $3m$ crystal, V_π depends on geometry of the sampler, because of the factor d/L_2. For LiTaO$_3$ at a wavelength of 1.2 μm, $r_{13} = 6.2 \times 10^{-12}$ m/V, $r_{33} = 26.7 \times 10^{-12}$ m/V, $n_0 \times 2.1305$, and $n_e = 2.1341$ (Kaminow, 1986). Therefore, $V_\pi = 6.0 d/L_2$ kV. For a 50-Ω microstrip transmission line on LiTaO$_3$ (with $\varepsilon = 45$), $d/L_2 \sim 8.5$ (Gupta et al., 1979), so $V_\pi \approx 50$ kV. The electro-optic effect in LiTaO$_3$ can be used to construct a modulator, as described previously for GaAs. For a bias $\Gamma_0 = \pi/2$, Eq. (21) is applicable, where V_π is given by Eq. (25).

In addition to hybrid samplers on which electrodes have been deposited, electro-optic sampling in which no electrical contacts are applied to the sampling crystal has been demonstrated (Valdmanis, 1987). In this technique, which is labeled external sampling and is discussed in detail in Chapter 4, the sampling crystal is immersed in the field lines emanating from the circuit node of interest (Valdmanis et al., 1983b).

B. Technical Details

1. Experimental Realizations

A generalized experimental schematic for measuring waveforms internal to an IC using electro-optic sampling and electrical excitation is shown in Fig. 14. Two frequency synthesizers are phase-locked together and generate signals at frequencies f_1 and f_2, which drive a pulsed laser and the IC, respectively. The probe beam from the laser is made circularly polarized by a quarterwave plate (QWP) to provide bias at $\Gamma_0 = \pi/2$, passes through a beam splitter, and is focused by a lens onto the DUT. In Fig. 14, the probe beam is incident on the DUT from the backside, is reflected by a contact back to the beam-splitter, and is directed from the beam-splitter through a polarizer to an optical receiver. Other optical sampling geometries are possible, and will be discussed in Section IV.B.2, as are other arrangements of optical retardation plates (Weingarten et al., 1988). The voltage on the contact being interrogated deforms the nearly circular polarization of the probe beam into slightly elliptical polarization, via the electro-optic effect, and this is converted into an amplitude change by passing the probe beam through the analyzer before detection. The analyzer is oriented

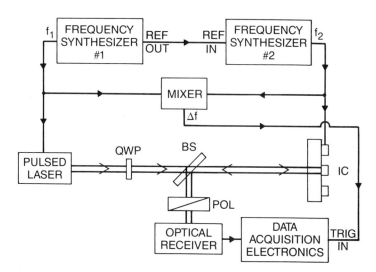

FIG. 14. General experimental schematic for electro-optic sampling. QWP, BS, and POL are quarterwave plate, beam-splitter, and polarizer, respectively.

along one of the principal axes of the elliptical polarization. The frequencies of the synthesizers are set so that $f_2 - mf_1 = \Delta f$, where $\Delta f \ll f_1, f_2$, and m is an integer. Successive pulses from the laser will sweep across the waveform on the IC at a rate Δf, which is the sampling rate. Depending on the particular experiment, Δf has a value between 10 Hz and 10 kHz (Taylor et al., 1986c). A microwave mixer generates a signal at Δf, which is used to trigger the data-acquisition electronics. A replica of the sampled waveform is reconstructed at the rate Δf when the data-acquisition device is an oscilloscope or signal averager, as shown schematically in Fig. 15.

When a spectrum analyzer, vector voltmeter, or lock-in amplifier is connected to the receiver, the power at the beat frequency Δf is measured, and the frequency response at the sampling point is measured as f_2 is varied (Rodwell et al., 1986c; Weingarten et al., 1988; Wiesenfeld and Heutmaker, 1988). Figure 16 shows, schematically, the microwave spectrum of the laser operating at frequency f_1, which, because the laser is pulsed, consists of a series of peaks at integral multiples of f_1 and the spectrum of the signal on the DUT, which has a fundamental at f_2, and may have harmonics, depending on the exact Fourier decomposition of the waveform. Because the electro-optic modulation is, from Eq. (21), $\Delta I = (\pi/V_\pi)\mathrm{I}(t)\mathrm{V}(t)$, the electro-optic effect will mix the frequency components of the laser and the DUT, producing sum and difference frequencies. For the case of $m = 1$, as shown, a signal at the difference frequency Δf can

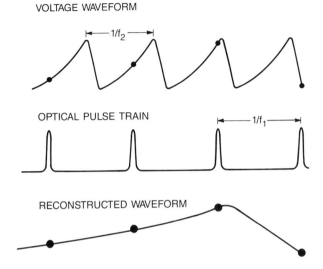

FIG. 15. Schematic of sampling in the time domain.

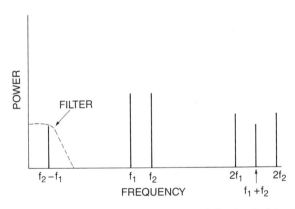

FIG. 16. Microwave spectrum of the optical pulse train (f_1), electrical signal (f_2), and the electro-optic signal ($f_2 \pm f_1$). A filter selects the beat frequency at $f_2 - f_1$.

readily be separated from all other mixing components using a filter (such as that found in a vector voltmeter or a spectrum analyzer). The magnitude of the signal at Δf depends on the response of the DUT at f_2, so as f_2 is varied (and in some arrangements f_1 is also varied; Wiesenfeld and Heutmaker, 1988), the measured signal at Δf will map out the response of the DUT in the frequency domain. Note that third-order nonlinearities,

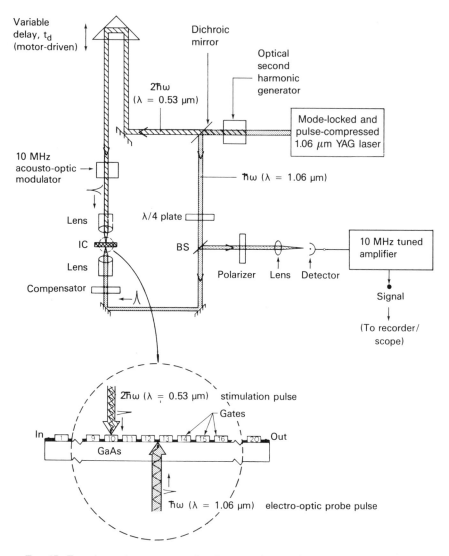

FIG. 17. Experimental arrangement for electro-optic sampling with optical excitation of the IC. The insert shows the positions of the optical beams for measurement of propagation delays in an inverter string.

such as the second term in Eq. (21), can also produce a signal at Δf $((f_1 + f_2) - 2f_1 = \Delta f$, for example), but this term is 10^{-8} times the linear term, and its effect is negligible.

Two types of laser sources have been used thus far for direct electro-optic sampling of ICs: pulse-compressed mode-locked Nd: YAG lasers

(Kolner and Bloom, 1986; Zhang and Jain, 1986) and gain-switched or mode-locked semiconductor injection lasers (Taylor et al., 1986a,c). Typically, the Nd:YAG laser pulse repetition rate is 100 MHz, so for circuits operating in the GHz regime, $m \sim 10$–50. The pulsed injection lasers can operate at repetition rates up to 20 GHz (Eisenstein et al., 1986), so $m = 1$ for circuits operating with clock rates of several GHz. Further details and considerations concerning choice of laser source are given in Section IV.B.6.

As shown in Fig. 14, excitation of the DUT is sinusoidal. A pulse input to the DUT can be obtained by inserting a step recovery diode between synthesizer #2 and the DUT (Wiesenfeld et al., 1987b).

An experimental schematic for electro-optic sampling of a DUT excited optically is shown in Fig. 17 (Zhang and Jain, 1986). Pulses from a pulse-compressed mode-locked Nd:YAG laser are frequency-doubled, and the second harmonic beam at 0.53 μm is separated from the residual fundamental beam at 1.06 μm by a dichroic mirror. The second harmonic beam traverses a variable optical delay and is focused into a photosensitive element in the IC, such as a GaAs MESFET. (See also Section III). This launches a current pulse in the DUT. The probe beam is the 1.06-μm beam, which is made circularly polarized by a QWP, and is incident on the DUT from the backside. As before, the probe is made elliptically polarized by the electro-optic effect in the GaAs substrate of the DUT, and this is converted to an amplitude change by an analyzer placed in front of the detector. To minimize the effects of low-frequency noise in the Nd:YAG laser, the pump beam is modulated at 10 MHz by an acousto-optic modulator, and the signal from the detector is measured using phase-sensitive electronics at 10 MHz. The waveform is reconstructed by varying the optical delay line in the excitation beam and recording the signal as a function of time delay between excitation and probe pulses.

2. Geometric Effects

In this section we discuss probing geometries and the geometric effects of a spatially inhomogenous electrical field, present in multielectrode DUTs, on the electro-optic sampling measurement. For the longitudinal electro-optic effect, the fractional modulation of the probe beam, as shown in the development leading to Eq. (21), depends on the difference in potential along the path of the probe beam in the electro-optic medium. The potential difference is the integral of the longitudinal component of the electric field along the path of the probe beam

$$\Delta V = \int_0^L \mathscr{E} \cdot d\hat{\mathbf{k}}, \qquad (26)$$

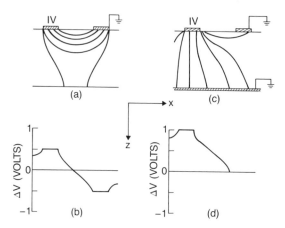

Fig. 18. Field lines (a) and (c) and differences in potential between front and back surfaces (b) and (d) for coplanar electrodes with 1V and ground applied potential. In (a) and (c) the back surface is floating and in (b) and (d) it is grounded.

where $d\hat{\mathbf{k}}$ is an infinitismal along the path of the probe beam. The value of ΔV will depend on the geometry and distribution of potentials applied to the DUT and on the precise path followed by the probe beam. Figure 18 shows the electric field lines and the difference of electrostatic potentials between the front surface and the back surface for a pair of coplanar electrodes, one of which is at 1-V potential and the other of which is at ground (Marcuse and Wiesenfeld, 1988). In Figs. 18a and 18b, the back surface of the substrate floats in potential, while in Figs. 18c and 18d, the back surface is grounded. The potential of the back surface is essentially independent of x when the substrate thickness is several times larger than the electrode spacing, so that the transverse spatial variations of ΔV in Figs. 18b and 18d are due to the front-surface variations in potential. When the back-surface potential floats, its value is the arithmetic mean of the potentials applied to the electrodes: 0.5 V for the example of Figs. 18a and 18b. For a probe-beam path normal to the substrate, the changes in ΔV along the transverse direction are the same regardless of the potential of the back surface, but the magnitude of ΔV is different and has a sign reversal when the back surface floats. The problem of establishing the potential of the back surface will be discussed in more detail later. Note also that because ΔV varies with x, the amplitude of the electro-optic signal will depend on the precise position of the probe beam.

In general, the probing geometry will also affect spatial resolution and crosstalk in the measurement. We now discuss the two commonly used probing geometries: backside probing and frontside probing. We shall then examine the effects of probing geometry on measured signal amplitude and crosstalk.

a. Probing Geometries

The commonly used probing geometries are shown in Fig. 19. The backside-probing geometry for a coplanar device (Freeman et al., 1985) is illustrated in Fig. 19a, and the frontside probing geometry is illustrated in Figs. 19b and 19c for coplanar (Heutmaker et al., 1988) and microstrip (Kolner and Bloom, 1984) devices, respectively. In the backside-probing geometry, the probe beam enters from the rear of the substrate, reflects off the contact, and exits through the rear of the substrate, as shown. The rear surface of the DUT must be accessible and of optical quality, so that the probe beam is not scattered there. The contact should be a good reflector.

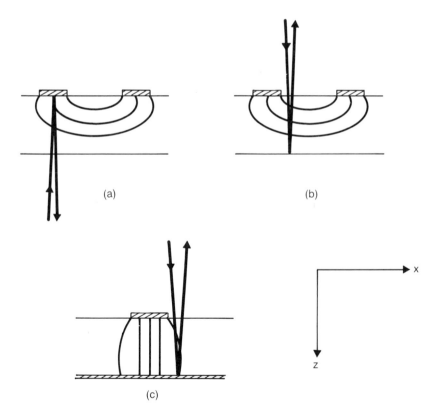

FIG. 19. Sampling geometries. (a) Backside for coplanar device. (b) Frontside for coplanar device. (c) Frontside for microstrip device.

A metallized gold contact is a good reflector for the infrared probe beam, while alloyed contacts are poorer reflectors because the substrate-contact interface is rough. In some devices with alloyed contacts, the ~30% reflectivity of the substrate–air interface is larger than that of the substrate–contact interface. The best electro-optic signal is then obtained by probing adjacent to the contact, even though, as seen in Fig. 18, the potential difference between back and front surfaces is smaller there than directly under the contact (Wiesenfeld et al., 1987b). In the backside geometry, the probe beam is focused onto the front surface, so that the spatial resolution is the diameter of the focused probe beam, which can be as small as 1 μm. The probe beam traverses through the electro-optic substrate twice, so that the fractional modulation of the probe-beam intensity is

$$\frac{\Delta I}{I} = \frac{2\pi \Delta V}{V_\pi} \tag{27}$$

where we have neglected the small nonlinear terms in Eq. (22). When the back surface of the substrate is fixed at ground potential (by a ground plane, for example) and the probe beam is reflected from the contact, $\Delta V = V$, the potential on the contact being probed, and

$$\frac{\Delta I}{I} = \frac{2\pi V}{V_\pi}. \tag{28}$$

The frontside-probing geometry (Figs. 19b and 19c) is used when the backside geometry is impossible to implement. The probe beam enters from the front of the DUT, reflects off the back surface of the surface of the substrate, and leaves the DUT at the front surface. Little probe-beam intensity will be lost when the back surface is metallized, as in the microstrip geometry of Fig. 19c. However, for an untreated back surface, the reflectivity will be only about 30%, reducing the reflected probe-beam intensity. Moreover, in the frontside geometry, the probe beam is focused on the back surface, so that its diameter is larger at the front surface. The best spatial resolution at the front surface is achieved when the confocal parameter of the laser beam is made equal to the substrate thickness (Heutmaker et al., 1988). In that case, the beam diameter at the front surface is $\sqrt{2}$ times that of the diameter of the focused spot on the rear surface, and the spatial resolution is $\sqrt{2}$ times poorer than in the backside-probing geometry. The fractional modulation of the probe beam is given again by Eq. (27). However, because the electrodes on the front surface are not transparent, the probe beam must enter the substrate adjacent to the electrode, where the magnitude of ΔV is smaller than that below the contact. (See Fig. 18.) Eq. (27) may be generalized to allow for probe paths that are not directly under the contact:

$$\frac{\Delta I}{I} = \frac{2\pi \, \Delta V_0 g(x)}{V_\pi} \tag{29}$$

where

$$g(x) = \frac{1}{\Delta V_0} \int_0^L \mathcal{E}_z(x) \, dz. \tag{30}$$

ΔV_0 is the potential difference between front and back surfaces directly under the contact. The integral in Eq. (30) is just that of Eq. (26) for propagation of the probe beam normal to the DUT and explicitly accounts for the x dependence of ΔV. $\mathcal{E}_z$ is the longitudinal component of the total electric field in the substrate. The plot of ΔV versus x in Fig. 18b gives the value of $g(x)$ for the particular electrode configuration shown, since $\Delta V(x) = \Delta V_0 g(x)$ and $\Delta V_0 = 1$ V.

b. Field Distribution and Crosstalk Considerations

There are several factors that influence the magnitude of the fractional modulation of the electro-optic signal. It is important that these factors be understood if the experimental measurement of $\Delta I/I$ is to be inverted to infer the amplitude V of the signal being probed, using Eq. (29). The magnitude of $g(x)$ at a particular probe position depends strongly on the potential of the back surface (Freeman et al., 1988). Contrary to early assumptions (Freeman et al., 1985), the potential of a floating back surface need not be at ground, even if the substrate is very thick. In the following discussion we consider only quasi-static electrical fields, so the discussion is valid in the limit of negligible propagation-time effects. The potential of a floating back surface will be essentially the arithmetic mean of the potential on the front surface (Marcuse and Wiesenfeld, 1988). Deposition of a good ground contact on the back surface will ensure that its potential is ground. If the substrate is placed in a package against a metallic box, the potential of the back surface will be somewhere between the preceding two extremes, since a Schottky barrier may form between the metal and the substrate, resulting in some of the potential drop occurring in the Schottky barrier (Rhoderick, 1978). Therefore, in many testing conditions, the potential of the back surface may not be known. If the back surface is not at ground potential, however, there should be an electro-optic signal near a ground line (Fig. 18d), which would have opposite phase to the electro-optic signal of a nearby signal-carrying line (Freeman et al., 1988, 1989). It may be possible, therefore, to estimate the potential of the back surface by such a measurement of the electro-optic signal at a ground line.

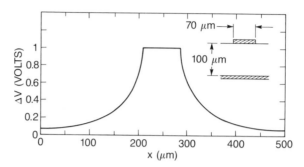

FIG. 20. Calculated transverse variation of potential difference between front and back surfaces for a 50-Ω microstrip device. Insert shows geometry of the device.

Because the value of g(x) depends on geometry, there is an uncertainty in relating $\Delta I/I$ to a signal voltage. In the backside-probing geometry, there is no uncertainty in the potential of the front surface since it is just the voltage V of the contact being probed. However, the potential changes as the probe moves transversely from the contact line, as seen in Fig. 18, and the magnitude of $\Delta I/I$ of a given value of V on the contact being probed will decrease. We show two examples from theoretical (Marcuse and Wiesenfeld, 1988) and experimental (Heutmaker et al., 1988) studies. The first example is a single-electrode structure. Figure 20 shows the transverse variation of g(x) at the front surface of a 50-Ω microstrip line on a 100-μm–thick GaAs substrate. Because the back surface is a ground plane, the potential at the front surface corresponds to $\Delta V_0 g(x)$. At a position 15 μm to the side of the electrode, g(x) drops to 0.5. Therefore, if no correction were made for this geometric factor and Eq. (28) were used, the magnitude of the voltage could be underestimated by a factor of two.

The second example is a multielectrode structure. In general, the field distribution in the substrate can become quite complex for a multielectrode structure. Several works have addressed this problem as it relates to electro-optic sampling (Freeman et al., 1987, 1988, 1989; Zhang and Jain, 1987; Lindemuth, 1987; Marcuse and Wiesenfeld, 1988). We propose a generalization of Eqs. (29) and (30) for the case of multiple-signal–carrying electrodes, based on superposition of the electric fields emanating from each of the electrodes (Heutmaker et al., 1988). If electrode i is located at position x_i, g_i for each electrode is given by

$$g_i(x) = \frac{1}{\Delta V_i} \int_0^L \mathscr{E}_z^{(i)}(x - x_i)\, dz \qquad (31)$$

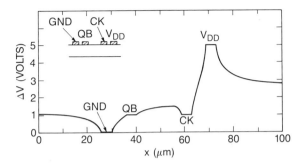

FIG. 21. Calculated transverse variation of potential difference for a multielectrode structure with grounded back surface. Schematic of the structure is shown in the insert.

where $\mathscr{E}_z^{(i)}$ and ΔV_i are the longitudinal component of the electric field and the potential difference between the back surface and the electrode, respectively, for electrode i. The generalization of Eq. (29) is

$$\frac{\Delta I}{I} = \frac{2\pi}{V_\pi} \sum_i \Delta V_i g_i(x - x_i). \qquad (32)$$

This is convenient, because $g_i(x)$ may be evaluated for individual electrodes.

Figure 21 shows $\Delta V(x)$ as a function of x for the four-electrode configuration of the second example, with electrode geometry as shown (Marcuse and Wiesenfeld, 1988; Heutmaker et al., 1988). For this example, the back surface of the substrate is grounded, and the electrodes, labeled GND, QB, CK, and V_{DD}, have defined potentials of 0, 1, 1, and 5V, from left to right. Notice that there is a region between the QB and CK electrodes where the potential is greater than 1V, because of the influence of the 5V V_{DD} line. Thus, if the probe were positioned 10 μm to the right of the QB electrode, the potential at the surface may be 50% higher than the potential on the electrode. However, the V_{DD} line does not have a time-varying voltage, and its effect will not be measured for a transient waveform. Figure 22 shows the modulation of $\Delta V(x)$ when the voltage on CK is varied from 0 to 1V, with all other voltages fixed. This is equivalent to displaying $g_{CK}(x)$, because $\Delta V_{CK} = 1$V. There is a substantial potential caused by CK as close to QB as 10 μm. Thus, if the probe beam were positioned 10 μm to the right of QB, it would experience the birefringence due not only to the electro-optic effect caused by the field of the signal on QB, but also additional birefringence due to the field from the signal on CK. This produces crosstalk in the measurement, even though there is no electrical crosstalk in the signals on CK and QB. The field distribution of

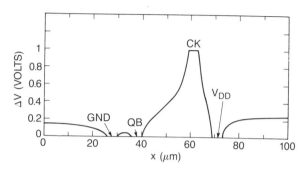

Fɪɢ. 22. Calculated modulation of ΔV for electrode configuration of Fig. 21, when potential on CK is varied from 1V to 0V.

even this relatively simple electrode configuration also makes it very difficult to predict, without calculation, the variation of $\Delta V(x)$. Hence, by measurement of $\Delta I/I$, it would be difficult to deduce accurately the magnitude of the voltage on the line being probed.

As just shown, crosstalk in the electro-optic sampling measurement can be a problem, especially for frontside probing. This has been demonstrated experimentally for the electrode configuration of Figs. 21 and 22. Figure 23 shows waveforms measured using the frontside-probing geometry at four positions for this electrode configuration, when the signal on the QB line is a 1-GHz sine wave and the signal on the CK line is a 2-GHz sine wave. GND is a ground line and V_{DD} is a 2V supply line. At positions b and c, the measured waveforms are those expected for the signals on QB and CK, respectively. However, when the probe is positioned at d or e, the measured waveform is a superposition of the waveforms on CK and QB, weighted most strongly by the waveform on the closer line. This demonstrates crosstalk and the need to properly position the probe beam. In Fig. 24, the fractional modulation of the electro-optic signal is measured as a function of x when the signal to QB is removed and only the CK line is active. This is an experimental measurement of $g_{CK}(x)$ for the CK line. The potential due to the CK signal extends across the 18-μm gap to the vicinity of the QB line and shows the origin of the crosstalk. Note the similarity between the measured $g_{CK}(x)$ and the calculated one, shown in Fig. 22. (In Fig. 24, however, $\Delta I/I$ becomes 0 on top of the CK electrode, because the experiment used the fronside-probing geometry. Also, the calculations for Fig. 22 used $V_{DD} = 5V$, while in the experimental circuit $V_{DD} = 2V$.) Figures 23 and 24 show that to minimize crosstalk in the frontside-probing geometry, the probe should be positioned between the active line to be interrogated and an inactive line, rather than between two active lines (Heutmaker et al., 1988).

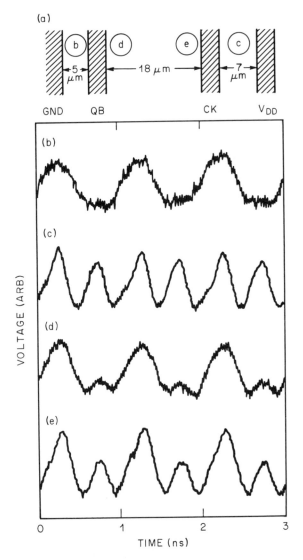

FIG. 23. Experimental measurement of crosstalk between active lines. (a) Layout of electrode configuration (same as Fig. 21). (b), (c), (d), and (e) are waveforms measured by probing at indicated positions in (a). 2-GHz signal and 1-GHz signal are applied to CK and QB, respectively. GND and V_{DD} are inactive ground and 2V bias lines, respectively. (Heutmaker et al., 1988)

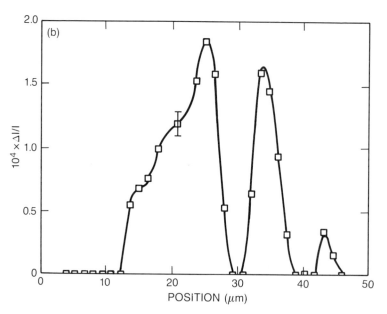

Fɪɢ. 24. Transverse variation of CK signal when 0V is applied to QB. This is an experimental measurement of $g_{CK}(x)$. Compare to Fig. 22. (Heutmaker et al., 1988)

Some practical considerations for calibrating the measurement of $\Delta I/I$ to the voltage on the contact being probed may be obtained by examination of Fig. 24. We note that in Fig. 24, the electro-optic signal falls off on either side of the *CK* line, but the magnitude of $\Delta I/I$ on one side is equal to that on the other side to within ~10% over the first 5 μm away from the line, in spite of the considerably different spacing of the adjacent circuit lines. Thus, if a consistent probe geometry is used (that is, constant spacing from the contact being probed), relatives values of *V* may be accurate to about 10%. An absolute calibration may be obtained, for electrical excitation of the DUT, by measurement of the electro-optic probe signal near an input line, where the input voltage is known. This will remove effects of nonideal orientation of the substrate crystal axes with respect to the analyzer and of optical bias other than $\pi/2$. Absolute calibration will be more difficult for optical excitation schemes.

Freeman et al. (1988, 1989) have studied the influence on the magnitude of the electro-optic signal of nonzero potential on the back surface of microwave coplanar transmission lines. For such structures, they found that the potential on the back surface was less than 10% of the potential of the signal electrode on the front surface if the substrate is thicker than the width of the signal electrode. The nonzero backside potential, if large, could lead directly to errors in inferring the magnitude of the signal being probed. Furthermore, the effect could produce crosstalk (Freeman et al., 1989) because the value of ΔV at a position away from the signal contact would be influenced by the potential on the signal contact via the nonzero potential on the back surface. The results obtained by Freeman et al. (1988, 1989) were explained quantitatively by their theoretical model (Freeman et al., 1987).

We conclude this section with additional comments on the effects of geometry on the measured signal and on the choice of probing geometry. For a general distribution of electrodes and, hence, of the electric field, the magnitude of $\Delta I/I$ will depend on the position of the probe beam. In the preceding discussion, we have ignored the effects of the finite diameter of the probe beam, which will change as a function of z inside the substrate, thereby integrating a spatially inhomogenous field. This is an additional complication in interpreting the measurements (Freeman et al., 1987). However, for coplanar devices, longitudinal variation of the potential occurs primarily in a depth comparable to the smallest spatial period of the electrode structure (Freeman et al., 1987; Marcuse and Wiesenfeld, 1988), which will generally be much smaller than the confocal parameter of the laser probe beam. Hence, in modelling the experiments, the probe-beam diameter can be taken equal to that at the front surface of the DUT. We have also ignored, in the previous discussion, the effects of propagation times of the electric field from the contact being probed to the position of the probe beam. The difference in optimal transverse positions of the probe beam for backside-versus frontside-probing geometries may be $\sim20~\mu m$, which would cause a time shift due to propagation delay of ~0.2 ps in a waveform measured by the two geometries, but no distortion. This time shift is unimportant, however, because it is insignificant compared to the propagation time of the probe beam through the thickness of the substrate. (See Section IV.B.3.) The amplitude of the electro-optic signal will depend, of course, on probing geometry. As shown already, crosstalk can became significant, especially for the frontside-probing geometry. We note that external electro-optic samplers used with ICs in a frontside-probing geometry (Valdmanis, 1987) also will suffer from cross-talk effects.

Finally, which probing geometry should be used for a particular measurement? If possible, the backside-probing geometry is preferable because (1) it has superior spatial resolution to the frontside geometry, (2) $g_i(x) = 1$ so there is no ambiguity concerning the potential of the front surface at the probe position (unless, due to poor reflectivity of the contact, the probe must be positioned adjacent to the contact where $g_i(x) < 1$), and (3) crosstalk will be minimized in this geometry. The frontside-probing geometry should be used when the backside-probing geometry cannot be implemented, for example, when the DUT is a microstrip device or a packaged IC. The frontside geometry, compared to the backside geometry has (1) poorer spatial resolution, (2) $g_i(x) < 1$, so sensitivity is reduced and an estimate of $g(x)$ is required to deduce voltage amplitude (See Section IV.B.4 for additional comments), and (3) possible crosstalk effects.

Nevertheless, the front-side-probing geometry has been used to analyze effectively monolithic microwave ICs (MMICs) with microstrip geometry (Rodwell et al., 1986c) and packaged coplanar ICs (Heutmaker et al., 1988).

c. Other Probing Geometries

To alleviate some of the problems associated with crosstalk described previously and to increase the sensitivity of the measurement, several other probing configurations have been proposed (Zhang and Jain, 1987). Two additional configurations are shown in Fig. 25. The first configuration (Fig. 25a) consists of probing the circuit from the back of the wafer and reflecting the infrared laser beam off the conductor of interest. This geometry differs from the backside-probing geometry of Fig. 19a in that the back has a treatment for establishing a ground potential, except for a hole for probing. Such a potential may be established by an n^+ implant on the bottom surface of the wafer or by depositing a conducting grid on the

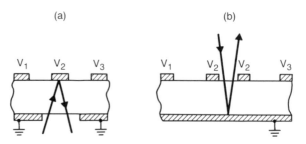

FIG. 25. Proposed configurations for electro-optic sampling. (Zhang and Jain, 1987)

back of the wafer and attaching it to ground. Alternatively, the back surface of the DUT may be established at ground potential using an ionic solution between the back surface and a transparent SnO support (Zhang and Jain, 1987). This permits optical access to the back of the substrate and reduces the magnitude of the backside potential (Zhang and Jain, 1987; Freeman et al., 1988). However, the ground plane may alter the impedance of the circuit (Freeman et al., 1988).

A second configuration (Fig. 25b) offers the advantages of probing from the top of the wafer, while reducing many of the accuracy or crosstalk problems. This configuration involves the design and fabrication of special probing windows at appropriate points within the integrated circuits to be investigated (Zhang and Jain, 1987). To implement this configuration, the designer of an IC mask set determines probe locations within the IC and opens up probe windows in the electrodes at the appropriate nodes. The major disadvantage of this configuration is that is requires circuits to be designed for testability via electro-optic probing. However, if testability is incorporated into the design of ICs, conventional frontside probing (Figs. 19b and 19c) could also be used, because probe positions between an active and an inactive line show minimal crosstalk, especially when the lines are well separated. Thus, an appropriate test point, which would be one where the contact to be probed is well separated from other active lines, could be incorporated into the layout of the IC.

3. Temporal Resolution

Several effects contribute to the overall temporal resolution of the electro-optic sampling experiment (Kolner and Bloom, 1986). Relevant effects include the intrinsic response time of the electro-optic effect, transit-time effects, laser-pulse duration, and timing jitter between the excitation signal and the probe pulse.

The electro-optic effect is an electronic optical nonlinearity, and, therefore, its response is expected to be in the femtosecond time domain. The lowest-energy region of dispersion in the dielectric response of GaAs is the restrahlen region, which begins at about 5 THz (Blakemore, 1982). Therefore, the response time of the intrinsic electro-optic effect should be below ~200 fs. This will be insignificant compared to the other sources of temporal broadening.

The relevant transit-time effects are the transit time of the optical probe pulse through the region containing the electrical fields inside the substrate and the transit time of the electrical signal across the diameter of the probe beam. These effects have been discussed in detail by Kolner and Bloom (1986), Weingarten et al. (1988), and Shibata et al. (1989). The optical transit-time effect arises from the propagation of the electrical waveform along the contact line, while the optical pulse is propagating through the electro-optic substrate. Because the electric field in the substrate changes as the electric signal propagates, the retardation caused by the electric signal will be averaged over the transit time of the pulse through the region containing the electric field inside the electro-optic substrate. Assuming that the field has large amplitude throughout the substrate, as, for example, in a microstrip transmission line (Fig. 19c), one may calculate an upper limit to the optical transit time. For substrate thickness L and index of refraction n, the optical transit time for a double pass through the substrate is

$$\Delta t_{\text{OTT}} = \frac{2nL}{c}. \tag{33}$$

For a 250-μm–thick sample of GaAs ($n = 3.5$) $\Delta t_{\text{OTT}} = 5.8$ ps. For coplanar devices, the geometry of the electric contacts may confine the electric field to a region much smaller than the thickness of the substrate, with a reduction of the effective L in Eq. (33) and, therefore, a reduction in Δt_{OTT}. The optical transit time (OTT) can also be reduced by a geometry in which the propagation velocity of the electrical signal is matched to a component of the propagation velocity of the optical probe beam (Valdmanis et al., 1982). This technique is useful for LiTaO$_3$ hybrid samplers, but not for GaAs hybrid samplers, because the microwave and optical velocities are nearly equal in GaAs (Kolner and Bloom, 1986).

The electrical transit time is due to the propagation of the electrical signal across the diameter of the optical beam. The retardation of the probe beam is due to the field produced by the electrical signal averaged over this propagation time. The electrical propagation time is (Kolner and Bloom, 1986)

$$\Delta t_{\text{ETT}} = \frac{w}{2} \sqrt{2 \ln 2 \varepsilon_{\text{eff}}}, \tag{34}$$

where w is the radius of the optical beam at the sampling point and ε_{eff} is the effective dielectric constant at the contact line of the DUT (Gupta et al., 1979). For a beam radius $w = 3$ μm and $\varepsilon_{\text{eff}} = 9$, $\Delta t_{\text{ETT}} = 40$ fs. In general, resolution is limited much more by the optical-transit-time effect than by the electrical-transit-time (ETT) effect.

The duration of the optical probe pulse can be a major limitation on temporal resolution, depending on the particular laser used. Pulses shorter than 100 fs are generated directly from mode-locked dye lasers (Fork et al., 1981), but dye lasers have not been used in direct electro-optic sampling measurements because their wavelengths are shorter than the absorption edge of GaAs or InP substrates. Direct electro-optic sampling measurements have been performed using either Nd:YAG laser systems operating at 1.06 μm or InGaAsP injection lasers operating at 1.3 or 1.55 μm. Typical pulse durations are 1–3 and 10–20 ps for the Nd:YAG laser systems and InGaAsP lasers, respectively. Further details of the laser sources are discussed in Section IV.B.6.

Timing jitter between the excitation source and the pulse train of the probe beam (which, in electro-optic sampling, arises with electrical excitation of the DUT) can degrade temporal resolution. Timing jitter is caused by random fluctuations of the pulse-repetition period with respect to the driving electronics. The absolute timing jitter is related to the phase noise of the probe laser pulse train and is determined from the measurement of the power spectrum of the laser, using a high-speed photodiode and an electronic spectrum analyzer (Kluge et al., 1984). The spectrum consists of a series of delta-function–like carrier peaks occurring at integral multiples, n, of the repetition rate of the pulse train, f, with each peak riding on top of one or more double-sided noisebands (Weingarten et al., 1988). The noisebands are caused by random fluctuations in the pulse energy (amplitude-noise sidebands) and/or pulse repetition rate (phase-noise sidebands). The ratio of the power of a phase-noise sideband to the power of the carrier increases as n^2, and the experimental verification of this relationship is the method for identifying phase-noise. The width of the phase-noise sideband in frequency δf is related to the correlation time for the frequency fluctuation τ by $\tau = 1/\delta f$. The magnitude of the rms timing jitter is given by (Robins, 1982)

$$\Delta t_J = \frac{1}{2\pi n f} \sqrt{P_n/P_c},$$ (35)

where P_n and P_c are the integrated powers in the phase-noise sidebands and carrier, respectively, measured at the nth harmonic of the laser repetition rate. Timing jitter is important only for measurement times longer than the frequency fluctuation correlation times τ. This will be the case for almost all measurements, however.

Timing jitter has been measured for mode-locked and gain-switched InGaAsP injection lasers, mode-locked Nd:YAG lasers, and synchronously mode-locked dye lasers. A power spectrum measurement at the fundamental of the repetition frequency ($n = 1$) for a 1.3-μm InGaAsP laser mode-locked at 2.021 GHz is shown in Fig. 26 (Taylor et al., 1986d). Two noise sidebands are evident, with widths (correlation times) of 600 Hz (1.7 ms) and 11 kHz (95 μs). The ratio of noise power to carrier power for measurements on this laser increases as n^2, indicating that they are phase-noise sidebands. The rms timing jitter calculated from Eq. (35) for the sum

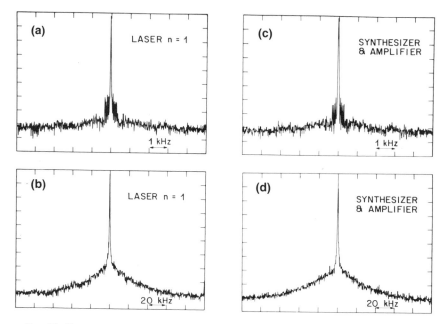

FIG. 26. Power spectra at the fundamental frequency ($n = 1$) for a 1.3-μm InGaAsP laser, mode-locked at 2.021 GHz: (a) and (b), and for the drive electronics: (c) and (d). The carrier peaks and noise sidebands are evident. Horizontal scales are one kHz/division for (a) and (c), and 20 kHz/division for (b) and (d). (Taylor et al., 1986d)

of the two sidebands is 0.55 ps. Also shown in Fig. 26 is the power spectrum for the synthesizer and amplifier used to drive the laser. The same two phase-noise sidebands are observed and the timing jitter in the driving electronics is calculated to be 0.54 ps. For all InGaAsP lasers studied by Taylor et al. (1986d), operating either mode-locked or gain-switched at rates between 1 and 4 GHz, the measured timing jitter was found to be identical to the jitter in the driving electronics. Furthermore, for all conditions studied, $\Delta t_J \lesssim 1$ ps, which is much shorter than the pulse duration. Therefore, timing jitter is not a serious problem for pulsed injection lasers. However, it is necessary to use low–phase-noise electronics to drive the pulsed injection lasers to have small timing jitter.

The timing jitter in mode-locked Nd:YAG lasers has been measured to be between 2 and 11 ps (Kolner and Bloom, 1986; Rodwell et al., 1986a) and may be caused by intensity fluctuations in the CW flashlamps pumping the laser. This jitter is comparable to or larger than the pulse duration. However, with a carefully designed external feedback circuit, the timing jitter has been reduced to 0.3 ps (Rodwell et al., 1986a; Rodwell et al.,

1989). Timing jitter in synchronously mode-locked dye lasers is about 20 ps (Kluge et al., 1984; Rodwell et al., 1986a).

Timing jitter also leads to amplitude noise (Weingarten et al. 1988). A detailed and careful analysis of timing jitter for a train of ultrashort laser pulses has been presented by Rodwell et al. (1986a, 1989). Timing jitter is not a problem for experiments using optical excitation when the excitation and probe pulses are derived from the same laser. In this case, all timing comes from an optical delay line, and only the micron-scale fluctuations of mechanical mounts lead to timing jitter (~fs) between excitation and probe.

The overall temporal resolution of the electro-optic sampling experiment is the convolution of the preceding effects. Since exact functional forms for all effects are not known, we assume a simple sum of square (Gaussian) convolution to arrive at a simple estimate of overall temporal resolution Δt. The most important effects are the optical transit time, the pulsewidth Δt_p, and the timing jitter. Hence,

$$\Delta t = (\Delta t_{OTT}^2 + \Delta t_P^2 + \Delta t_J^2)^{1/2}. \tag{36}$$

Taking $\Delta t_{OTT} \approx 4$ ps and the most favorable values for the other factors for the two types of laser used for direct electro-optic sampling gives values of Δt of 5 and 11 ps for the mode-locked Nd:YAG source and the mode-locked InGaAsP source, respectively.

Finally, note that the temporal resolution just calculated gives the limitation for resolving the transition time of an electrical signal. It is possible, however, to measure propagation times for a pulse or other well-defined shape to better accuracy, if no significant dispersion of the waveform occurs (Wiesenfeld et al., 1987b). In this case, the limitation on temporal resolution is imposed by signal-to-noise considerations.

4. Voltage-Sensitivity

In this section we consider the voltage-sensitivity and calibration issues of the electro-optic probing technique. For the signal-to-noise calculation, we proceed analogously to Kolner and Bloom (1986). The photocurrent in the detector is $i = RI_0$, where R is the responsivity of the photodetector (typically, $R \approx 0.6$ A/W). The signal and noise powers from the photodetector (P_S and P_N, respectively) are

$$P_S = \overline{i_s^2} R_L \tag{37}$$

$$P_N = 2ei_{avg}R_L B_f + 4kTB_f \tag{38}$$

where $\overline{i_s}$ is the time-averaged signal photocurrent, i_{avg} is the average photocurrent, e is the electron charge, R_L is the load resistor on the photodetector, B_f is the electronic bandwidth of the measurement, k is Boltzmann's constant, and T is the temperature. The first term in Eq. (38) is the shot-noise and the second term is the Johnson noise. We do not consider other noise terms here. From Eq. (20), making the small signal approximation that $\pi V/V_\pi \ll 1$, the photocurrent can be written

$$i = \frac{RI_0}{2}\left(1 + \frac{\pi V(t)}{V_\pi}\right).$$ (39)

For the time-dependent signal $V(t)$ on the DUT we choose a form which is positive only (Kolner and Bloom, 1986):

$$V(t) = \frac{V_0}{2}(1 + \cos \omega t)$$ (40)

where the signal frequency is ω. Eq. (39) can then be written

$$i = \frac{RI_0}{2}\left[\left(1 + \frac{\pi V_0}{2V_\pi}\right) + \frac{\pi V_0}{2V_\pi}\cos \omega t\right].$$ (41)

The first term on the right-hand side of Eq. (41) is the average photocurrent and the second term is the signal photocurrent. To a good approximation, $i_{avg} = RI_0/2$. Then, the ratio of signal power to noise power is

$$S/N = \frac{i_{avg}^2\left(\dfrac{\pi V_0}{2V_\pi}\right)^2 R_L}{4ei_{avg}R_L B_f + 8kTB_f},$$ (42)

where we have taken $\overline{\cos^2 \omega t} = 1/2$. When i_{avg} is large enough so the first term in the denominator in Eq. (42) is much larger than the second, the signal is shot-noise–limited, and the signal-to-noise ratio is

$$S/N = \frac{i_{avg}}{4eB_f}\left(\frac{\pi V_0}{2V_\pi}\right)^2.$$ (43)

The minimum detectable voltage, V_{min}, is the value of V_0 for which, in a one-Hz bandwidth, the signal-to-noise ratio is unity. From Eq. (43),

$$V_{min} = \frac{4V_\pi}{\pi}\sqrt{e/i_{avg}} \text{ V}/\sqrt{\text{Hz}}.$$ (44)

Consider as a first example, shot-noise–limited operation for 10-μW average power incident on the photodetector. Then, for $V_\pi = 8$ kV and $i_{avg} = 6$ μA, $V_{min} = 1.7$ mV/$\sqrt{\text{Hz}}$. As a second example, let 10 mW be incident on the photodetector, which produces $i_{avg} = 6$ mA. Then, for shot-noise–limited operation, $V_{min} = 53$ μV/$\sqrt{\text{Hz}}$. From Eq. (38), shot-noise–limited operation depends on both i_{avg} and R_L. For a high-speed photodetector with $R_L = 50\,\Omega$, shot-noise–limited operation requires $i_{avg} \gg 1.2$ mA ($I_0/2 \gg 2$ mW). For a higher-impedance photoreceiver (for example, $R_L = 1$ MΩ), shot-noise–limited operation can be achieved for $i_{avg} \gg 60$ nA ($I_0/2 \gg 100$ nW). Thus, it is important to use a high-impedance receiver in order to operate in the shot-noise–limited regime. Preferably, the receiver should also have high gain to overcome other noise sources in the signal-acquisition electronics. In addition to shot-noise and Johnson noise, $1/f$ noise and excess laser noise also may be important.

Kolner and Bloom (1986) have performed more extensive calculations of signal-to-noise ratio, including the effects of varying the optical bias Γ_0 and the load resistor R_L. The reader is referred to this excellent paper for details.

From Eq. (42) it can be seen that reduction of the measurement bandwidth B_f increases the signal-to-noise ratio. One way to accomplish this is to modulate the excitation to the DUT and to detect changes in probe-beam intensity using narrow-bandwidth, synchronous detection techniques. By choosing a high enough modulation frequency, the magnitude of excess laser noise can be reduced. This method has been used in conjunction with Nd:YAG probe lasers, with a modulation frequency of 10 MHz. For electrical excitation, the DUT is amplitude-modulated for an analog IC (Weingarten et al., 1985) and phase-modulated for a digital IC (Rodwell et al., 1986b). Phase modulation is necessary for the digital IC if the operation of the device is to be unperturbed by the modulation, but it causes a reduction in the sensitivity of the measurement (Rodwell et al., 1986b). For DUTs excited optically, the optical excitation beam is amplitude-modulated (Valdmanis et al., 1982; Kolner and Bloom, 1984; Zhang and Jain, 1986; Taylor et al., 1986b).

A second bandwidth-reduction scheme relies upon acquiring data directly at the frequency offset Δf (Fig. 14) and reducing the bandwidth of the signal around Δf using a variable bandpass AC filter (Taylor et al., 1986c). The bandpass of the filter must be sufficiently wide to pass all relevant harmonics necessary to faithfully reconstruct the sampled waveform and, therefore, may be several times larger than Δf. The signal is acquired at a rate Δf, which is typically about one kHz. In general, when signal averaging is used and the data is averaged at Δf, the final signal-to-noise ratio does not depend on whether high-frequency modulation or averaging at Δf is used, but, rather, depends only on the total data-acquisition time. It is difficult, however, to average signals at a rate faster than one kHz, using commercial signal averagers. This places an upper limit on Δf for direct signal averaging. If it is necessary to operate at a higher frequency due to laser $1/f$ noise, the modulation techniques are preferable. Excess laser noise is a serious problem for Nd:YAG lasers up to about 5 MHz (Kolner and Bloom, 1986) but does not appear to be a problem for InGaAsP injection lasers at a frequency of a few kHz (Taylor et al., 1986c; Wiesenfeld and Heutmaker, 1987).

The linearity of the electro-optic probe measurement has been discussed in detail by Kolner and Bloom (1986) and Kolner (1987). The linearity depends somewhat on the bias Γ_0 and also on the measurement-sensitivity. Kolner and Bloom (1986) have shown that for shot-noise–limited operation with a sensitivity of 20 $\mu V/\sqrt{Hz}$, the electro-optic probe signal is linear to within a deviation of 10^{-3} over a range of 140 dB, at the quarter-wave bias point. For Γ_0 as small as $\pi/64$, the dynamic range for linearity within 10^{-3} is still 90 dB. The dynamic range scales with sensitivity, and for a sensitivity of 1 $mV/\sqrt{Hz}$, the dynamic ranges previously mentioned are reduced by 34 dB. In an alternative description of dynamic range, Kolner (1987) has exploited the analogy between an electro-optic modulator and a microwave mixer. Distortion in the electro-optic signal in this description is caused by gain compression and intermodulation distortion (IMD). A conservative estimate of the dynamic range of the electro-optic measurement is given by the spurious-free dynamic range (SFDR). The SFDR is the ratio of the input voltage at which the third-order IMD products have power equal to the shot-noise–limited minimum detectable voltage, V_{min} (Eq. 44), to V_{min}. For $\Gamma_0 = \pi/2$, the SFDR is

$$SFDR = \frac{20}{3} \log(2i_{avg}/eB_f).$$ (45)

For V_{min} of 20 $\mu V/\sqrt{Hz}$ and 1 $mV/\sqrt{Hz}$, as before, the SFDR are 112 and 89 dB, respectively.

5. Noninvasiveness

A major advantage of optical-probing techniques over conventional electronic-probing techniques is their noninvasiveness. This arises not only from the lack of mechanical contact to the nodes being probed, but also by avoidance of parasitics associated with the electrical probe needles or wires used for direct electrical measurements. Also, the generation of extraneous fields due to the charging effects associated with electron beam probes do not exist. Finally, because of the relatively low powers needed for optical probing or optical carrier generation, the occasional damage problems associated with mechanical and electron-beam probes are avoided. However, carrier generation and the inverse electro-optic effect, described shortly, are two factors that may cause the optical probe beam to affect the DUT.

When an optical beam is used for excitation of the DUT, the photon energy must be above the bandgap and chosen so that the absorption coefficient (and thus the absorption depth) results in optimal carrier generation and transport in the chosen device. The electrical signals generated will then have the appropriate magnitude, risetime, and duration for excitation of the DUT. However, sufficient precautions must be taken to ensure that no spurious carriers are generated inside the DUT, either directly or by carrier diffusion from the photoexcited region, which could modify the performance of the device or features of the device being tested. In practice, this is relatively easy to achieve, particularly when one uses photoconductive gaps or field-effect transistors external to the nodes of interest.

For the optical probe beam, however, the photon energy must be selected to be sufficiently below the bandgap of the material being probed to avoid significant photogeneration of carriers. Although in principle, the longer the probe wavelength the better this requirement can be met, the presence of impurities and impurity transitions, such as those from acceptor and deep levels, can lead to photocarrier generation even at wavelengths longer than the band edge. The absorption spectra for GaAs (Spitzer and Whelan, 1959; Blakemore, 1982) is shown in Fig. 27 for several dopant concentrations. A probe wavelength between 1.3 and 3.5 μm would be optimal for use in most of these GaAs samples, although a wavelength of 1.06 μm is also acceptable. For a sample with a doping level of 1×10^{17} cm^{-3} (sample #2), the absorption for a 200-μm–thick substrate is ~1% at 1.3-μm wavelength. For highly doped samples, ($n_e = 5.4 \times 10^{18}$ cm^{-3}, sample #4) free carrier absorption can cause a strong reduction in the electro-optically sampled signal because of the strong attenuation of the probe beam. Even with the very small absorption coefficients in GaAs in the 1–4-μm range, a long optical path inside the DUT can cause absorption that is possibly significant. Such a long path may be produced by scattering of the pump or probe beam caused by roughness or defects in the substrate, or it could be produced by multiple internal reflections (Freeman, 1988).

For intense pump and probe beams, direct carrier generation by band-to-band two-photon absorption is possible. The two-photon absorption coefficient for GaAs is ~20 cm/GW at 1.06-μm wavelength (Van Stryland et al., 1985). Therefore, to avoid carrier generation by two-photon absorption, the peak intensity of the pump and probe beams must be limited below ~100 MW/cm^2.

The second factor to consider, to ensure noninvasiveness, is the role of the inverse electro-optic effect (Bass et al. 1962; Morozov et al., 1980). In electro-optic materials, the inverse electro-optic effect can lead to optical

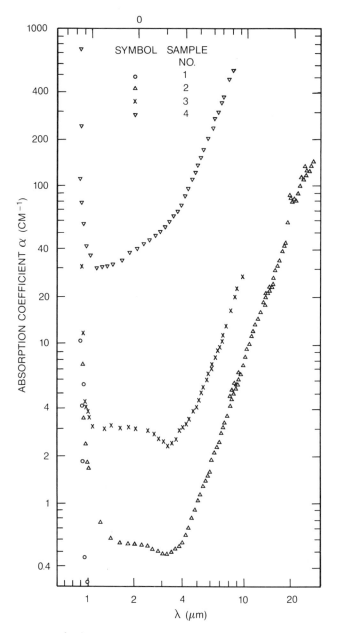

FIG. 27. Absorption spectra for samples of GaAs with different free carrier densities, caused by n-type doping. Doping densities for samples 1–4 are 0, 1.3, 4.9, and 54×10^{17} cm^{-3}, respectively. (Blakemore, 1982)

rectification, that is, the generation of a DC polarization field that is proportional to the amplitude of the optical wave. Optical rectification has been used as a spectroscopic tool for sampling of fast behavior within electro-optic materials, as described by Auston et al. (1984). For the electro-optic sampling geometries considered here for GaAs and related compounds, Kolner and Bloom (1986) have estimated the magnitude of the voltage induced by this effect to be given by

$$V = \frac{n^3 r_{14} L}{\varepsilon c} I_0\left(\frac{\tau'}{\tau}\right)[1 - \exp(-\tau/\tau')],$$

where τ', is an RC time constant characteristic of the transmission line on the DUT, τ is the laser pulsewidth, L is the thickness of the GaAs sample, and I_0 is the peak intensity of the optical pulse. For a 10-μm–wide transmission line on a GaAs wafer of $L = 100$-μm thickness, with a laser pulsewidth of $\tau = 5$ ps, and a peak intensity I_0 of 6×10^7 W/cm^2 (which is a typical intensity used with a Nd:YAG laser probe beam), the estimated peak voltage is of the order of 40 μV. Clearly, the optical rectification signals are very small compared to typical voltages to be measured in the DUTs, and they will be discriminated against by the use of phase-sensitive electronics in the detection system.

6. Laser Sources

The laser is a key component of the electro-optic sampling system. The ideal laser source would have the following properties: (1) pulse duration as short as possible to enhance temporal resolution; (2) average power as high as possible maximize sensitivity; (3) low timing jitter with respect to an electrical signal to enhance overall temporal resolution; (4) variable repetition rate to allow precise synchronization with the drive signal to the DUT; (5) wavelength longer than about 1 μm so that the substrate (GaAs or InP) is transparent to the laser probe; and (6) minimal excess noise (that is, noise above shot-noise). No one laser source maximizes all these attributes. Two types of laser sources have been used for direct electro-optic sampling: pulse-compressed Nd:YAG lasers (Kolner and Bloom, 1986; Zhang and Jain, 1986) and pulsed InGaAsP injection lasers (Taylor et al., 1986a; Wiesenfeld et al., 1987b). Some properties of these lasers are summarized in Table II.

The properties of the actively mode-locked Nd:YAG laser source, especially as it relates to electro-optic sampling, have been discussed in detail by Weingarten et al. (1988). The strongest output of the Nd:YAG laser is at the wavelength 1.06 μm, and this is the wavelength that has been used for electro-optic sampling. The repetition rate for commercially available

TABLE II

Laser Sources

Laser source	Pulse duration	Rep. rate	Avg. power	Wavelength	Comments
Nd:YAG (pulse-compressed)	2 ps	100 MHz	350 mW[a]	1.06 μm	Can be doubled for optical excitation; requires active feedback for jitter reduction
InGaAsP diode (mode-locked)	10 ps	1–20 GHz	1 mW	1.3, 1.55 μm	Compact, low jitter
InGaAsP diode (gain switched)	18 ps	1–20 GHz	1 mW	1.3, 1.55 μm	Compact, low jitter
Color-center	10–15	100 MHz	100 mW	1.45–1.70 μm	Not yet used; expected jitter comparable to Nd:YAG

[a] Should be limited to 100 mW to prevent two-photon absorption.

mode-locked Nd:YAG lasers is around 100 MHz. Thus, for measurements of DUTs in the GHz range, it is necessary to lock the driving signal to the DUT to a harmonic of the laser repetition rate. The duration of the pulse directly from the Nd:YAG laser is typically 90 ps. The pulse duration is shortened using a fiber-grating-pair pulse compressor (Grischkowsky and Balant, 1982; Kafka et al., 1984). In this arrangement, the pulse from the laser is coupled into an optical fiber of about one-km length. The pulse is frequency-chirped in the fiber because of self-phase modulation, producing extra optical bandwidth and a pulse that has the red-shifted frequency components emerge first and the blue-shifted frequency components emerge last. The emerging pulse is sent through a compressor consisting of a pair of diffraction gratings that acts as a dispersive delay line with anomalous dispersion (Treacy, 1969). Thus, in the grating pair, the red frequency components travel slower than the blue frequency components. By proper adjustment of the separation of the gratings, all frequency components of the chirped pulse emerge from the grating pair at the same time, producing a pulse of 1 to 3 ps, which is then delivered to the experiment.

The average power obtained directly from the Nd:YAG laser can be as high as 20 W. However, to avoid stimulated Raman scattering in the optical fiber within the pulse compressor, and two-photon absorption within the GaAs (or InP) electro-optic substrate, the average power should be limited to less than 400 to 700 mW and 100 mW, respectively (Weingarten et al., 1988).

Timing jitter, which can degrade both the temporal resolution and the sensitivity, has a value of 2 to 10 ps for the Nd:YAG laser. With an external feedback loop for stablizing the frequency of the laser to a high-stability electronic reference oscillator, the timing jitter has been reduced to 0.3 ps (Rodwell et al., 1986a, 1989).

There are several sources of excess noise in the pulse-compressed Nd: YAG laser system. For frequencies below 100 kHz, $1/f$ noise is large. For frequencies above 100 kHz, amplitude-noise from the pulse-compressor, caused by stimulated Raman scattering (SRS) or polarization fluctuations, dominates. SRS noise can occur when the 4% reflections from the ends of the optical fiber are sufficient to cause the fiber to become a synchronous resonator for the pulses generated by Raman scattering of the Nd:YAG pulse in the glass fiber. This problem is avoided by changing the fiber length to make it asynchronous with the Nd:YAG laser pulse train, and by lowering the input power to the fiber. Polarization fluctuations in the fiber translate into intensity fluctuations after the grating-pair compressor because the transmission of the grating pair depends on the input polarization. The fluctuations may arise from temperature fluctuations and/or guided acoustic wave Brillouin scattering (Shelby et al., 1985). This polarization-noise can be reduced by placing the optical fiber in a temperature-controlled environment, or by using polarization-maintaining fiber. The polarization-noise is difficult to eliminate and increases the noise floor to 10 dB above the shot-noise limit (Weingarten et al., 1988). The combination of the Nd:YAG laser system and fiber-grating pair pulse compressor, however, is a workhorse system, as the number of results obtained with its use indicates. The high average power available gives good sensitivity, even though the shot-noise limit is not achieved.

Semiconductor diode lasers of the alloy InGaAsP have been developed primarily for optical communications purposes (Agrawal and Dutta, 1986). The best developed stoichiometries of this alloy system are $In_{0.70}Ga_{0.30}As_{0.66}P_{0.34}$ and $In_{0.58}Ga_{0.42}As_{0.93}P_{0.07}$, which produce lasers operating at 1.30 and 1.55 μm, respectively. These wavelengths are ideal for electro-optic sampling of GaAs or InP. A key feature of the InGaAsP lasers is their ability to convert a high-speed electrical-drive signal into direct high-speed modulation of the laser output. Indeed, small-signal modulation bandwidths up to 22 GHz have been reported (Bowers et al., 1986).

Short pulses from InGaAsP diode lasers can be generated by mode-locking or gain-switching. In mode-locking, the diode is integrated into an external cavity. The ends of the cavity are formed by one facet of the diode (typically, 30% reflecting) and an external feedback element, such as a mirror. The other facet of the diode is antireflection-coated and the diode is coupled to the mirror by a high-numerical-aperture lens. The diode is driven by a superposition of DC and RF current, where the frequency of the RF matches the resonant frequency of the external cavity. The repetition rate of the laser may be increased to a harmonic of the fundamental frequency of the external cavity by driving it with the appropriate RF frequency. A pulse duration of 10 ps is easily achievable with such an arrangement. Pulses of 5-ps duration at repetition rates up to 20 GHz (Eisenstein et al., 1986) and of 0.6 ps duration at 16 GHz (Corzine et al., 1988) have been reported. For gain-switching, the bare laser diode is driven by a combination of DC and RF current (AuYeung, 1981). By suitable adjustment of the relative current levels, the laser diode will generate single pulses at a repetition rate equal to that of the RF frequency; hence, the repetition rate is continuously tunable. The pulses from a gain-switched laser correspond to the first pulse in the relaxation oscillation of the laser diode (Agrawal and Dutta, 1986). Pulses as short as 11 ps directly (White et al., 1985) and 6 ps by compression using an optical fiber (Takada et al., 1986) have been produced from gain-switched InGaAsP lasers. Pulse durations of 18 to 20 ps are more generally achievable. The pulse duration from diode lasers is the main limitation to the temporal resolution in electro-optic sampling: 10 ps → 33 GHz and 18 ps → 18 GHz.

Approximately one-mW average power can be produced from the pulsed diode lasers, which becomes tens of microwatts incident on the receiver. This produces a shot-noise–limited sensitivity of ~ 2 mV/$\sqrt{\text{Hz}}$. However, for pulsed diode lasers, $1/f$ noise is not significant above 1 kHz (Wiesenfeld and Heutmaker, 1987). No other sources of excess noise exist for the injection lasers if they are driven by high-purity electrical sources. Furthermore, the timing jitter for diode lasers is less than 1 ps (Taylor et al., 1986d).

The advantage to using diode lasers for electro-optic sampling is simplicity. The lasers are compact and have high repetition rate and small jitter, so there is no need for feedback and/or synchronization loops. The disadvantage in using diode lasers, compared to Nd:YAG lasers, is their longer pulse duration and lower power, leading to poorer temporal resolution and shot-noise–limited sensitivity, respectively. Nevertheless, InGaAsP diode lasers have been successfully applied to electro-optic sampling measurements. Furthermore, the technology of InGaAsP injection lasers is developing, and improvements in pulse duration (bandwidth) and average power can be expected.

Other laser sources for direct electro-optic sampling—such as mode-locked color-center lasers, long-wavelength dye lasers, and soliton lasers (Mollenauer and Stolen, 1984)—could be used (but have not been, as of the late 1980s) for direct electro–optic sampling. For comparison, relevant properties for a commercially availabe mode-locked color-center laser are included in Table II.

7. Discrete Hybrid Samplers

In some applications, the DUT is a small, discrete device and an electro-optic sampling measurement is used for its high temporal resolution. This is the case for high-speed photodiodes, with bandwidths that exceed the measurement capabilities of conventional electronic-test instrumentation. To measure the response of such a DUT, it is connected to a hybrid sampler, which may be a microstrip transmission line on $LiNbO_3$ (Valdmanis et al., 1982), $LiTaO_3$ (Valdmanis et al., 1983a; Kolner et al., 1983a,b), or GaAs (Kolner and Bloom, 1984; Taylor et al., 1986a). The cross sections of such hybrid samplers are shown in Fig. 13, and their operating principles have been discussed in Section IV.A. A discussion of $LiTaO_3$ hybrid samplers is also found in Chapter 4.

The main limitation in using hybrid samplers is the electrical connection to the DUT, which may have a bandwidth smaller that that of the DUT itself. The highest-speed connections can be obtained when the DUT is bonded directly to the hybrid sampler. Furthermore, to achieve the highest bandwidth, it is necessary to sample near the input of the microstrip transmission line because of dispersion in the microstrip line. (See Chapter 4.) For example, a microstrip transmission line that originates on the DUT and continues across the interface between the DUT and the sampler (Valdmanis et al., 1982) has been used to measure the subpicosecond risetime of the pulse response of a Cr-doped GaAs photoconductive detector (Valdmanis et al., 1983a). In that experiment, great care was taken to minimize transit-time effects and dispersion on the microstrip transmission line. Wang and Bloom (1983) used a gold-wire mesh (Kolner et al., 1983b) to bond a very-high–speed GaAs Schottky photodiode directly to a microstrip transmission line on a $LiTaO_3$ hybrid sampler.With this arrangement, Wang and Bloom (1983) measured a 100-GHz bandwidth for the detector.

Packaged photodiodes with bandwidths much greater than the response of conventional sampling oscilloscopes have been fabricated (Bowers et al., 1985; Bowers and Burrus, 1987). Thus, to measure their response, a hybrid sampler with a high-speed coax-microstrip transition is required (Taylor et al., 1986b; Tucker et al., 1986). The limitation to the measurement bandwidth may be the coax-microstrip transition. However, the advantage of such hybrid samplers, compared to directly bonded samplers, is that they may be reused to study many devices. Furthermore, for packaged devices, the connector is an integral part of the device and, thus, contributes in principle to the device bandwidth. Using a GaAs microstrip hybrid sampler with K connectors, a bandwidth of 67 GHz has been measured for a high-speed InGaAs photodiode (Tucker et al., 1986).

C. RESULTS

In this section we survey experiments in which the electro-optic sampling technique has been applied for direct probing of operating digital and analog ICs, and of field distributions of microwave waveguide structures. We also discuss experiments in which properties of discrete devices have been probed using electro-optic sampling with hybrid samplers.

1. Digital ICs

Table III lists in chronological order several measurements performed on digital ICs. A brief description of the key results follows.

a. Measurement of On-Chip Waveforms and Propagation Delays

The first electro-optic sampling measurements of on-chip waveforms and propagation delays in inverter strings were reported almost simultaneously by two groups (Jain, 1985; Zhang and Jain, 1986; Rodwell et al., 1986b) on separate samples of the same test IC. The test IC consisted of a string of 20 inverter gates of standard buffered-FET logic (BFL) design (Swierkowski et al., 1985); this string is illustrated schematically in the inset (cross-sectional view) of Fig. 17, with each labeled block representing a single inverter gate.

In the first report (Jain, 1985; Zhang and Jain, 1986), optical excitation was used to excite logic-level electrical pulses with relatively fast risetimes for both the 1-to-0 and the 0-to-1 transitions, by addressing appropriate field-effect transistors within the circuit with ~2-ps, 532-nm pulses of ~6-pJ energy. Representative data are shown in Fig. 28. For the data shown in Fig. 28a, the logic level at the input of the inverter string was initially set at logic 0 (-3V), and the logic state at the output of the gate 10 was optically switched from logic 0 to logic 1 ($\approx$1V). The leading edges of the logic-switched waveforms at the output of several downstream gates (12 through 15) are shown in Fig. 28a. Similarly, leading edges of output waveforms for transitions of both polarities (0-to-1 and 1-to-0) at the outputs of gates 11 and 12 are shown in Fig. 28b. Similar results were obtained by the measurements performed with electrical excitation by Rodwell et al. (1986b). From the measurements, individual gate propagation delays, τ_g, of 72 ± 3 to 78 ± 3 ps were deduced for various gates in this

TABLE III

DIRECT- E-O PROBING OF DIGITAL ICs

Circuit	Key results	Reference
2.6-GHz 8:1 multiplexer	Observation of serial word at output of MUX	Freeman et al., 1985
BFL inverter string	Measurement of propagation delays by optical excitation	Zhang and Jain, 1986
BFL inverter string	Measurement of propagation delays by electrical excitation	Rodwell et al., 1986b
2.7-GHz 8:1 multiplexer	Measurement of phases in eight-phase counter	Rodwell et al., 1986b
1.6-GHz MUX (8:1) and DEMUX (1:8)	Measurement of strobe pulses and timing patterns at internal nodes of counters and flip-flops	Jain and Zhang, 1986
7-GHz frequency divider (SDHT)	Measured effect of gate bias on waveforms at input buffer	Taylor et al., 1986c
18-GHz frequency dividers (CEL and BFL)	Verification of operation by measuring CL, Q, $\bar{Q}$	Jensen et al., 1987
5-GHz InGaAs/InP MISFET inverters	Measurement of propagation delays in individual MISFETs	Wiesenfeld et al., 1987b
1.6-GHz MUX and DEMUX	Detailed analysis of waveforms in internal flip-flops and counters	Zhang et al., 1987
1.7-GHz decision circuit	Measured waveforms in D-FF and propagation delays; packaged devices	Heutmaker et al., 1988
Inverter chain and frequency divider (HEMT)	Measurement of averaged propagation delays and output waveforms	Joshin et al., 1988

string. These measured values were consistent with values obtained from less precise measurements performed by conventional electronic techniques (Swierkowski et al., 1985). More importantly, these measurements represented the first clear experimental observation of on-chip pulse propagation in high-speed GaAs ICs.

The electro-optic sampling experiments were also extended by both groups to measure temporal delays between devices internal to each inverter. As seen in Fig. 29, temporal delays as small as 4 picoseconds were measured between waveforms observed on either side of level-shifting diodes in the inverters; measurements with such high temporal resolution are performed readily with optical excitation because of the subpicosecond timing accuracy allowed by the nearly jitter-free nature of the highly synchronized optical excitation-and-probe method.

In more recent work, Wiesenfeld et al. (1987b) have measured propagation delays in a two-stage InGaAs/InP metal-insulator-semiconductor FET (MISFET) inverter circuit (Antreasyan et al., 1986, 1987) in which the

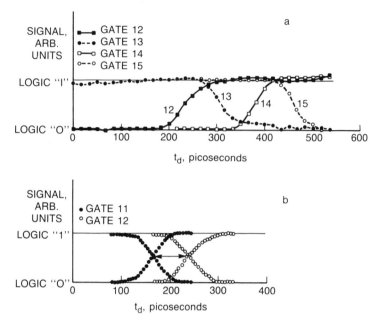

FIG. 28. Waveforms of switching transitions of downstream gates in an inverter string. (a) Output of gate 10 is optically switched from 0 to 1. (b) Superposition of waveforms on gates 11 and 12 when the output of gate 10 is optically switched from 0 to 1 and from 1 to 0. (Zhang and Jain, 1986)

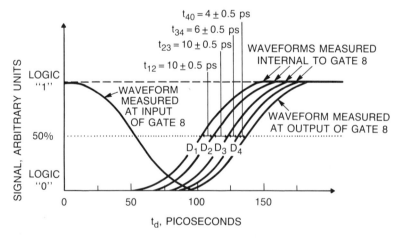

FIG. 29. Details of propagation delays inside a single gate. D_1–D_4 are waveforms measured after internal diodes inside the gate. (Zhang and Jain, 1986)

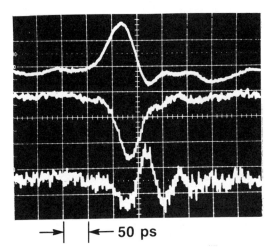

FIG. 30. Propagation delays in a two-stage InGaAs/InP MISFET inverter circuit. From top to bottom traces are input of comb generator pulse to the gate of first FET, output of first FET, and output of second FET. (Wiesenfeld et al., 1987b)

drain of the first FET was connected to the gate of the second FET. The circuits were driven electrically and sampled at the gate and drain electrodes of the first FET and at the output of the second FET, using a gain-switched InGaAsP semiconductor laser (~19-ps temporal resolution). To measure propagation delays, the circuit was driven at 2 GHz by the output of a comb generator (2V, 43-ps FWHM). As depicted in the traces of Fig. 30, propagation delays of 15 ± 4 ps and 33 ± 4 ps were measured for the first and second FETs, respectively. Note that the accuracy of the propagation delay measurement exceeds the nominal temporal resolution because pulse positions (rather than transition times) can be measured to an accuracy that depends on signal-to-noise ratio (Wiesenfeld et al., 1987b). The electro-optic probe works as well in this InP-based IC as in GaAs ICs because the electro-optic r_{41} coefficient of InP (Suzuki and Tada, 1984) is equal, to within 10%, to that of GaAs. More recently, Joshin et al. (1988) have used a gain-switched InGaAsP laser to measure the average propagation delay (54 ps) for a string of 32 high electron mobility transistor (HEMT) inverters.

b. High-Speed GaAs Multiplexers

The first digital IC of MSI complexity to be probed by direct electro-optic sampling was a GaAs multiplexer. In the first measurements, shown in Fig. 31, Freeman et al. (1985) measured the serial output waveform at the output buffer of a Triquint 2.7-GHz 8:1 time-domain multiplexer

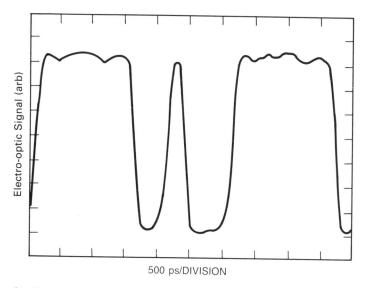

500 ps/DIVISION

FIG. 31. Serial output waveform of 2.7-GHz, eight-bit multiplexer. Output word is 111101000. (Freeman et al. 1985)

(MUX) to which a static parallel eight-bit word (11110100) was input by simply setting bit switches at the desired binary levels. Subsequently, Rodwell et al. (1986b) reported a preliminary measurement of all eight phases at the output of the eight-phase counter (used to control data combination in this 8:1 multiplexer).

In later work, Zhang et al. (1987) performed a more detailed analysis of on-chip waveforms at several internal nodes in 1.6-gigabit–rate eight-bit multiplexers–demultiplexers (Gigabit Logic #10G040 and 10G041). The results provided accurate timing information between the internal nodes, including such timing parameters as set-up and hold timing margins in flip-flops, and gate-propagation delays in normally loaded and normally exercised gates. The electro-optic sampling measurement also revealed nonideal logic-circuit operation (including effects such as clock feed-through at internal nodes), even in circuits operating at gigabit speeds. The measurements were performed directly on wafers with conventional probe cards used for biasing and feeding of the gigabit rate clock inputs and using the backside-probing geometry. A clock frequency of 1.31 GHz (period = 762 ps), synchronized to within a few Hz of a precise integer multiple of the 82-MHz repetition rate of the mode-locked Nd:YAG laser, was used for the timing analysis. The frequencies were deliberately offset by a few Hz to sweep the phase of the sampling pulses linearly with respect

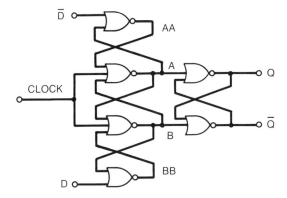

FIG. 32. Circuit diagram of D-flip-flop in 1.6-Gbit eight-bit multiplexer–demultiplexer.

to the sampled waveform. (See Section IV.B.1.) The input to the IC was phase-modulated at 1.5 MHz and the electro-optic signal was detected in a narrowband receiver to reduce the effects of $1/f$ noise from the Nd:YAG laser source. No method of jitter reduction was necessary for this experiment, since timing accuracy of less than 10 ps was not necessary.

Figure 32 shows the circuit diagram and the names (A, AA, etc.) used to designate nodes at which waveforms were measured in one of the four falling-edge triggered D-flip-flops (D-FF) that comprise the four-stage counter in the multiplexer–demultiplexer studied. Figure 33 shows the waveforms measured at the various internal nodes in this flip-flop, along with the measured phase of the clock input to this flip-flop. From the timing relationships between such waveforms, the timing margins for set-up and hold of this flip-flop and the propagation delays of each of the six NOR gates comprising the D-FF can be inferred very accurately. The propagation delays for gates Q (fan-out = 2) and A (fan-out = 3), τ_Q, and τ_A, respectively, were measured to be 90 and 97 ps, respectively. The maximum frequency of operation (f_{max}) of each of the flip-flops can then be inferred from the timing margins with any test clock frequency $f < f_{max}$. For example, the set-up and hold timing margins are given by the amount of time by which the falling edges in the outputs of AA and BB precede and follow the corresponding clock falling edge, respectively. From the smallest timing margin (190 ps) measured for set-up of this D-FF when operated at $f = 1.31$ GHz, Zhang et al. (1987) deduced its maximum operating frequency to be f_{max} equal to 1.74 GHz. This was verified experimentally by gradually increasing the clock frequency until the multiplexer ceased to operate. A value of $f_{max} = 1.7$ GHz was obtained by the latter technique, in good agreement with that deduced from the flip-flop set-up time measurements.

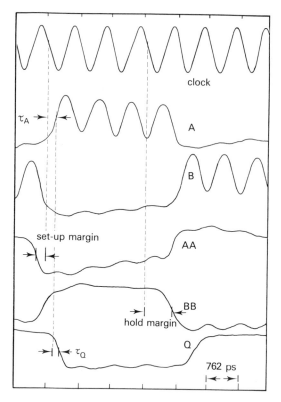

Fig. 33. Waveforms measured at internal nodes of D-flip-flop of Fig. 32. Timing parameters are indicated. (Zhang et al., 1987)

Timing patterns measured by electro-optic sampling for all eight phases of a four-stage synchronous counter in the 1.6-GHz multiplexer (constructed with D-FFs similar to that just described) are shown in Fig. 34, and signals measured for the eight-phase strobe pulses that control data combining and separation are depicted in Fig. 35. (The strobe pulses are decoded from synchronous counter outputs as $S_i = \overline{Q_i + \overline{Q}_{i+1}}$ by a simple combinational logic circuit.) Note that the relatively large ripples and apparent glitches observed in the zero state on several of the waveforms depicted in Figs. 33–35 are not random noise, but are a manifestation of intrinsic operational characteristics of the logic circuits used. Two 1 inputs to the NOR gates used result in a "hard zero" (0'), whose actual voltage is lower than the "zero" (0) normally obtained with a single 1 input, because of the extra current drawn by the second logic FET in the gate and the consequent increased potential drop $(V_0 = V_{DDL} - (I_A + I_B)R_{\text{eff}})$ (Zhang

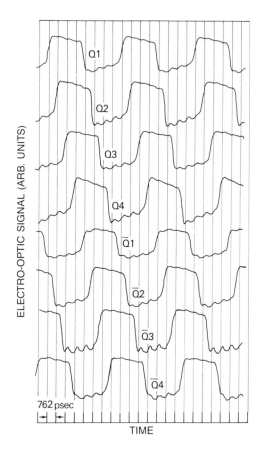

FIG. 34. Waveforms measured at the outputs of the four-stage synchronous counter for the 1.6-Gbit multiplexer. (Zhang et al., 1987)

and Jain, 1987). The difference between the levels of the single-1-input zero (0) and the dual-1-input zero (0') is the source of several of the apparent glitches in the observed data. For example, in the output latch of edge-triggered flip-flops, simple analysis (timing diagram and truth table) shows that a pair of 1's will appear alternating with a single 1 at the inputs of one of the NOR gates (AA or BB) for exactly half of a clock period, manifesting itself at the flip-flop output waveforms (Q, $\bar{Q}$) as a ripple at precisely the clock period. (This effect of "clock feedthrough" manifests itself on the zero state of each of the traces of Figs. 33–35; see Q_3 and $\bar{Q}_3$ in Fig. 34 for a clear illustration.) The situation is similar when Q_i is represented by the data string (00001111). Then, $\bar{Q}_{i+1}$, nominally represented by the data string (10000000), is more accurately given by (10000'000).

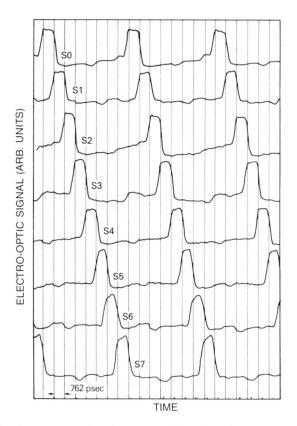

ELECTRO-OPTIC SIGNAL (ARB. UNITS)

S0
S1
S2
S3
S4
S5
S6
S7

762 psec

TIME

Fɪɢ. 35. Waveforms measured at the outputs of the eight-phase strobe decoder for the 1.6-Gbit multiplexer. (Zhang et al., 1987)

The hard zero (0') manifests itself as an apparent negative glitch of one clock-period width, exactly midway between the 1 pulses in the strobe waveforms. (See S7 in Fig. 35 for the clearest illustration of this effect.)

All the timing data observed previously, as well as the observation of clock feedthrough effects, show good agreement with computer modelling, viz. SPICE simulations (Nagel, 1975). However, detailed examination of the data also shows evidence of apparent crosstalk in the measurement (Section IV.B.2). The crosstalk is significant when the bottom surface of the substrate is allowed to float and was manifested in the measurements on the multiplexer circuit as slow ramps (rising in the 0 state, falling in the 1 state) in much of the data. The observed waveforms were attributed to a convolution of signals of varying magnitude from several adjacent signal-bearing lines and device pads (Zhang and Jain, 1987). The ramps were

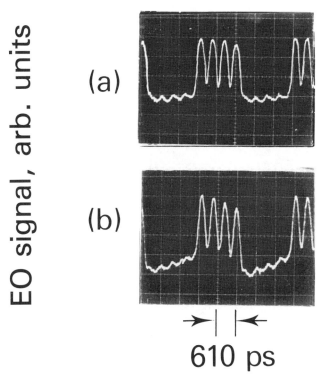

EO signal, arb. units

(a)

(b)

→| |←

610 ps

FIG. 36. Waveforms observed under node A of the D-flip-flop (Fig. 32) (a) with and (b) without use of SnO grounding plane at the back surface of the GaAs substrate. (Zhang and Jain, 1987)

observed to disappear (Fig. 36) when the bottom surface of the substrate was well grounded, by using a transparent, highly conducting ionic solution between the GaAs substrate and a grounded SnO substrate support.

c. Frequency Dividers

Taylor et al. (1986c) probed points internal to a 7-GHz frequency divider circuit (Pei et al., 1984) using an electro-optic sampling system based on a gain-switched 1.3-μm InGaAsP injection laser. The system sensitivity was 2 mV/$\sqrt{\text{Hz}}$ and the temporal resolution was ~18 ps. The frequency dividers, which were based on GaAs/AlGaAs selectively-doped heterojunction transistor technology, were driven at 2.4 and 5 GHz in separate experiments. In the 2.4-GHz experiments, saturation and pulse-shaping effects on the waveforms at the output (drain) of an input buffer stage were

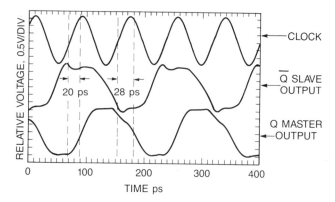

FIG. 37. Measured waveforms for CEL frequency divider circuit operating at 12.1 GHz. Timing delays are indicated. (Jensen et al., 1987)

observed as a function of gate bias when the circuit was driven with a 1-V peak-to-peak sinusoidal input.

Subsequently, Jensen et al. (1987) reported elegant electro-optic sampling experiments on 18-GHz static frequency dividers. The frequency dividers (Jensen et al., 1986) were implemented by either capacitively enhanced logic (CEL) or buffered FET logic, and were fabricated with 0.2-μm gate length MESFETs, molecular beam epitaxy (MBE)-grown active channels, and air bridge interconnects. Electro-optic sampling was used both to characterize signal timing and, thus, measure internal gate delays, as well as to verify divider operation. An example of such measurements, performed at 12.1 GHz on the CEL divider, is shown in Fig. 37. In this frequency divider circuit (Jensen et al., 1986), implemented with a biphase clock master–slave D-flip-flop design, the $\bar{Q}$ output of the slave is fed back to the D input of the master. Note that in this design, the maximum toggle rate is $(2\tau_{\mathrm{g}})^{-1}$, where τ_{g} is the average propagation delay of the loaded (NOR) gates. Similar to the discussion of set-up and hold margins described in the preceding multiplexer timing analysis (Section IV.C.1.b), the delay that determines the maximum toggle rate of this flip-flop is the longer of the two delays between the rising edge of the clock (CL) and the rising and falling edges of the master Q waveforms. From the 28-ps delay measured between CL and the falling edge of Q, a maximum toggle frequency of 17.9 GHz inferred, consistent with separate conventional measurements of $f_{\max} = 17$ GHz (at 170-mW power dissipation). Similar electro-optic sampling measurements on the BFL divider implied an $f_{\max} \sim 19.5$ GHz (with 658-mW power dissipation), which is in reasonable agreement with the value of 18.1 GHz observed by these workers in standard electrical measurements.

Joshin et al. (1988) have verified the operation of a 2-GHz frequency divider circuit fabricated with HEMT flip-flops and a flip chip circuit design, using an InGaAsP diode laser-based electro-optic sampling system.

d. Decision Circuit

A 1.7-GHz planar integrated decision circuit (AT&T 494A; O'Connor et al., 1984), whose function is to amplify input data with respect to a logic threshold and to synchronize it to a reference clock signal, was probed electro-optically by Heutmaker et al. (1988). Because the sampling was performed on packaged devices (with opaque bottoms), these authors employed frontside probing for their study (Fig. 19b). A gain-switched 1.3-μm InGaAsP injection laser, with a pulse duration of ~22 ps, was used in this work. The measurement-sensitivity was $20 \text{ mV}/\sqrt{\text{Hz}}$, rather than the shot-noise–limited value of $2 \text{ mV}/\sqrt{\text{Hz}}$. because of the reduced sensitivity (smaller fractional modulation of the probe) in the frontside-probing geometry, as discussed in Section IV.B.2.

A schematic of the decision circuit is shown in Fig. 38. Data is amplified in the input amplifier and synchronized to the clock in the D-flip-flop. For the data reported, the circuit was driven with a 1-GHz sinusoid (simulating a periodic 1–0 data pattern), and clocked at 2 GHz (generated by frequency doubling and phase-shifting part of the 1-GHz drive signal). The measured waveforms in an R–S flip-flop, internal to the D-flip-flop, are shown in Fig. 39. When the reset (R) pulse is high, the flip-flop output (Q) switches low; when the set pulse (S) is high, Q switches high. The set and reset pulses are synchronized to the clock, so Q has a fixed phase with respect to the clock. The signal at Q is subsequently amplified (and squared up) to produce the output Z (Heutmaker et al., 1988). Propagation delays in each of the three FETs in the input amplifier were measured by driving the circuit with pulses from a comb generator at 1 GHz; the results are shown in Fig. 40. The comb generator actually produces two pulses. For the data shown, the amplitude of the input was adjusted so that only the

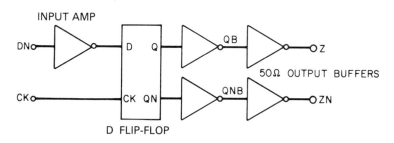

FIG. 38. Block diagram of 494A decision circuit.

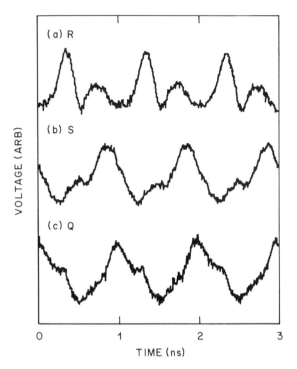

Fɪɢ. 39. Measured waveforms inside R–S flip-flop in 494A decision circuit. (Heutmaker, et al., 1988)

second pulse saturated in the amplifier. The measured propagation delays for the three FETs were 29 ± 5, 18 ± 2, and 29 ± 3 ps, respectively, which agree well with the average delay, 25 ps, measured for similarly fabricated FETs using a ring oscillator (O'Connor et al., 1984).

Frontside probing suffers from apparent crosstalk problems, as discussed in Section IV.B.2, and care must be taken to properly position the probe beam. Crosstalk arising in the measurement of the decision circuit is shown in Fig. 23. The causes of crosstalk in the front-side probing geometry and some methods to avoid its influence were discussed by Heutmaker et al. (1988) and in Section IV.B.2. Even with the reduced sensitivity and cross-talk problems caused by frontside probing, it was possible to evaluate effectively the packaged IC.

2. Analog Circuits (MMICs)

As summarized in Table IV, the two major classes of analog devices studied by direct electro-optic sampling are GaAs microwave amplifiers

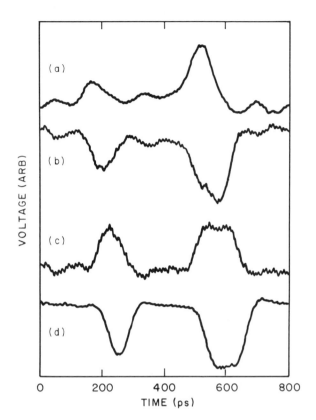

FIG. 40. Measurement of propagation delays in input amplifier section of 494A decision circuit. (a) is the input pulse from a comb generator, and (b), (c), and (d) are the measured waveforms after the first, second, and third FETs in the amplifier, respectively. (Heutmaker et al., 1988)

TABLE IV

DIRECT E-O PROBING OF ANALOG (MM) ICs

Circuits	Key results	Reference
12-GHz travelling-wave amplifier (TWA)	Distortion of internal waveforms with overdrive	Weingarten et al., 1985
Microwave transmission lines	Lateral voltage distribution	Weingarten et al., 1986
~18-GHz TWA	Internal signal propagation, line attenuation, frequency response, large-signal saturation characteristics	Rodwell et al., 1986c
Microwave transmission lines	S-parameter measurements	Weingarten et al., 1987
GaAs coplanar waveguide	Use of harmonic wave mixing to measure standing waves	Zhu et al., 1987
4.5-GHz FET amplifier	Frequency response	Wiesenfeld et al., 1988
Nonlinear transmission line	Pulse compression	Rodwell et al., 1988
Nonlinear transmission line	3.5-ps falltime	Madden et al., 1988
GaAs coplanar waveguide	Measurement of standing waves to 20.1 GHz	Zhu et al., 1988

and transmission lines. These have been tested at frequencies as high as 40 GHz, with signals launched either by bond wires to individual devices, or, for wafer-level testing, by means of high-bandwidth coplanar probes (Cascade Microtech) in a microwave probe station, modified to permit bottom illumination (that is, backside probing) (Weingarten et al., 1988). Except for the experiments reported by Wiesenfeld and Heutmaker (1988), all the work described in this section was performed using a pulse-compressed 1.06-μm Nd:YAG laser system with pulses of ~1.5 to 3-ps duration (at an 82-MHz repetition rate) and pulse-to-pulse rms timing fluctuations as low as ~0.3 ps (Rodwell et al., 1986a, 1989). As usual, the multigigahertz drive frequencies chosen for testing were selected to have values that corresponded to precise integer multiples of the laser repetition rate plus a small frequency offset (of ~1–100 Hz) and were phase- or amplitude-modulated at ~1 MHz to improve signal-to-noise ratios by synchronous detection. Because of the ultrashort durations of the pulses used in most of the reported measurements and relatively large optical transit times (~3 ps in a 125-μm–thick GaAs sample), the system temporal resolution was significantly affected by the OTT and was typically between 4 and 5 picoseconds. The relatively high average powers of the Nd:YAG lasers resulted in a voltage-sensitivity of ~100 $\mu V/\sqrt{Hz}$. For analog circuits, frequency-domain characterization (as well as time-domain characterization) is useful and has been performed directly using electro-optic sampling. A summary of the key results reported for amplifiers and transmission lines characterized by direct electro-optic sampling is described next.

a. Amplifiers

The first work using direct electro-optic sampling of GaAs MMICs was reported by Weingarten et al. (1985), who measured the effect of bias voltages and overdrive on the linearity of 4-GHz signals at internal nodes of a Varian 12-GHz travelling-wave amplifier (TWA). Clipping effects, caused by variation of the drain bias, as well as more severe distortions caused by overdrive of the input sinusoid, were observed on the drain electrodes probed; however, these effects were not well characterized in this preliminary report.

In much more extensive subsequent work, Rodwell et al. (1986c) characterized the signal propagation, frequency response, and distortion characteristics of multistage TWAs, using time-domain and frequency-domain electro-optic sampling measurements. These authors attempted to analyze two different five-stage Varian TWAs, one constructed with microstrip transmission lines and the other constructed with coplanar transmission lines (Riaziat et al., 1986), and having upper frequency cutoffs of 18 and

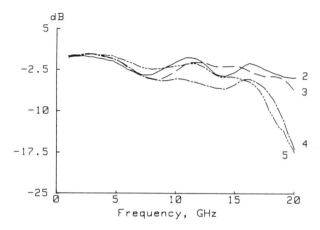

FIG. 41. Small-signal frequency response for successive gates in a five-stage distributed TWA. Devices are numbered in order of propagation for forward wave. (Rodwell et al., 1986c; © 1986, IEEE)

20 GHz, respectively. However, because of packaging problems (large package-ground interference), verified by optical probing between the chip and package ground planes, testing of the coplanar TWA was not possible. On the other hand, the microstrip TWA was characterized relatively completely, as illustrated by its measured small-signal frequency response in Fig. 41. For these measurements, the TWA input was driven with a sinusoid of low amplitude, and the small-signal transfer function from the input to the FET gate and drain terminals was measured by electro-optic sampling at each of these nodes. (The gate and drain signals correspond to relative drive and output levels, respectively.) The electro-optic sampling signals shown in Fig. 41 were obtained at the gate of each stage (labeled) of the TWA. The measured frequency characteristics were compared to simulations in which the process-dependent circuit parameters were varied to obtain the best fit. The results of this fitting procedure implied changes in values of the various parameters from their design values: gate termination resistance (from the 50-Ω design to 80 Ω); gate-source capacitance, C_{GS} (from 1.0 to 1.14 pf/mm); gate-drain capacitance, C_{GD} (from 0.03 to 0.06 pf/mm); source resistance, R_S (from 0.58 to 0.72 Ω); and source inductance, L_S (from 0.14 to 0.10 nH). Such changes were within the range expected due to normal process variations. However, the electro-optic sampling measurement has enabled direct evaluation of the circuit parameters, which could not have been obtained otherwise.

In addition, Rodwell et al. (1986c) observed a strong frequency dependence of the signals on each of the drain electrodes (see Fig. 42); this

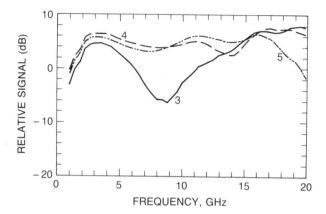

FIG. 42. Small-signal frequency response for successive drains in five-stage TWA. (Rodwell et al., 1986c; © 1986, IEEE)

dependence was attributed to interference between the forward and reverse waves, as confirmed by a simple analytical model that included the interference terms. Finally, the large-signal operation of these amplifiers was characterized by making a series of large-signal temporal measurements on the drain terminals of the FETs in the microstrip TWA. Depending upon the propagation delay between successive drains, the reverse propagating harmonic currents were observed to either increase or decrease peak voltages at other drains, thus modifying the normal saturation characteristics of the preceding FETs. It is clear that the study of such interesting effects caused by the propagation and interference characteristics of on-chip signals has been facilitated significantly by direct electro-optic sampling techniques.

Wiesenfeld and Heutmaker (1988) have reported the measurement of the frequency response of a FET amplifier by electro-optic sampling using a gain-switched InGaAsP laser and frontside probing. The amplifier studied was the first FET within the input amplifier section of the decision circuit shown in Fig. 38. The frequency response was measured to 8 GHz and showed a -3-dB frequency between 4 and 4.5 GHz.

b. Microwave Transmission Lines and S-Parameter Measurements

Electro-optic sampling exhibits great potential for the measurement of scattering (or S-) parameters in microwave circuits with potential measurement bandwidths of several hundred gigahertz. S-parameters describe the relationship between transmitted and reflected waves at the input and output ports of a microwave device or network, and they are related to fundamental device parameters (Sze, 1981). Unlike standard microwave test instruments, which use directional couplers and bridges to isolate and measure the forward and reverse waves on a transmission line, the electro-optic sampling measurement involves measuring the voltage standing wave as a function of position along the transmission lines (similar to a slotted-line measurement), from which the forward and reverse waves and the scattering parameters of microwave devices connected to these transmission lines can be inferred.

Although no measurements of S-parameters of active devices, such as microwave FETs, have been reported by this technique as of the late 1980s, Weingarten et al. (1988) have reported on efforts in this direction using a pulse-compressed Nd:YAG laser-based system. These researchers reported vector standing-wave measurements for 40-GHz excitation of an unterminated coplanar waveguide transmission line on GaAs (Fig. 43) and for 20-GHz excitation of the same transmission line terminated with an

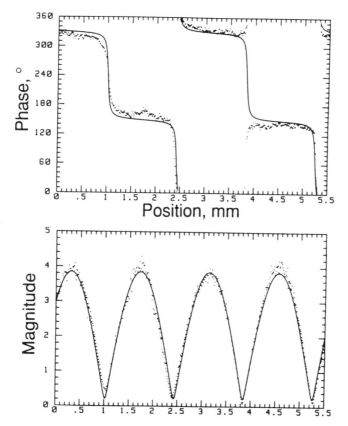

FIG. 43. Phase and magnitude of a 40-GHz standing wave on an open-terminated GaAs coplanar waveguide transmission line, as a function of position along the waveguide. Points are the data and the solid line is a fit. (from Weingarten et al., 1988; © 1988 IEEE)

unmatched load. By curve-fitting to the basic transmission-line equation for standing waves (Ramo et al., 1965), reflection coefficients (which correspond to the S-parameter S_{11}) of 0.90 and 0.06 were inferred from the measured data at 40 and 20 GHz, respectively.

As explained by Weingarten et al. (1987, 1988), one advantage of the measurement of S-parameters with electro-optic sampling is that this technique defines a reference plane directly on the IC, thus eliminating the relatively complex de-embedding requirement for test fixtures in conventional network analyzers. Weingarten et al. (1987) have also described a highly simplified error model (relating the measured S-parameter values to actual values) for the electro-optic sampling probe. Majidi-Ahy et al. (1987) have measured standing waves for even and odd guided modes of coplanar waveguides.

In independent experiments reported prior to the work of Weingarten et al. (1987), Zhu et al. (1987) demonstrated measurement of standing-wave patterns at frequencies of 8.2 and 12.3 GHz in GaAs coplanar waveguides (CPW). In this work, the technique was named harmonic-mixing electro-optic probing (HMEOP). HMEOP and the frequency-domain technique used by Weingarten et al., (1987, 1988) are essentially identical. In HMEOP, the GaAs CPW transmission-line modulator combined with the photodetector is viewed as a mixer, such that the mixing current is given by (Kolner, 1985)

$$
i_{mix} = i_{avg} \pi \frac{V_m}{V} \sum_{n=1}^{\infty} \frac{\sin(\pi n f_0 \tau_p)}{\pi n f_0 \tau_p}
$$
$$
\times [\sin 2\pi(n f_0 + f_m)t - \sin 2\pi(n f_0 - f_m)t], \qquad (46)
$$

where i_{avg} is the average photocurrent, V_m is the amplitude of the modulating voltage, V_π is the halfwave voltage of the GaAs modulator, n is the order of harmonic component, τ_p is the pulsewidth of the mode-locked laser, f_0 is the repetition rate of the mode-locked laser, and f_m is the frequency of modulation of the CPW. The electro-optic signal at the intermediate mixing frequencies $f_i = n f_0 \pm f_m$ is proportional to V_m, that is, to the amplitude of the modulating voltage. Using this technique, Zhu et al. (1987) measured the mixing signal of an open-terminal CPW corresponding to the mixing of an 8.2107 GHz (f_m) microwave signal with the 100th harmonic of the 82 MHz (f_0) mode-locked laser, at an intermediate mixing frequency of 10.7 MHz (f_i). The measured standing-wave patterns at 8.2 GHz (and in a similar experiment at 12.3 GHz) were in excellent agreement with theory. The work has been extended to 20.1-GHz frequency (Zhu et al., 1988).

Electro-optic sampling has also been applied to the measurement of electrical shockwaves generated on nonlinear microwave transmission lines (Rodwell et al., 1988; Madden et al., 1988). The purpose of these specially designed transmission lines (Rodwell et al., 1987) is the production of very fast electrical transients by compression of input waveforms, for potential use in wide-bandwidth electronic systems, such as waveform samplers. Madden et al. (1989) measured a waveform with a 1–6 ps falltime. This measurement could not have been performed without electro-optic sampling.

3. CW Electro-optic Probing

The use of electro-optic probing was first proposed for the characterization of very-high–speed devices via picosecond and subpicosecond resolution sampling. Interestingly enough, a related application of the direct electro-optic sampling probe, as proposed and demonstrated by Zhu, Lo, and coworkers (Zhu et al., 1986; Lo et al., 1987a,b) is in the utilization of this high–spatial-resolution noninvasive probe for CW probing of low-frequency electrical fields. These authors have reported the use of the CW electro-optic probe (CWEOP) for (a) the measurement of transverse surface potentials (and field distributions) in coplanar waveguides (Zhu et al., 1986) and for (b) inferring insulating behavior in special GaAs test samples, by studying space charge boundaries (inferred from the second derivative of the electro-optic signal) in $n^+ - i - n^+$ and $n^+ - p - n^+$ test structures (Lo et al., 1987a,b). The test structures were made by different annealing treatments of structures derived from the same semi-insulating substrate. These authors have also proposed using CWEOP with suitably oriented beams and electrode geometries for the three-dimensional mapping of internal fields in GaAs and GaAs-like materials (Lo et al., 1987b). However, the overall utility of this relatively complex mapping procedure remains to be demonstrated. A summary of the key results obtained with CWEOP appears in Table V.

4. Discrete Devices

For the measurement of the response of discrete devices, the ultrafast temporal response of the electro-optic effect is important. In this section, we discuss electro-optic sampling measurements of FETs, photodetectors, and electro-optic waveguide optical modulators. In the FET measurements, the substrate on which the FET is fabricated is used as the sampling medium; for the other devices, a hybrid electro-optic sampler is used. We note also that the pulse shape produced by a 2-GHz comb generator has been measured using a hybrid sampler (Taylor et al., 1986a,c).

TABLE V

Applications of CW Electro-Optic Probing (CWEOP)

Measurement	Reference
Lateral field distribution in a coplanar waveguide	Zhu et al., 1986
Measurement of surface properties including measurement of the space charge boundary of semi-insulating GaAs substrates	Lo et al., 1987a
Proposal for use of CWEOP to measure the internal static field distribution in planar GaAs devices	Lo et al., 1987b

a. FETs

There has been considerable interest and experimental activity since about 1980 in the study of the photodetection (Sugeta and Mizushima, 1980; Gammel and Ballantyne, 1980) and electrical switching characteristics of high-speed transistors, as well as in the development of new techniques for their characterization (Hammond et al., 1986; Smith et al., 1981; Cooper and Moss, 1986). Zhang and Jain (1987) performed studies of photodetection and electrical switching in GaAs MESFETs with gatewidths between 10 and 50 μm. Photodetection experiments were performed by illuminating the gate-to-drain region with pulses at above-bandgap photon energy from a frequency-doubled (0.53-μm) and pulse-compressed ($\sim$3-ps pulse-width) Nd:YAG laser. Electro-optic probing was performed under the gate, source, and drain electrodes with synchronized pulses at subbandgap (1.06-μm) photon energy from the fundamental of the same pulse-compressed Nd:YAG laser. (See Fig. 17.) Studies of electrical switching characteristics of the FETs were performed by optically switching a photo-conductive gap adjacent to the gate electrode with the 0.53-μm pulses, followed by electro-optic sampling under the gate and drain electrodes to monitor the input and output waveforms.

Figure 44 shows typical transients observed under the gate and drain electrodes in the photodetection experiments (Zhang and Jain, 1987). The gate signal is primarily due to a fast Schottky photodiode-type behavior in the depletion region around the gate, whereas the drain signal shows clear evidence of photoconductive behavior (and long decay times corresponding to slow recombination) and of photoconductive gain. Zhang and Jain (1987) also measured the propagation delay of a GaAs FET with a 50-μm gatewidth.

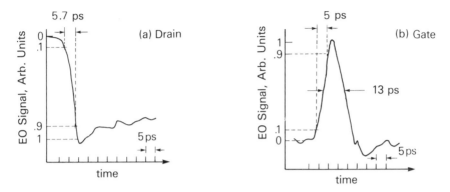

FIG. 44. Transient response of optically excited GaAs MESFET under the drain (a) and gate (b) electrodes. (Zhang and Jain, 1987)

b. Photodetectors

The impulse response of photodetectors is measured using an optical excitation scheme and a hybrid sampler. A representative experimental schematic for the measurement of the impulse response of a photodiode, using an injection laser and a hybrid sampler, is shown in Fig. 45 (Taylor et al., 1986b). This is similar to the arrangement for optical excitation and

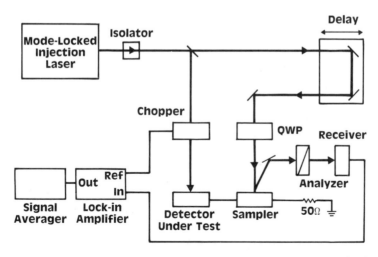

FIG. 45. Experimental schematic for measurement of the impulse response of a photodetector using electro-optic sampling and a hybrid sampler.

direct electro-optic sampling (Fig. 17). Ultrashort pulses from a mode locked laser are divided into two beams. One beam excites the photodetector under test. The voltage pulse produced by the excitation is coupled onto the hybrid sampler, where the voltage pulse is sampled electro-optically by time-delayed pulses in the second beam. The excitation beam is chopped, and the electro-optic signal is acquired using phase-sensitive electronics. The time delay between excitation and probe beams is produced using a mechanical optical delay line. The voltage pulse produced by the photodiode is the convolution of the laser intensity $I(t)$ with the impulse response of the photodetector $h_D(t)$ (Kolner et al., 1983a):

$$V_D(t) = I(t)*h_D(t). \tag{47}$$

On the hybrid sampler, $V_D(t)$ is convolved with the impulse response of the sampler, $h_s(t)$, and the measurement of the voltage on the sampler by a delayed optical pulse is a cross-correlation operation. Since the operations of correlation and convolution are associative, the measured response for the experimental arrangement of Fig. 45 is

$$V(t) = I(t) \star \{h_s(t)*[I(t)*h_D(t)]\} = [I(t) \star I(t)]*[h_s(t)*h_D(t)]. \tag{48}$$

The laser pulse enters twice, once for excitation and once for probing, and appears on the right-hand side of Eq. (48) as the autocorrelation of the laser-pulse intensity. Because the laser-pulse autocorrelation can be measured accurately and independently by time-delayed second-harmonic generation in a nonlinear optical crystal (Ippen and Shank, 1977), the effects of the laser-pulse duration can be eliminated from the measured signal $V(t)$ by deconvolution. To the extent that the impulse response of the sampler has a much wider bandwidth than that of the photodiode, deconvolution of the autocorrelation of the laser pulse from the signal produces the impulse response of the photodiode. The impulse response is measured in the time domain, and the deconvolution is performed in the frequency domain. Subsequently, the deconvolved impulse response is calculated by Fourier transformation of the frequency-domain response. Thus, the deconvolved frequency response is one step closer to the original data, and, where signal-to-noise ratio is a limitation, may be more accurate. Note that, in principle, a photodiode impulse response that is shorter than the laser pulsewidth can be measured. Note also that in the optical excitation scheme of Fig. 45 there is no timing jitter between excitation and probing, since both beams are derived from the same laser pulse.

The power available from the laser source influences the signal-to-noise ratio of the experiment, since the laser pulse is involved twice in the measurement. Let the peak power of the excitation pulse be I_P, and let the average power in the probe beam be I_{avg}. The signal voltage produced by excitation of the photodiode (the DUT) is $V = R I_P R_L$, where R is the responsivity of the detector and R_L is its load resistance. The photocurrent in the receiver used in the electro-optic measurement system is, from Eq. (39),

$$i = \frac{R I_{avg}}{2} \left(1 + \frac{\pi R I_P R_L}{V_\pi} \right). \tag{49}$$

The first term on the right-hand side of Eq. (49) is the average photocurrent and the second term is the signal photocurrent. Then, from Eqs. (37) and (38), and taking $i_{avg} = R I_{avg}/2$, the signal-to-noise power ratio for shot-noise–limited operation is

$$S/N = \frac{R R_L}{4 e B_f} \left(\frac{\pi R}{V_\pi} \right)^2 I_P^2 I_{avg}. \tag{50}$$

Thus, the S/N depends on the square of the peak power in the excitation beam times the average power in the probe beam (Wiesenfeld et al., 1987a). Clearly, for measurements of photodiodes, high-power laser sources are advantageous. The limitation on the maximum usable power from the laser is imposed by nonlinearity in and possible damage to the detector.

The impulse response of a very-high–speed planar GaAs Schottky photodiode was measured by Wang and Bloom (1983) using 5-ps pulses from a mode-locked CW dye laser and a LiTaO$_3$ hybrid sampler. To ensure a high-speed connection to the sampler, the transition from the photodiode to the microstrip transmission line on the hybrid sampler was accomplished using a gold-mesh bond wire (Kolner et al., 1983b). Results are shown in Figs. 46 and 47. Figure 46 shows the impulse response measured and calculated after deconvolution of the autocorrelation of the laser pulses used. The full-width at half-maximum (FWHM) of the deconvolved impulse response is 5.4 ps. In the frequency domain, the -3-dB bandwidth is 100 GHz, as shown in Fig. 47. This extremely high bandwidth is achieved by including series inductance in the device, which is trimmed to produce maximum bandwidth. In this measurement, the extremely large bandwidth of the electro-optic effect was used to advantage to measure frequency response from DC to >100 GHz. It is possible to use conventional electronics to measure frequency response at high frequencies, but only in discrete bands. A very high–speed indium tin oxide/GaAs photodetector, with -3-dB bandwidth of 110 GHz, has been reported by Parker et al. (1987). To measure this bandwidth, the response of the diode to 0.1-ps pulses from a mode-locked CW dye laser was measured using a microwave waveguide mixer diode. Therefore, the bandwidth of the

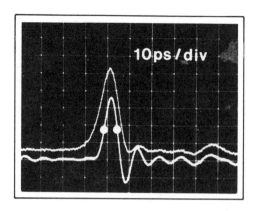

FIG. 46. Impulse response of GaAs Schottky photodiode. The upper trace is the measured response and the lower trace is the deconvolved response. The FWHM are 8.5 and 5.4 ps, respectively. (Wang and Bloom, 1983)

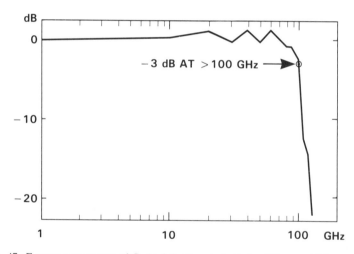

FIG. 47. Frequency response of GaAs Schottky photodiode. (Wang and Bloom, 1983)

measurement was limited to the 75–110 GHz range, due to the finite range of the mixer diode. Because the conventional electronic measurement is performed in a discrete waveguide band, no phase information is available to reconstruct the temporal waveform. Electro-optic sampling, however, provides the temporal waveform directly and the frequency and phase responses by Fourier transformation.

For optical-communications applications, InGaAs photodiodes (which

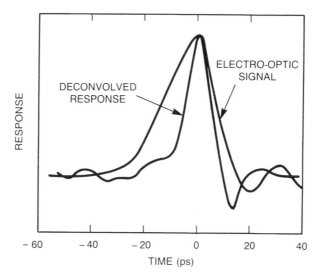

FIG. 48. Measured and deconvolved impulse response for back-illuminated InGaAs/InP punch-through photodiode. (Tucker et al., 1986)

are responsive at 1.3 and 1.55 μm) are important. The impulse responses of very-high–speed InGaAs photodiodes and avalanche photodiodes have been measured using electro-optic sampling (Taylor et al., 1986b,e; Tucker et al., 1986). For these measurements, the source was a mode-locked 1.3-μm InGaAsP injection laser. The sampler was a 50-Ω microstrip line deposited on a semi-insulating GaAs substrate and mounted in a package with K connectors. The photodiodes studied were back-illuminated InGaAs/InP punch-through p-i-n devices (Lee et al., 1981). Results for the widest-bandwidth measurement for such devices are shown in Figs. 48 and 49 (Tucker et al., 1986). The device used in this measurement had an active area of 150 μm^2 and an intrinsic layer thickness of 0.5 μm (Bowers et al., 1985). The chip was mounted on a small, low-inductance L-shaped bracket and connected via a short length of gold-mesh wire to the center pin of a K connector. This device and mount are nominally the same as those described by Bowers et al. (1985). The photodiode package is connected via the K connectors to the hybrid electro-optic sampler. The sampler, in turn, is connected between the photodiode package and the bias T for the photodiode, so that the measured impulse response is that due only to the convolution of the photodiode and the sampler, and it does not include the bandwidth limitation of the bias T. The measured electro-optic response of the photodiode and its deconvolved impulse response are shown in Fig. 48.

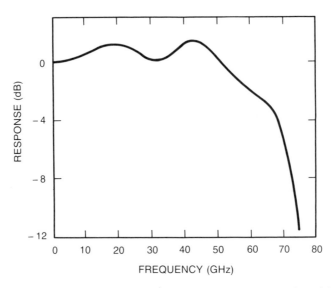

Fig. 49. Frequency response of InGaAs/InP photodiode. Bandwidth (-3 dB) is 67 GHz. (Tucker et al., 1986)

The width (FWHM) of the impulse response is less than 9 ps. The deconvolved frequency response is shown in Fig. 49. There are ripples of magnitude 2 dB and the -3-dB bandwidth is 67 GHz. The measured bandwidth of 67 GHz is larger than the nominal 47-GHz bandwidth of the K connector. It is likely that the K connector is overmoded in the high-frequency range, but this does not appear to degrade the response. The factors that affect the ultimate speed of the photodiode are intrinsic layer thickness and active area of the chip, and parasitics in the chip and mount (Bowers et al., 1985). From a circuit analysis of the mount used in these measurements, it appears that parasitic inductance and capacitance limit the bandwidth to 67 GHz (Tucker et al., 1986).

The impulse and frequency responses of a different back-illuminated InGaAs/InP punch-through photodiode were reported by Taylor et al. (1986b). This device (Burrus et al., 1985) had different intrinsic layer thickness and active area from the device just described and was mounted in a modified commercial package used for microwave detectors. The impulse response for this device was 22-ps FWHM (Taylor et al., 1986b). Also, the impulse response of an InGaAs avalanche photodiode (Campbell et al., 1983) has been measured as a function of its gain using electro-optic sampling (Taylor et al., 1986e).

c. Optical Modulators

A technique for measuring the response of waveguide electro-optic modulators that predates the conventional electro-optic sampling technique uses the DUT as the hybrid sampler (Alferness et al., 1980). In this technique, a train of optical pulses is synchronized to a train of short electrical pulses (Alferness et al., 1980) or to a high-frequency sinusoidal electrical drive signal (Korotky et al., 1987) by phase-locking a frequency synthesizer driving the mode-locked laser with a frequency source for the electrical signal. The optical pulses are coupled through the waveguide modulator, and the electrical drive pulses or sinusoid are coupled to the electrodes of the modulator. Transmission of the modulator is measured using a slow optical detector. As the phase between the optical and electrical signals is varied, the signal on the slow optical detector reconstructs the response of the modulator DUT. The experiment is analogous to electro-optic sampling using electrical excitation.

In the first demonstration of this technique, the switching time was measured for an optical directional coupler switch made with Ti-diffused waveguides on a LiNbO$_3$ substrate, using 5-ps pulses from a mode-locked CW dye laser (Alferness et al., 1980). The switching 10–90% rise time was measured to be 110 ps. In an experiment using 15-ps pulses from a mode-locked InGaAsP laser, the frequency response of a Ti-diffused LiNbO$_3$ waveguide modulator was measured to a frequency of 40 GHz (Korotky et al., 1987). The -3-dB bandwidth for this device was 22.5 GHz, and the response at 40 GHz was 8 dB below the low-frequency response.

V. Direct Probing by Charge-Density Modulation

In this section we consider the charge-density probing technique. This technique is applicable to all substrate materials and, hence, may be important for probing silicon as well as GaAs ICs and devices. We first present the general principles of the technique. Following that, we present technical details of its implementation and, finally, applications of the technique.

A. PRINCIPLES

An optical probing technique based on charge-density variation in active devices has been proposed and demonstrated by Heinrich, et al. (1968a). The technique is based on the dependence of the optical index of refraction of the device on the local charge density, and on the detection of variations of the local index of refraction by interferometric techniques. The carrier-density dependence of the index of refraction is present in all semi-conductor materials and, in particular, in non–electro-optic materials. Thus, the technique can be used to probe internal nodes in Si integrated circuits.

Free carrier plasma refraction makes the dominant contribution to the carrier-dependence of the index of refraction, but electro-refraction and band-filling effects also contribute (Keller et al., 1988; Koskowich and Soma, 1988a,b). The index of refraction of a semiconductor is perturbed by the presence of a free carrier plasma according to (Jackson, 1962)

$$n = n_0 \sqrt{1 - \omega_p^2/\omega^2}, \tag{51}$$

where n_0 is the unperturbed index of refraction, in the absence of carriers, and ω is the angular frequency of the optical probe beam. The plasma frequency, ω_p is given by (Ichimura, 1973)

$$\omega_p^2 = \frac{e^2}{\varepsilon} \left(\frac{n_e}{m_e^*} + \frac{r_h}{m_h^*} \right), \tag{52}$$

where e is the electron charge; n_e and n_h are the electron and hole concentrations, respectively; m_e^* and m_h^* are the electron and hole effective masses, respectively; and ε is the dielectric constant. Combining Eqs. (51) and (52), the change in refractive index due to free carriers, $\Delta n = n(n_e, n_p) - n_0$ is given by

$$\Delta n = -\frac{n_0 e^2 \lambda^2}{8\pi^2 c^2 \varepsilon} \left(\frac{n_e}{m_e^*} + \frac{n_h}{m_h^*} \right), \tag{53}$$

where $\lambda = 2\pi c/\omega$ is the probe wavelength. For a single carrier plasma (or a plasma with equal electron and hole densities) with density n and effective mass m^*,

$$\Delta n = n_p' n$$

where

$$n_P' = -\frac{n_0 e \lambda^2}{8\pi^2 c^2 \varepsilon m^*} \tag{54}$$

Because the coefficient $n_p' \approx 3 \times 10^{-29} \text{ m}^3$, Δn will be small. It is possible to detect small changes of refractive index interferometrically, however. Consider Fig. 50, which shows a Michaelson interferometer. A probe beam is divided by a 50%-reflecting–50%-transmitting beam-splitter (BS). In one arm of the interferometer is a sample of length L, in which the index is modulated by free carriers. In the other arm is a bias plate, which intro-

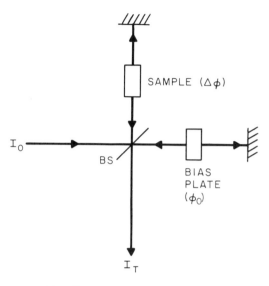

FIG. 50. Michaelson interferometer.

duces a fixed retardation between the two arms of the interferometer. After reflection from two mirrors, the two beams are recombined by the beam-splitter and detected by a photodiode. The phase change produced by free carriers after two passes through the sample is given by

$$\Delta\phi = \frac{4\pi}{\lambda}\Delta nL$$

$$= -\frac{n_0 e^2 \lambda}{2\pi c^2 \varepsilon}\left(\frac{n_e}{m_e^*} + \frac{n_h}{m_h^*}\right)L. \tag{55}$$

The phase change caused by the bias plate is ϕ_0 for two passes. Assume the two arms of the interferometer are exactly equal in length except for the bias plate and free carriers in the sample. If the electric field incident on the entrance to the interferometer is $\mathscr{E} = \mathscr{E}_0 e^{-j\omega t}$, the field at the photodiode is

$$\mathscr{E}_T = \frac{\mathscr{E}_0}{2} e^{-j\omega t}(e^{j\Delta\phi} + e^{j\phi_0}). \tag{56}$$

The total detected intensity is $I_T = (1/2)\sqrt{\varepsilon/\mu}\,\mathscr{E}_T\mathscr{E}_T^*$, and is given by

$$I_T = I_0 \cos^2\left(\frac{\phi_0 - \Delta\phi}{2}\right), \tag{57}$$

where $I_0 = (1/2) \sqrt{\varepsilon/\mu} \, \mathcal{E}_0^2$. The transfer function for the interferometer, Eq. (57), is the same as the transfer function for the electro-optic modulator, Eq. (16), with a 90° phase shift; hence, Fig. 12 shows the form of the transfer function. Maximum sensitivity and linearity is achieved at the point of maximum slope of the transfer function. This occurs for the interferometer when ϕ_0 is an integral multiple of $\pi/2$. In this case, Eq. (57) becomes

$$I_T = \frac{I_0}{2} (1 + \sin \Delta\phi). \tag{58}$$

Because $\Delta\phi$ is small, the fractional modulation caused by charge-density modulation is

$$\frac{\Delta I}{I} = \frac{I(n_e, n_h) - I(0)}{I(0)} = \Delta\phi - \frac{1}{6}(\Delta\phi)^3 + \cdots, \tag{59}$$

or, to first order using Eq. (55),

$$\frac{\Delta I}{I} = -\frac{n_0 e^2 \lambda}{2\pi c^2 \varepsilon} \left(\frac{n_e}{m_e^*} + \frac{n_h}{m_h^*} \right) L. \tag{60}$$

The products $n_e L$ and $n_h L$ are sheet carrier densities, in units of carriers/m^2.

To obtain an expression relating modulation of the probe beam to voltage, we assume a linear relationship between carrier density in the active region of a device and voltage applied to the device, that is, a constant capacitance (Heinrich et al., 1986a)

$$V = enL/C_a, \tag{61}$$

where n is the dominant carrier density (effective mass m^*) and C_a is the capacitance per unit area of the device. Combining Eqs. (60) and (61),

$$\frac{\Delta I}{I} = -\frac{n_0 e \lambda}{2\pi c^2 \varepsilon m^*} C_a V. \tag{62}$$

Consider, as an example, a forward-biased Si diode with a junction capacitance of $C_a = 8 \times 10^{-3}$ F/m^2 (Heinrich et al., 1986a). Assume $V = 1$ V and λ is 1.3 μm. For Si, $n_0 = 3.5$, $m^* = 0.13 m_e$, and $\varepsilon = 12\varepsilon_0$. Substitution of these numbers into Eq. (62) gives $\Delta I/I = 8 \times 10^{-4}$. Fractional modulation two orders of magnitude larger (9.8%) has been reported when probing the base of a planar silicon n-p-n transistor, upon application of 2V to the base (Hemenway et al., 1987).

In the preceding analysis, we have considered carrier-induced changes in the refractive index caused by a free carrier plasma. This effect will occur in all semiconductor materials. In direct-gap semiconductors such as GaAs, however, the Burstein-Moss effect will also produce a carrier-density dependence of the refractive index. The Burstein-Moss effect is due to bandfilling, which causes the absorption edge of the semiconductor to be shifted to higher energy compared to the same material with no free carriers. By a Kramers-Kronig transformation, this shift of the absorption edge will translate into a change in the index of refraction. Keller et al. (1988) have calculated that in GaAs, for a probe wavelength of 1.32 μm, the change in refractive index due to the Burstein-Moss effect depends linearly on carrier density (for small enough density) and is of the same sign and about 30% larger than the shift due to plasma refraction. Application of an electric field across the semiconductor, as in a biased diode, will also cause a change in the shape of the absorption edge, because of the Franz-Keldysh effect. Again by Kramers-Kronig transformation, the Franz-Keldysh effect will produce a carrier-density dependence of the index of refraction, which is electrorefraction (Keller et al., 1988; Koskowich and Soma, 1988a,b). In GaAs, the magnitude of this change, however, will be less than 10% of that due to free carrier plasma and Burstein-Moss effects (Keller et al., 1988). In Si devices, the Burstein-Moss effect will not be important, but the magnitude of the index changes due to electrorefraction can be as large as 50% of that caused by free carrier refraction, depending on the device parameters (Koskowich and Soma, 1988b). For interferometric detection of refractive index changes, Eq. (58) is valid with $\Delta\phi = 4\pi\Delta n L/\lambda$. However, in a detailed analysis, Δn has several components:

$$\Delta n = n'_\text{P} n + n'_B n + \Delta n_\text{FK}, \qquad (63)$$

where n'_P is given by Eq. (54), n'_B is the coefficient for the Burstein-Moss effect, and Δn_FK is the refractive index change caused by the Franz-Keldysh effect. In what follows we consider free carrier plasma refraction to be the dominant mechanism relating Δn to n, which will be the case for silicon devices.

B. TECHNICAL ASPECTS

An experimental schematic for the measurement of waveforms internal to a silicon integrated circuit is shown in Fig. 51 (Black et al., 1987; Hemenway et al., 1987). To probe through a silicon substrate, it is necessary to use a laser wavelength below the absorption edge of silicon. Probe pulses are generated at such a wavelength from a gain-switched, 1.3 μm InGaAsP injection laser. The probe beam passes through a polarizing beam-splitter (PBS # 1), a 45° Faraday rotator, and a second polarizing

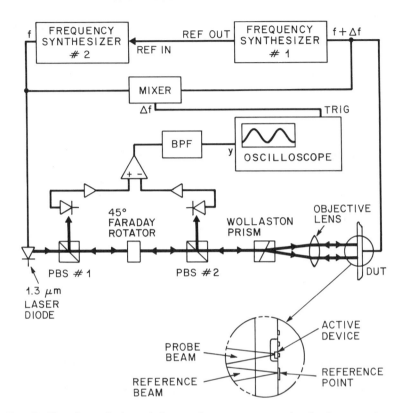

FIG. 51. Experimental schematic for sampling measurement using the charge-sensing technique. PBS, BPF, and TRIG are polarizing beam-splitter, bandpass filter, and trigger, respectively.

beam-splitter (PBS #2) before passing through a Wollaston prism. The two polarizing beam-splitters and the Faraday rotator comprise a Faraday isolator which isolates the laser from reflections. The Wollaston prism divides the incoming beam into two orthogonally polarized beams, which are then focused onto the DUT from the backside, which must be polished. One of the beams is the probe beam, which is focused through the active device, and the other is a reference beam, which is focused onto an inactive region of the DUT: either a contact or simply the substrate surface. The beams are reflected from the front surface of the DUT and are recombined in the Wollaston prism and sent back to the Faraday isolator. The two beams impinging on and reflecting from the DUT form an interferometer, similar in principle to that shown in Fig. 50, and the intensity of the beam recombined in the Wollaston prism is given by Eq. (57). Maximum sensitivity will occur when there is a relative phase shift of an integral multiple of $\pi/2$ between the probe and reference beams in the absence of signal applied to the active device. Such a phase shift may be produced by adjustment of the Wollaston prism and, therefore, the position of the reference beam. In this case, the recombined, reflected beam will be circularly polarized, and a phase change caused by carriers in the active device will cause it to become elliptically polarized. The axes of the carrier-induced elliptical polarization will be oriented at 45° with respect to the axes of the Wollaston prism. PBS #2 will reflect the polarization parallel to one of the axes of the carrier-induced elliptical polarization into a photodetector. After passage through the Faraday rotator, the second axis of the elliptical carrier-induced polarization will be reflected by PBS #1 into another photodetector. Because the signals in the photodetectors correspond to different axes of the carrier-induced elliptical polarization, the signals will be 180° out of phase and equal in magnitude (given by Eq. (60)). Hence, if the signals are detected differentially, as in Fig. 51, the electrical signal will be twice the magnitude of that given by Eq. (60).

In spite of difficulties, it is possible to use a multi-longitudinal-mode laser source with the experimental arrangement of Fig. 51. In a multimode laser, mode-partition noise and spectral variations in the laser, due to thermal instabilities or optical feedback, might cause noise at the detector, because each mode has a different wavelength and will, therefore, have a different phase shift through the interferometer. The Wollaston prism can be adjusted so that only one mode of the laser has exactly a multiple of $\pi/2$ phase shift in the interferometer. However, because both polarizations travel the same optical path length through glass in the Wollaston prism, the phase shifts for the modes will not be much different than the multiple of $\pi/2$. Thus, excess noise due to mode-partition noise and thermal instabilities should be small. To remove the excess noise, however, a single-frequency laser source, such as a distributed feedback (DFB) diode laser, could be used. Pulses as short as 26 ps have been generated from a gain-switched DFB laser, without external compression (Takada et al., 1986), and this would be an appropriate source.

For circuits operating at frequencies below about one GHz, the laser can be operated CW (Hemenway et al., 1987), and the time resolution in the measurement system is achieved by using fast photodetectors. For circuits operating at higher frequencies, however, it is necessary to use ultrashort laser pulses and achieve time resolution by varying the delay between the arrival of the probe pulse and the waveform being interrogated (Black et al., 1987). The delay between laser pulse and waveform can be achieved electronically, as shown in Fig. 51. As in the electro-optic sampling experiments, two frequency synthesizers are phase-locked and offset in frequency by Δf. One synthesizer drives the circuit and the second drives the injection laser. The pulses from the laser sample the waveform at the point being probed at a rate Δf, and the resulting signal can be displayed on an oscilloscope for a "real-time" reconstruction of the waveform or sent to a signal averager for further processing.

Next, consider the sensitivity of the technique. For this calculation, we take $n_e = n_h = n$ and $m^* = (m_e^{*-1} + m_h^{*-1})^{-1}$. Furthermore, let the sheet charge density in the active device be $\delta n = nL$. We consider only one of the detectors in Fig. 51. The signal and noise powers at the photodetector are given by Eqs. (37) and (38), respectively, and, as before, we consider only shot-noise and Johnson noise. The photocurrent in the photodetector is RI_0, and from Eqs. (54) and (57), making the valid small-signal approximation, the photocurrent is

$$i = \frac{RI_0}{2}\left(1 - \frac{n_0 e^2 \lambda}{2\pi c^2 \varepsilon m^*}\delta n\right). \tag{64}$$

The first term on the right in Eq. (64) is the average photocurrent ($RI_0/2 = i_{avg}$) and the second term is the signal photocurrent. The variation in δn is assumed linear in time so that $\langle \delta n \rangle = \delta n_0/2$, where δn_0 is the maximum sheet charge density injected into the active region. Then, from Eqs. (37), (38), and (64), the signal-to-noise power ratio is

$$S/N = \frac{\dfrac{i_{avg}^2}{4}\left(\dfrac{n_0 e^2 \lambda}{2\pi c^2 \varepsilon m^*}\right)^2 \delta n_0^2}{(2e i_{avg} + 4kT/R_L)B_f}. \tag{65}$$

The minimum detectable sheet charge density is the value of δn_0 for which $S/N = 1$. We now assume shot-noise–limited operation for which the first term in the denominator of Eq. (65) is much larger than the second term. From Eq. (65), with $S/N = 1$ (Heinrich et al., 1986a),

$$\delta n_0 = \frac{4\pi c^2 \varepsilon m^*}{n_0 e^2 \lambda}\sqrt{2e/i_{avg}} \quad \text{(carriers/m}^2 \cdot \sqrt{\text{Hz}}\text{)}, \tag{66}$$

where we have taken $B_f = 1$ Hz. Using Eq. (61), Eq. (66) can be written in terms of the minimum detectable voltage in a one-Hz bandwidth:

$$V_{min} = \frac{4\pi c^2 \varepsilon m^* C_a}{n_0 e \lambda}\sqrt{2e/i_{avg}} \quad \text{(V/}\sqrt{\text{Hz}}\text{)}. \tag{67}$$

For silicon at a probe wavelength of 1.3 μm, the minimum detectable sheet charge density when $i_{avg} = 1$ mA is 2.2×10^{12} carriers/m^2, from Eq. (66). For a forward-biased diode with $C_a = 8 \times 10^{-3}$ F/m^2, $V_{min} = 44\ \mu$V/$\sqrt{\text{Hz}}$, from Eq. (67).

For Eqs. (66) and (67), we have neglected Johnson noise compared to shot-noise. This is valid when $i_{avg} \gg 4kT/2eR_L$, which for a 50-Ω load resistor, requires $i_{avg} \gg 1$ mA. This is of some consequence for systems using a CW laser diode, because most high-speed photodetectors use a 50-Ω load resistor. To be shot-noise–limited would require about 3 mA of average photocurrent, which means about 5 mW of optical power imping-ing on the detector (R = 0.6 A/W). Thus, 10-mW CW power is required from the injection laser, assuming the only loss is the 50% loss in the polarizing beam-splitter. While this amount of power is possible from an injection laser, other losses in the optical system will increase the power required from the injection laser to a level that may be unachievable. The use of two photodetectors, as in Fig. 51, increases the signal by 6 dB, while increasing the noise by 1.3 dB (Hemenway et al., 1987). It is clear, how-ever, that to operate the system in the shot-noise limit using a CW laser source and high-speed, low-impedance photodetectors requires at least 10 mW of optical power, which is near the limit of what can be generated by a 1.3-μm InGaAsP laser. If a high-impedance preamplifier can be used with the photodetector, the optical power required for shot-noise–limited operation is reduced. Indeed, if the time resolution is achieved by using a pulsed injection laser, a "slow" photodiode with a high-impedance pream-plifier can be used, and the shot-noise limit is more easily achieved.

The S/N ratio is inversely proportional to the electronic bandwidth of the measurement system, as seen from Eq. (65). Therefore, as in the electro-optic sampling experiments, some scheme of narrow-band detec-tion is employed. In the arrangement of Fig. 51, the signal is passed through a bandpass filter at Δf, with the bandwidth adjusted to the mini-mum width consistent with distortionless acquisition of the signal. It is necessary for Δf to be large enough so that the $1/f$ noise (or excess laser noise) can be reduced. In the scheme of Black et al. (1987), the clock signal to the DUT is chopped at 20 kHz, to overcome the $1/f$ noise, and synchro-nous detection of the signal on the photodiodes is employed. In that arrangement, Δf is chosen to be 33 Hz, to provide a "real-time" recon-struction of the signal on an oscilloscope. In any case, the waveform is measured by the charge-sensing technique at a rate Δf and can be averaged on a signal averager at that rate.

Calibration of the magnitude of the signal measured by the charge-sensing technique from one active region of an IC to another may be difficult. The experiment measures sheet-charge densities. Considering only one type of charge carrier, Eq. (60) can be written as

$$\frac{\Delta I}{I} = -\frac{n_0 e^2 \lambda}{2\pi c^2 \varepsilon m^*} \delta n. \tag{68}$$

In order to determine the value of δn, it is necessary to measure both ΔI and I. For conversion from δn to voltage, using Eq. (61), it is necessary also to know the value of C_a for the active region being probed. From the design parameters of the IC, C_a can be estimated, pessimistically, to within 50% and, optimistically, to within 20%. However, for different active devices of the same type on the same chip, relative values of C_a should be consistent to within ~10%. Moreover, to compare the absolute magnitudes of the voltage waveforms measured at different active regions of the IC, it will also be necessary to ensure that the interferometer remains biased at a relative phase delay of $\pi/2$ between the probe and reference beams and that the amplitudes of the probe and reference beams are adjusted to be equal, so that in the absence of signal, the recombined beam in the Wollaston prism is circularly polarized. This will require care, because the reference beam must be focused to a point near the probe beam and the reflectivity of the DUT at a given distance from a particular probe site may vary from site to site due to the diversity of the circuit arrangement. (For example, the reference point may be a surface, contact, alloyed contact, etc.) Perhaps, with care, the accuracy of absolute voltage determination at different points on the IC should be comparable to the accuracy of C_a, about 10%.

As with electro-optic sampling, the spatial resolution is limited by the diameter of a diffraction-limited focal spot because, in the back-side probing geometry of Fig. 51, the probe beam is focused on the contact just above the active region to a diffraction-limited diameter as small as 1 to 2 μm. For same ICs, the dimensions of the active region may be comparable to or smaller than that of the focal spot (Black et al., 1987), resulting in a reduction of signal and nonuniform phase change across the profile of the probe beam. Because the reference beam is focused on an inactive contact or the surface of the substrate, its diameter is less critical than that of the probe beam. However, as just mentioned, it is necessary that the phase and amplitude of the probe beam be well controlled to maintain the proper bias of the interferometer. Frontside probing is not effective because the charge in the active region is necessarily below a contact that is opaque to the probe radiation.

The probe beam is modulated by changes in the index of refraction of the substrate which can be caused by carrier effects in the active region and also by thermal lattice effects (Hemenway et al., 1987). The carrier contribution to the index of refraction arises from carrier-density changes in the operating device and follows the waveform propagating through the device. Carrier thermal relaxation processes might be slower than the waveform and will heat the lattice in the vicinity of the active region, causing changes in the refractive index. The thermal contribution to the index of refraction will be a low-frequency effect, and a computer model by

Hemenway et al. (1987) shows that the thermal contribution has a corner frequency of one MHz and rolls off at 30 dB/decade at higher frequencies. Hence, the thermal effect should not be significant for measurement of high-speed repetitive waveforms.

The other factors affecting the temporal resolution of the charge-sensing technique are (1) the speed of the photodetector for a CW probe or the pulse duration for a pulsed laser, (2) timing jitter between the laser and the clock driving the IC for a pulsed laser source, and (3) the optical transit time for the probe beam through the active region. Points (2) and (3) have been elaborated in Section IV.B.3. We consider point (1). For a CW laser, the system response is determined primarily by the bandwidth of the photodetector. A long wavelength InGaAs p-i-n photodiode with a bandwidth as high as 67 GHz has been reported (Tucker et al., 1986). InGaAs p-i-n photodiodes with a bandwidth around 20 GHz are more common (Burrus et al., 1985; Taylor et al., 1986b; Wang et al., 1987), however, and provide adequate bandwidth for measurement of most existing ICs. Generally, these photodiodes are used with 50-Ω electronics and therefore many mW of optical power at the photodiode is required for shot-noise–limited operation, as discussed already. Among pulsed laser sources, gain-switched 1.3-μm InGaAsP injection lasers have produced pulses below 20 ps (White et al., 1986; Wiesenfeld et al., 1987b), which corresponds to a bandwidth of >16 GHz. Details concerning laser sources have been discussed in Section IV.B.6. With a pulsed laser source, a high-impedance photodetector can be used, with the result that shot-noise–limited operation is more easily achievable.

The overall temporal resolution of the measurement is determined by the convolution of mechanisms (1)–(3), but is dominated by the duration of the laser pulse or the bandwidth of the photodiode. Black et al. (1987) have reported a temporal resolution of 42 ps, limited by the duration of the laser pulse, for a system using a gain-switched laser. Using shorter laser pulses that can be generated by other gain-switched InGaAsP lasers, temporal resolution below 20 ps is possible.

The charge-sensing technique is noninvasive because the probe wavelength of 1.3 μm is below the bandgap of silicon or GaAs. In practice, there is some absorption even at 1.3 μm, but because the absorption coefficient is small, it leads to only small changes in the operation of the DUT. With 20 mW incident on a bipolar transistor, Heinrich et al. (1986b) deduce an effective base current of 6.6 nA caused by absorption of the laser beam. For a different bipolar junction transistor, Black et al. (1987) observe a change of 0.1% in the collector current due to 5-mW incident optical power. These changes are small enough to qualify the technique as noninvasive.

In summary for this section, the charge-sensing technique is noninvasive and is capable of detecting sheet-charge densities with a theoretical sensitivity in silicon of 2.2×10^{12} carriers/m^2-$\sqrt{Hz}$ for one mA of detected photocurrent. A spatial resolution of 1 to 2 μm and a temporal resolution below 20 ps (using a gain-switched laser) are possible.

C. APPLICATIONS

As of the late 1980s, the technique of optical probing by charge-density modulation has not yet been used extensively. All the applications of the technique reported to date have involved discrete devices rather than ICs. The applications can be divided into real-time probing and sampling of waveforms.

1. Real-Time Probing

The first demonstration of the technique used a large area (0.5 mm × 0.5 mm) abrupt junction Si p-n diode (Heinrich et al., 1986a). For this demonstration, a 1-MHz signal was applied to the forward-biased diode, and the optically measured signal was displayed on a 20-MHz bandwidth oscilloscope. Rise-and falltimes of ~250 ns were attributed to the capacitance and series resistance of the diode.

Measurement in real time of a 100-MHz signal was reported for a vertical Si n–p–n bipolar transistor with a cutoff frequency f_T of 1 GHz (Heinrich et al., 1986b). The emitter area was 11 μm × 11 μm, and the probe beam was focused through the collector, base, and emitter of the device. Using a CW Nd:YAG laser operating at 1.3 μm, an "emitter-coupled logic" (ECL) level signal (0.8V) at 100 MHz was probed and the optically measured signal was observed directly on an oscilloscope. In this measurement, the average photocurrent on the detector was 10 mA, corresponding to ~30 to 40 mW optical power incident on the DUT. With 3 mW from a CW, single-mode semiconductor laser, the signal could not be measured with acceptable signal-to-noise ratio at 100 MHz, but could be measured at 25 MHz. This variation of signal-to-noise ratio with bandwidth and optical power follows from Eq. (65), which shows that for a given S/N ratio, the average photocurrent must be increased if the bandwidth is increased.

The charge-sensing technique has been applied to GaAs reverse-biased Schottky diodes. Koskowich and Soma (1988a) have performed calculations suggesting that measurement of a voltage in the Schottky barrier by plasma refraction may be more sensitive than measurement by electro-optic sampling. They also measured a waveform at 20 MHz. Keller et al. (1988) have measured the charge density and applied voltage in a GaAs Schottky diode by carrier-induced refraction changes and electro-optic sampling, respectively. They have been able to calibrate the absolute magnitudes of both the charge and voltage in the diode by careful measurement of both ΔI and I. They have also resolved spatially the charge distribution under the Schottky barrier by observation of a voltage-dependent standing-wave effect in the charge-sensing measurement.

In the preceding examples, a Nomarski wedge, rather than a Wollaston prism, was used to produce orthogonally polarized probe and reference beams, which resulted in significant mode-partition noise and/or noise due to thermal fluctuations in the laser unless a single-mode laser source was used. In the work of Hemenway et al. (1987), the experimental arrangement of Fig. 51 was introduced, which allowed use of a multimode laser source. Using the arrangement of Fig. 51 with a CW, multimode semiconductor laser, Hemenway et al. (1987) measured in real time the response of a planar Si n-p-n bipolar transistor to a 25 Mbit/s return-to-zero (RZ) data stream. The emitter for this device had a cross section 11 μm × 11 μm, and the probe beam was directed through the emitter. The result of the measurement for a 2V input signal is shown in Fig. 52. The reported fractional modulation $\Delta I/I$ is 9.8%, which means that the capacitance per unit area for this device, C_a, is large; from Eq. (62), C_a is about 0.5 F/m². The fractional modulation is 3–4% for an ECL level signal. The baseline wander in the optically detected signal in Fig. 52 is due to the one-MHz low-frequency cutoff in the amplifier used in the experiment and to the lattice heating effects discussed already. The noise floor for the measurement of Fig. 52 is 2 dB above the Johnson noise limit and 9 dB above the shot-noise limit. Even so, the signal-to-noise ratio is good, which suggests that the effect could be used as an optical modulator or an optical interconnection with a low bit error rate (Heinrich et al., 1986a,b; Hemenway et al., 1987). In order to achieve a low bit error rate, however, it is necessary to remove the low-frequency components, which are responsible

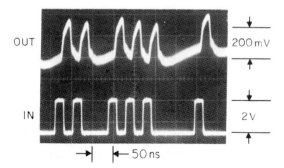

FIG. 52. Real-time measurement of 25-Mbit/s RZ data stream in a planar Si n-p-n bipolar transistor. Upper trace is measured signal and lower trace is input voltage. Horizontal scale is 50 ns/division. (Hemenway et al., 1987; © 1987 IEEE)

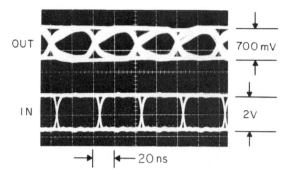

Fɪɢ. 53. Eye diagram of Manchester-coded pseudorandom bit stream at 25 Mbit/s, using charge-sensing measurement technique and a planar Si n-p-n bipolar transistor. Lower trace is input data stream and upper trace is received optical signal. (Hemenway et al., 1987; © 1987 IEEE)

for the baseline wander in Fig. 52, from the signal. This can be achieved using Manchester coding which enables the data to be encoded with no frequency component below 5 MHz. The penalty for such a coding scheme is a reduction of the data-transmission rate by a factor of 2 below the nominal bit rate. Figure 53 shows an eye diagram for a Manchester coded pseudorandom bit stream using the Si n-p-n bipolar transistor as a modulator. The eye of the received signal has a signal-to-noise ratio of six, suggesting that data transmission with a bit error rate below 10^{-9} is possible. In order to achieve this signal-to-noise ratio at higher bit rates, it will be necessary to have more optical power incident on the device, so that i_{avg} is larger.

2. Sampling

Black et al. (1987) have measured the response of a high-speed silicon bipolar junction transistor using a gain-switched InGaAsP injection laser and a system similar to that in Fig. 51. The device studied had $f_T = 5$ GHz and an emitter area of $1.5 \ \mu m \times 5 \ \mu m$. The smaller dimension of the emitter is comparable to the focused diameter of the laser probe beam, causing some reduction in signal. The laser is gain-switched at a 100 MHz rate, using a comb generator driven by frequency synthesizer #2 in Fig. 51. Frequency synthesizer #1, which drives the device, is locked to a harmonic of synthesizer #2. As mentioned previously, in this experiment the signal to the DUT is chopped at 20 kHz and detected synchronously. The harmonic of synthesizer #2 and the frequency of synthesizer #1 are offset by 33 Hz, which is the rate at which the signal on the DUT is sampled. The laser pulse duration is 42 ps, which limits the bandwidth of the measure-

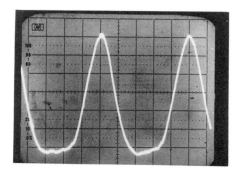

FIG. 54. Sampled one-GHz waveform measured in a Si bipolar junction transistor. Horizontal scale is 200 ps/division. (Black et al., 1987)

ment system to 8 GHz. With this laser source, the system operation is shot-noise–limited, although the Johnson noise power is only 2 dB below the shot-noise power. A "real-time" reconstruction of the response of the bipolar transistor to a one-GHz sine-wave input, with bias level adjusted to show clipping, is presented in Fig. 54. Waveforms for frequencies between 10 MHz and 2.5 GHz were measured for this device (Black et al., 1987).

Thus far, only discrete devices and not ICs have been studied by the charge-sensing technique. While, in principle, real-time probing could be extended to DUTs operating at GHz frequencies, this will be difficult because required optical power for photodiodes followed by 50-Ω electronics will be tens of mW or because photoreceivers with higher than 50-Ω impedance will have to be designed. The sampling technique has been demonstrated into the GHz range and with shorter laser pulses could be extended beyond 10 GHz. If the difficulties for GHz operation of the real-time probing technique can be overcome, however, it is possible that with suitably designed devices it could provide the basis for optical modulators and, perhaps, optical interconnects.

VI. Phase-Space Absorption Quenching

Phase-space absorption quenching (PAQ), a novel technique for optically probing the logic state of a quantum-well (QW) device, has been reported by Chemla et al. (1987). The technique was applied to a modulation-doped (MD) QW field-effect transistor, and could be used with high-speed MD QW devices, such as the selectively-doped heterojunction transistor (Drummond et al., 1986). The technique relies on the large absorption cross section for the exciton transitions in QW structures, even at room temperature (Chemla and Miller, 1985) and the sensitivity of the exciton absorption to free carriers (Schmitt-Rink et al., 1985). If the channel for carriers in an FET contains the QW structure, the one-component free carrier plasma due to current in the channel significantly modifies the absorption spectrum of the QW (Chemla et al., 1988). Transitions into the states occupied by the free carriers are inhibited because of the Pauli exclusion principle and absorption at the corresponding wavelengths is therefore quenched. The magnitude of the change in absorption coefficient near the first exciton peak can be larger than 10^4 cm^{-1}. Thus, if an optical probe beam with a wavelength within the region of quenched absorption passes through the channel, the transmitted intensity of the probe beam will depend on the carrier density in the channel and, hence, on the logic state of the device.

A differential absorption spectrum for an InGaAs MD QW FET is shown in Fig. 55 (Chemla et al., 1987). The sample contains a single InGaAs QW 100 Å thick, surrounded by barrier layers of InAlAs and grown on an InP substrate. Only the InGaAs QW absorbs probe wavelengths longer than 1 μm. The FET is made with a Schottky gate, and the carrier density in the gate is controlled by the gate-source voltage. The probe beam is incident from the rear of the substrate and reflects off

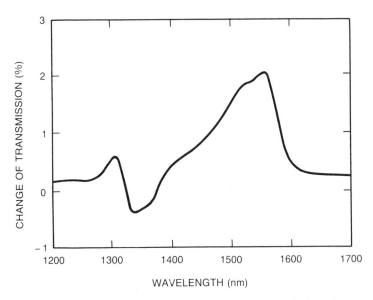

FIG. 55. Differential absorption spectrum for a MD InGaAs QW FET as the gate-source voltage is changed between −0.5 and +1.5V. Transmission is increased near the $n = 1$ exciton absorption peak at 1.55 μm. Note the derivative lineshape near the position of the $n = 2$ exciton absorption at 1.32 μm. (Chemla et al., 1987)

the metallized gate contact so that it passes through the QW twice. The spectrum in Fig. 55 is the difference in probe transmission as the gate-source voltage changes from -0.5 to + 1.5V. At the gate-source voltage of -0.5V, there are no free carriers in the QW in the gate (Chemla et al., 1987). Even though the QW absorbing layer is only 100 Å thick, changes in transmission as large as 2% are observed. The broad positive peak (bleaching) observed around 1.55 μm is due to quenching of the absorption in the region of the first exciton transition caused by the presence of carriers at 1.5 V gate-source voltage. The free carriers, electrons in the present example, fill phase-space in the conduction band and occupy states normally available for absorption, thus bleaching the absorption at energies up to the Fermi level; hence, the name phase-space absorption quenching. PAQ is a generalization of the Burstein-Moss shift. Concurrently with PAQ, many-body effects cause bandgap renormalization, shifting the band edge to lower energy (Chemla et al. 1988). Thus, the second exciton peak is shifted to longer wavelength in the presence of carriers, producing the derivative lineshape near the second exciton peak at 1.32 μm. Because there are not enough free carriers to fill phase-space up to the level of the second exciton peak, PAQ is not observed there.

From the shape of the differential absorption spectrum near the first exciton peak, the carrier density and temperature in the gate can be deduced (Bar-Joseph et al. 1987). Figure 56 shows differential absorption

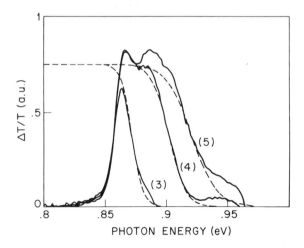

FIG. 56. Differential spectra of the $n = 1$ exciton absorption of 10 K. Gate-source voltage is modulated from $-0.5V$ to 0, 0.5, and 1.0V for curves (3), (4), and (5), respectively. Electron temperatures T_e and densities n deduced from the data are (3) $T_e = 50$ K, $n = 8 \times 10^{10}$ cm^{-2}, (4) $T_e = 80$ K, $n = 4.3 \times 10^{11}$ cm^{-2}, and (5) $T_e = 120$ K, $n = 6.4 \times 10^{11}$ cm^{-2}. (Bar-Joseph et al., 1987)

spectra for the InGaAs MD QW FET at 10 K for gate-source voltage modulation between -0.5V and 0, 0.5, and 1.0V for curves (3)–(5), respectively (Bar-Joseph et al., 1987). The high-energy tails of the spectra are fit to Fermi distributions, enabling determination of carrier density and temperature in the QW, with the results given in the caption to Fig. 56. It is clear that the absorption of a probe beam at a wavelength near the exciton peak will be sensitive to the carrier density in the gate and, hence, to the logic state of the device. However, at a particular probe wavelength, the variation in probe transmission with carrier density will be linear over only a limited range of carrier densities, since the bleaching at a given wavelength saturates, as seen in Fig. 56.

Invasiveness of the technique as a probe of devices or ICs requires some elaboration. The probe beam is absorbed in the QW, with an absorption coefficient as large as 2×10^4 cm^{-1} in the absence of carriers (Bar-Joseph et al., 1987). For a probe pulse energy of 0.1 pJ, this will produce $\sim 2 \times 10^4$ electron–hole pairs in a layer of 100 Å thick. If the gate area is 1 μm × 10 μm, this will produce a carrier sheet charge density of 2×10^{11} cm^{-2}, which is similar to the density produced by a gate-source voltage of 0.5V. Therefore, for noninvasiveness, the probe pulse energy must be small, perhaps ~ 0.01 pJ.

The PAQ sampling technique has been applied to the measurement of multigigahertz waveforms in a depletion-mode InGaAs/InAlAs MODFET configured as a 50-Ω common-source inverter (Wiesenfeld et al., 1989). The experimental arrangement was similar to that used for electro-optic sampling measurements with InGaAsP injection lasers (Taylor et al., 1986c; Wiesenfeld et al., 1987b) and is similar to the arrangement depicted in Fig. 14. In the PAQ sampling measurement, the quarterwave plate and polarizer are unnecessary because the technique is based on intensity changes in the probe beam rather than polarization or phase changes. For the PAQ sampling measurement, the laser source was a mode-locked external-cavity semiconductor laser, with a grating feedback element (Bessenov et al., 1983), which produced pulses of 10 to 15-ps duration over the wavelength range from 1.48 to 1.56 μm. The MODFET was sampled under the gate channel from the back surface of the substrate, as described previously. Waveforms at 2 GHz were measured for several values of gate modulation and DC gate bias. As the gate bias became more negative, the PAQ signal showed clipping and then was extinguished, because the channel was pinched off at the most negative gate bias.

The PAQ sampling technique, which is charge-sensitive, was combined with electro-optic sampling, which is voltage-sensitive, to measure propagation delays internal to the MODFET (Wiesenfeld et al., 1989). In this measurement, a ~ 50-ps electrical pulse produced by a comb generator was

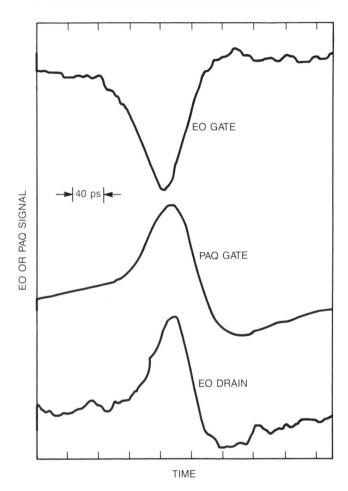

FIG. 57. Measured waveforms following input of ∼50-ps electrical pulses to InGaAs/ InAlAs MODFET. Waveforms are, from top to bottom, measured by electro-optic sampling under the gate pad, measured by PAQ sampling under the gate channel, and measured by electro-optic sampling under the drain pad, respectively. Horizontal scale is 40 ps/ division. (Wiesenfeld et al., 1989)

coupled into the gate of the device. Signals were measured under the gate and drain pads by electro-optic sampling and under the gate channel by PAQ sampling. The results of this measurement are shown in Fig. 57. Propagation delays are extracted from the arrival times of the peaks of the electrical waveforms. The delay for appearance of the PAQ signal is 12 ± 4 ps and the propagation delay to the drain electrode is 15 ± 4 ps. Because the PAQ signal is sensitive to carriers, the 12-ps delay in its appearance is due to the dynamics of the carriers in charging the gate channel.

The experiments just reported are an initial use of the PAQ effect for high-speed sampling. Another particularly promising aspect of the PAQ effect is the large magnitude of the modulation of the probe beam. For this reason, use of the effect as an optical interconnect has been suggested (Chemla et al., 1987).

In related work on quantum-well materials, Knox et al. (1986) have shown that electric fields applied in the plane of quantum wells, in a GaAs/AlGaAs material, broaden the exciton transitions. Therefore, transmission of a probe beam at the wavelength of the exciton peak will be sensitive to the field applied to the material. This effect is excitonic electroabsorption. Excitonic electroabsorption is a field-sensitive probing technique, unlike PAQ, which is charge-sensitive. The response of the excitonic electroabsorption was measured to be faster than 330 fs (Knox et al., 1986). The utility of this effect for optical probing of devices or ICs remains to be studied.

VII. Summary and Outlook

Despite their relative infancy, techniques for direct optical probing of high-speed devices and ICs already show enormous potential. Not only are a large variety of optical techniques manifesting themselves for the solution of a variety of measurement problems, but also the implementation of some of these in real-world test equipment has begun. Most notable among the techniques is electro-optic sampling, which not only appears to allow a large variety of methods for its implementation, but also has demonstrated excellent versatility for solving a large variety of test problems requiring information from a few nodes.

In this chapter we have reviewed a variety of techniques that possess only partially overlapping sets of requirements, conditions, and capabilities. Thus, there is no one general technique that will be optimal for all measurement problems associated with high-speed ICs and devices. In the following sections, we shall compare the probing techniques and present a few anticipated needs for probing systems. We distinguish between probing of discrete devices, packaged devices, and wafers. We also distinguish between characterization and screening applications. For characterization applications, a relatively small number of devices need to be studied in detail in order to aid in design and/or development of ICs and devices. For screening, a large number of (production) devices must be measured, but less thoroughly, to ensure their reliability.

A. COMPARISON OF PROBING TECHNIQUES

There are several features common to all the techniques presented in this chapter. All are direct optical techniques and are, therefore, noncontact. No mechanical probe need be applied to the DUT (except for electrical inputs), and, thus, the electrical parasitics and possible mechanical-damage problems of the probe are avoided. All techniques achieve high temporal resolution, by virtue of the ultrashort optical pulses employed. However, the modulation of the probe beam produced by the DUT is small, and (as of the late 1980s) high temporal resolution has required a sampling measurement. A real-time measurement of devices operating in the multigigahertz frequency range requires high-speed, low-impedance photodetectors. With the low-impedance photodetectors, Johnson noise becomes significant, and shot-noise–limited operation requires increasingly higher optical power at shorter temporal resolution. Finally, all probing techniques afford excellent spatial resolution, because the optical probe (and/or excitation) beams can be focused to near diffraction limited spot sizes of about one μm. We now consider the individual techniques.

Photocarrier generation. This set of techniques relies on optical radiation to stimulate particular nodes in an IC. The effect of the stimulus is manifested in the IC output. The techniques are applicable to device, packaged-device, or wafer-level testing. However, they might require special fabrication to include appropriate photosensitive elements at specific locations in the DUT.

Electro-optic sampling. This technique has been the most widely applied to date and may remain the most important. The technique responds to the electric field produced by a voltage-carrying contact on the DUT, and, thus, any node on a device or IC can be probed. The major deficiency of direct electro-optic sampling is the requirement that the substrate of the DUT be an electro-optic material (for example, GaAs or InP, but not Si). However, the highest-speed ICs, at present, are being fabricated on electro-optic substrates, which mitigates the deficiency. Fractional probe modulation for ~1V signals is about 10^{-4} for the electro-optic effect; so (as of the late 1980s) high-speed measurements have been sampling measurements. Electro-optic sampling has been applied to discrete devices, as well as packaged devices (Heutmaker et al., 1988) and to ICs on wafer (Weingarten et al., 1988). As required temporal resolution approaches one ps, direct electro-optic sampling will require the DUT to have a thin substrate, to reduce the optical transit-time limitation.

Charge-density probing. This technique is newer and less well studied than those just presented. Because it measures charge density, the technique is applicable to all substrate materials. For GaAs and other electro-optic substrates, the electro-optic effect will be a complication to charge-density probing, but may be accounted for or used to advantage (Keller et al., 1988; Koskowich and Soma, 1988a). The technique has been applied to the real-time measurement of waveforms at up to 100 MHz, because it can have large fractional modulation of the probe beam (up to 9%; Hemenway et al., 1987). Sampling measurements by charge-density probing in the gigahertz range are feasible, and a result at one GHz has been reported (Black et al., 1987). A backside-probing geometry is required, and discrete-device or wafer-level (but not packaged-device) testing is possible. The interferometric optical-detection technique necessary for charge-density probing requires precise alignment and may be a drawback.

Phase-space absorption quenching. This new and interesting technique that requires a quantum-well structure can produce a relatively large fractional probe-beam modulation (2%). The first application of the technique to a sampling measurement of multigigahertz waveforms in discrete InGaAs/InAlAs MODFETs has been reported (Wiesenfeld et al., 1989).

B. FUTURE NEEDS

Measurement of waveforms (frequency response) at single nodes internal to ICs or devices have been performed using direct optical-probing techniques (electro-optic sampling, in particular). The achievement of high-temporal resolution has required a sampling measurement, which requires repetitive input waveforms. In the future, it will be desirable to be able to probe the DUT as it is exercised with random or pseudorandom inputs. This will require real-time probing. To achieve this requires increased sensitivity, which will mean either an increase in usable optical-probe power or a reduction of the frequency at which the DUT can be tested. Possibly a real-time sampling configuration, in which only selected portions of the waveform, rather than the entire waveform, are measured, can be developed. It will also be desirable to be able to probe several nodes of the DUT simultaneously.

Calibration issues, both signal (voltage and/or charge density) and spatial, are important. To translate a measurement of the fractional modulation of the probe beam into voltage on a contact line or charge density in the active region of a device will require a model of the device. It may not be desirable to perform detailed field or charge-density calculations for all devices, so it will be important to develop rules of thumb that can convert general geometric features of the DUT into calibration factors. As discussed in Section IV.B.2 and elsewhere, some first steps in this direction have been taken. In the spatial domain, it will be necessary to position accurately the probe beam to within about one μm to study a desired node of an IC. For characterization, this can be done manually, but for screening applications, this will require a computer-controlled automatic image recognition system, which can find test points reproducibly on devices or ICs. In general, screening applications will require considerably more computer interfacing and control of experiments than has been the norm to date.

Finally, let us conclude with comments on temporal resolution. For current high-speed ICs, and the next-generation research devices, temporal resolution of existing systems is already adequate. Indeed, the bandwidth of systems using gain-switched InGaAsP injection lasers of ~20-ps duration is about 16 GHz, and very few ICs operate near this frequency. However, discrete devices with bandwidths approaching 100 GHz are being fabricated, which will require temporal resolution approaching one ps. This resolution approaches the limit of ultrashort pulses generated by Nd:YAG laser technology (at present), but other lasers in the 1–2-μm wavelength range with femtosecond pulses exist—notably, the soliton laser (Mollenauer and Stolen, 1984). However, to attain one-ps resolution, the degradation of the temporal resolution by the optical transit-time effects must be minimized. Extension of direct optical-probing techniques to the subpicosecond time domain will indeed be challenging.

References

Agrawal, G.P. and Dutta, N.K. (1986). *Long-Wavelength Semiconductor Lasers*. Van Nostrand Reinhold, New York.

Alferness, R.C., Economou, N.P., and Buhl, L.L. (1980). "Picosecond Optical Sampling Technique for Measuring the Speed of Fast Electro-Optic Switch/Modulators," *Appl. Phys. Lett.* **37**, 597–599.

Antreasyan, A., Garbinski, P.A., Mattera, V.D., Shah, N.J., and Temkin, H. (1986). "Monlithically Integrated Enhancement-Mode InP MISFET Inverter," *Electron. Lett.* **22**, 1014–1016.

Antreasyan, A., Garbinski, P.A., Mattera, V.D., Wiesenfeld, J.M., Tucker, R.S., Shah, N.J., and DiGiuseppe, M.A. (1987). "Gigahertz Logic Based on InP Metal-Insulator–Semiconductor Field-Effect Transistors by Vapor Phase Epitaxy," *IEEE Trans. Electron Dev.* **ED-34**, 1897–1901.

Auston, D.H. (1988). "Ultrafast Optoelectronics," in *Ultrashort Laser Pulses and Applications, W.* Kaiser, Springer–Verlag, Berlin, 183–233.

Auston, D.H., and Glass, A.M. (1972). "Optical Generation of Intense Picosecond Electrical Pulses," *Appl. Phys. Lett.* **20**, 398–399.

Auston, D.H. Cheung, K.P., Valdmanis, J.A., and Kleinman, D.A. (1984). "Cherenkov Radiation from Femtosecond Optical Pulses in Electro-optic Media," *Phys. Rev. Lett.* **53**, 1555–1558.

AuYeung, J. (1981). "Picosecond Optical Pulse Generation at Gigahertz Rates by Direct-Modulation of a Semiconductor Laser," *Appl. Phys. Lett.* **38**, 308–310.

Bar-Joseph, I., Kuo, J.M., Klingshern, C., Livescu, G., Chang, T.Y., Miller, D.A.B., and

Chemla, D.S. (1987). "Absorption Spectroscopy of the Continuous Transition from Low to High Electron Density in a Single Modulation-Doped InGaAs Quantum well," *Phys. Rev. Lett.* **59**, 1357–1360.

Bass, M., Franken, P.A., Ward, J.F., and Weinreich, G. (1962), "Optical Rectification," *Phys. Rev. Lett.* **9**, 446–448.

Becker, P.C., Fork, R.L., Brito Cruz, C.H., Gordon, J.P., and Shank, C. V. (1988). "Optical Stark Effect in Organic Dyes Probed with Optical Pulses of 6-fs Duration," *Phys. Rev. Lett.* **60**, 2462–2464.

Bessenov, Y.L., Bogotov, A.P., Vasilev, P.P, Morozov, V.N., and Sergeev, A.B. (1983). "Generation of Picosecond Pulses in an Injection Laser with an External Selective Resonator," *Sov. J. Quantum Electron.* **12**, 1510–1512.

Black, A., Courville, C., Schultheis, G., and Heinrich, H. (1987). "Optical Sampling of GHz Charge Density Modulation in Silicon Bipolar Junction Transistors," *Electron. Lett.* **23**, 783–784.

Blakemore, J.S. (1982). "Semiconducting and Other Major Properties of Gallium Arsenide," *J. Appl. Phys.* **53**, R123–R181.

Bowers, J.E., and Burrus, C.A. (1987). "Ultrawide-Band Long-Wavelength p-i-n Photodetectors," *J. Lightwave Tech.* **LT-5**, 1339–1350.

Bowers, J.E., Burrus, C.A., and McCoy, R.J. (1985). "InGaAs PIN Photodetectors with Modulation Response to Millimeter Wavelengths," *Electron. Lett.* **21**, 812–814.

Bowers, J.E., Hemenway, B.R., Gnauck, A.H., and Wilt, D.P. (1986). "High-Speed InGaAsP Constricted-Mesa Lasers," *IEEE J. Quantum Electron.* **QE-22**, 833–844.

Burrus, C.A., Bowers, J.E., and Tucker, R.S. (1985). "Improved Very-High-Speed Packaged InGaAs Punch-Through Photodiode," *Electron. Lett.* **21**, 262–263.

Campbell, J.C., Dentai, A.G., Holden, W.S., and Kasper, B.L. (1983). "High-Performance Avalanche Photodiode with Separate Absorption 'Grading' and Multiplication Regions," *Electron. Lett.* **19**, 818–820.

Chemla, D.S., and Miller, D.A.B. (1985). "Room-Temperature Excitonic Nonlinear-Optical Effects in Semiconductor Quantum Well Structures," *J. Opt. Soc. Am.* **B2**, 1155–1173.

Chemla, D.S., Bar-Joseph, I., Klingshern, C., Miller, D.A.B., Kuo, J.M., and Chang, T.Y. (1987). "Optical Reading of Field-Effect Transistors by Phase-Space Absorption Quenching in a Single InGaAs Quantum Well Conducting Channel," *Appl. Phys. Lett.* **50**, 585–587.

Chemla, D.S., Bar-Joseph, I., Kuo, J.M., Chang, T.Y., Klingshern, C., Livescu, G., and Miller, D.A.B. (1988). "Modulation of Absorption in Field-Effect Quantum Well Structures," *IEEE J. Quantum Electron.* **QE-24**, 1664–1676.

Cooper, D.E., and Moss, S.C. (1986). "Picosecond Optoelectronic Measurement of the High-Frequency Scattering Parameters of a GaAs FET," *IEEE J. Quantum Electron.* **QE-22**, 94–100.

Corzine, S.W., Bowers, J.E., Przybylek, G., Koren, U., Miller, B.I., and Soccolich, C.E. (1988). "Actively Mode-Locked GaInAsP Laser with Subpicosecond Output," *Appl. Phys. Lett.* **52**, 348–350.

Drummond, T.J., Masselink, W.T., and Morkoc, H. (1986). "Modulation-doped GaAs/(Al,Ga)As Heterojunction Field-Effect Transistors: MODFETs," *Proc. IEEE* **74**, 773–822.

Eisenstein, G., Tucker, R.S., Koren, U., and Korotky, S.K. (1986). "Active Modelocking Characteristics of InGaAsP-Single Mode Fiber Composite Cavity Lasers," *IEEE J. Quantum Electron.* **QE-22**, 142–148.

Fork, R.L., Greene, B.I., and Shank, C.V. (1981). "Generation of Optical Pulses Shorter than 0.1 ps by Colliding Pulse Modelocking," *Appl. Phys. Lett.* **38**, 671–672.

Fork, R.L., Brito Cruz, C.H., Becker, P.C., and Shank, C.V. (1987). "Compression of

Optical Pulses to Six Femtoseconds by Using Cubic Phase Compensation," *Opt. Lett.* **12**, 483–485.

Freeman, J.L. (1988). Ph.D. thesis, Stanford University.

Freeman, J.L., Diamond, S.K., Fong, H., and Bloom, D.M. (1985). "Electro-Optic Sampling of Planar Digital GaAs Integrated Circuits," *Appl. Phys. Lett.* **47**, 1083–1084.

Freeman, J.L. Jefferies, S.R., and Auld, B.A. (1987). "Full-Field Modeling of the Longitudinal Electro-Optic Probe," *Opt. Lett.* **12**, 795–797.

Freeman, J.L., Bloom, D.M., Jefferies, S.R., and Auld, B.A. (1988). "Accuracy of Electro-Optic Measurements of Coplanar Waveguide Transmission Lines," *Appl. Phys. Lett.* **53**, 7–9.

Freeman, J.L., Bloom, D.M., Jefferies, S.R., and Auld, B.A. (1989). "Sensitivity of Direct Electro-Optic Sampling to Adjacent Signal Lines," *Appl. Phys. Lett.* **54**, 476–478.

Gammel, J.C., and Ballantyne, J.M. (1980). "Integrated Photoconductive Detector and Waveguide Structure," *Appl. Phys. Lett.* **36**, 149–151.

Gheewala, T. (1987). "Survey of High Speed Test Techniques," in *Characterization of Very High Speed Semiconductor Devices and ICs*, R.K., Jain, ed., *SPIE Proc.* **795**, 2–9.

Grischkowsky, D., and Balant, A.C. (1982). "Optical Pulse Compression Based on Enhanced Frequency Chirping," *Appl. Phys. Lett.* **41**, 1–3.

Gupta, K.C., Garg, R., and Bahl, I.J. (1979). *Microstrip Lines and Slotlines*. Artech, Dedham, Massachusetts.

Hammond, R.B., Paulter, N.G., and Gibbs, A.J. (1986). "GaAs Photoconductors to Characterize Picosecond Response in GaAs Integrated Devices and Circuits," *Proceedings of the International Conference on High-Speed Electronics, Stockholm*. Springer-Verlag, New York, 223–225.

Heinrich, H.K., Bloom, D.M., and Hemenway, B.R. (1986a). "Noninvasive Sheet Charge Density Probe for Integrated Silicon Devices," *Appl. Phys. Lett.* **48**, 1066–1068, and *erratum, ibid*, **48**, 1811.

Heinrich, H.K., Hemenway, B.R., McGroddy, K.A., and Bloom, D.M. (1986b). "Measurement of Real-Time Digital Signals in a Silicon Bipolar Junction Transistor Using a Noninvasive Optical Probe," *Electron. Lett.* **22**, 650–652.

Hemenway, B.R., Heinrich, H.K., Goll, J.H., Xu, Z., and Bloom, D.M. (1987). "Optical Detection of Charge Modulation is Silicon Integrated Circuits Using a Multimode Laser-Diode Probe," *IEEE Electron Dev. Lett.* **EDL-8**, 344–346.

Henderson, T., Aksun, M.I., Peng, C.K., Morkoc, H., Chao, P.C., Smith, P.M., Duh, K.-H.G., and Lester, L.F. (1986). "Microwave Performance of a Quarter-Micrometer Gate Low-Noise Pseudomorphic InGaAs/AlGaAs Modulation-Doped Field Effect Transistor," *IEEE Electron Dev. Lett.* **EDL-7**, 649–651.

Henley, F.J. (1984). "An Automated Laser Prober to Determine VLSI Internal Node Logic States," *Proc. IEEE International Test Conf.*, 536–542.

Heutmaker, M.S., Cook. T.B., Bosacchi, B., Wiesenfeld, J.M., and Tucker, R.S. (1988). "Electro-Optic Sampling of a Packaged High-Speed GaAs Integrated Circuit," *IEEE J. Quant. Electron.* **QE-24**, 226–233.

Ichimura, S. (1973). *Basic Principles of Plasma Physics*. Benjamin, Reading, Massachussetts.

Ippen, E.P., and Shank, C.V. (1977). "Techniques for Measurement," in *Ultrashort Light Pulses*, S.L. Shapiro, ed. Springer-Verlag, Berlin, 83–122.

Jackson, J.D. (1962). *Classical Electrodynamics*. Wiley, New York.

Jain, R.K. (1982). "Synchronously Modelocked CW Dye Laser: Recent Advances and Applications," *Picosecond Laser Applications*, L. Goldberg, ed., *SPIE Proc.* **322**, 2–12.

Jain, R.K. (1984). "Picosecond Optical Techniques Offer a New Dimension for Microelectronics Test," *Test and Measurement World* **4** (6), 40–53.

Jain, R.K. (1985). "New Optical and Electro-Optical Techniques for the Measurement of

Very High Speed Integrated Circuits," *Microelectronics and Photonics*, R.P. Bajpai, ed. Siddharth, New Delhi, 155–182.

Jain, R.K., and Snyder, D.E. (1983a). "Addressing and Control of High-Speed Logic Circuits with Picosecond Light Pulses," *Opt. Lett.* **8**, 85–87.

Jain, R.K., and Snyder, D.E. (1983b). "Switching Characteristics of Logic Gates Addressed Directly by Picosecond Light Pulses," *IEEE J. Quantum Electron.* **QE-19**, 658–663.

Jain, R.K., and Zhang, X.-C. (1986). "New Picosecond Techniques for the Control and Measurement of High-Speed GaAs ICs," *Tech. Digest IEEE GaAs IC Symp.*, 141–144.

Jain, R.K., Snyder, D.E., and Stenersen, K. (1984). "A New Technique for the Measurement of Speeds of Gigahertz Digital ICs," *IEEE Electron Dev. Lett.* **EDL-5**, 371–373.

Jensen, J.F., Salmon, L.G., Deakin, D.S., and Delaney, M.J. (1986). "Ultra-High Speed GaAs Static Frequency Dividers," *Tech. Dig. 1986 Int. Electron Device Meeting*, 476–479.

Jensen, J.F., Weingarten, K.J., and Bloom, D.M. (1987). "Development of 18 GHz Static Frequency Dividers and Their Evaluation by Electro-optic Sampling," in *Picosecond Electronics and Optoelectronics* **II**, F.J. Leonberger, C.H. Lee, F. Capasso, and H. Morkoc, eds. Springer-Verlag, Berlin, 184–187.

Joshin, K., Kamite, K., Mimura, T., and Abe, M. (1988). "Electro-optic Sampling of a Flip-Chip with a Distributed Feedback Laser Diode," in *Ultrafast Phenomena* **VI**, T. Yajima, K. Yoshihara, C.B. Harris, and S. Shinoya, eds. Springer-Verlag, New York, 189–191.

Kafka, J.D., Kolner, B.H., Baer, T., and Bloom, D.M. (1984). "Compression of Pulses from a Continuous-Wave Modelocked Nd:YAG Laser," *Opt. Lett.* **9**, 505–507.

Kaminow, I.P. (1974). *An Introduction to Electrooptic Devices*. Academic Press, New York.

Kaminow, I.P. (1986). In *CRC Handbook of Laser Science and Technology*, **IV**, Part 2, M.J. Weber, ed. CRC Press, Boca Raton, Florida. 253–278.

Keller, U., Diamond, S.K., Auld, B.A., and Bloom, D.M. (1988). "Noninvasive Optical Probe of Free Charge and Applied Voltage in GaAs Devices," *Appl. Phys. Lett.* **53**, 388–390.

Kluge, J., Wiechert, D., and Von der Linde, D. (1984). "Fluctuations in Synchronously Mode-Locked Dye Lasers," *Opt. Comm.* **51**, 271–277.

Knox, W.H., Miller, D.A.B., Damen, T.C., Chemla, D.S., Shank, C.V., and Gossard, A.C. (1986). "Subpicosecond Excitonic Electroabsorption in Room-Temperature Quantum Wells," *Appl. Phys. Lett.* **48**, 864–866.

Kolner, B.H. (1985). Ph.D. thesis, Stanford University.

Kolner, B.H. (1987). "Internal Electro-optic Sampling in GaAs," in *Characterization of Very High Speed Semiconductor Devices and ICs*," R.K. Jain, ed., *SPIE Proc.* **795**, 310–316.

Kolner, B.H., and Bloom, D.M. (1984). "Direct Electro-optic Sampling of Transmission Line Signals Propagating on a GaAs Substrate," *Electron. Lett.* **20**, 818–819.

Kolner, B.H., and Bloom, D.M. (1986). "Electrooptical Sampling in GaAs Integrated Circuits," *IEEE J. Quantum Electron.* **QE-22**, 79–93.

Kolner, B.H., Bloom, D.M., and Cross, P.S. (1983a). "Electro-optic Sampling with Picosecond Resolution," *Electron. Lett.* **19**, 574–575.

Kolner, B.H., Bloom, D.M., and Cross, P.S. (1983b). "Picosecond Optical Electronic Measurements," in *Picosecond Optoelectronics*, G. Mourou, ed., *Proc. SPIE* **439**, 149–152.

Korotky, S.K., Eisenstein, G., Tucker, R.S., Veselka, J.J., and Raybon, G. (1987). "Optical Intensity Modulation to 40 GHz Using a Waveguide Electro-optic Switch," *Appl. Phys. Lett.* **50**, 1631–1633.

Koskowich, G.N., and Soma, M. (1988a). "Voltage Measurement in GaAs Schottky Barriers Using Optical Phase Modulation," *IEEE Electron Dev. Lett.* **EDL-9**, 433–435.

Koskowich, G.N., and Soma, M. (1988b). "Optical Charge Modulation as an Internal Voltage Probe for CMOS ICs," *IEEE J. Quantum Electron.* **QE-24**, 1981–1984.

Lee, T.P., Burrus, C.A., Ogawa, K., and Dentai, A.G. (1981). "Very-High-Speed Back-Illuminated InGaAs/InP PIN Punch-Through Photodiode," *Electron Lett.* **17**, 431–432.

LeFur, P., and Auston, D.H. (1976). "A Kilovolt Picosecond Optoelectronic Switch and Pockel's Cell," *Appl. Phys. Lett.* **28**, 21–23.

Lindemuth, J. (1987). "Optimal Performance of the Electro-optical Sampler," *SPIE Proc.* **793**, 120–124.

Lo, Y.H., Zhu, Z.H., Pan, C.L., Wang, S.Y., and Wang., S. (1987a). "New Technique to Detect the GaAs Semi-Insulating Surface Property—CW Electro-optic Probing," *Appl. Phys. Lett.* **50**, 1125–1127.

Lo, Y.H., Wu, M.C., Zhu, Z.H., Wang, S.Y., and Wang, S. (1987b). "Proposal for Three-Dimensional Internal Field Mapping by CW Electro-optic Probing," *Appl. Phys. Lett.* **50**, 1791–1793.

Madden, C.J., Rodwell, M.J.W., Marsland, R.A., Bloom, D.M., and Pao, Y.C. (1988). "Generation of 3.5-ps Fall-Time Shock Waves on a Monolithic GaAs Nonlinear Transmission Line," *IEEE Electron Dev. Lett.* **EDL-9**, 303–305.

Madden, C.J., Marsland, R.A., Rodwell, M.J.W., Bloom, D.M., and Pao, Y.C. (1989). "Hyperabrupt-Doped GaAs Nonlinear Transmission Line for Picosecond Shock-Wave Generation," *Appl. Phys. Lett.* **54**, 1019–1021.

Majidi-Ahy, R., Weingarten, K.J., Riaziat, M., Auld, B.A., and Bloom, D.M. (1987). "Electro-optic Sampling Measurement of Coplanar Waveguide (Coupled Slot Line) Modes," *Electron. Lett.* **23**, 1262–1263.

Marcuse, D., and Wiesenfeld, J.M. (1988). Unpublished results.

Mishra, U.K., Jensen, J.F., Brown, A.S., Thompson, M.A., Jelloian, L.M., and Beaubien, R.S. (1988). "Ultra-High-Speed Digital Circuit Performance in 0.2-μm gate-length AlInAs/GaInAs HEMT Technology," *IEEE Electron. Dev. Lett.* **EDL-9**, 482–484.

Mollenauer, L.F. and Stolen, R.H. (1984). "The Soliton Laser," *Opt. Lett.* **9**, 13–15; and *erratum, ibid*, 105.

Morozov, B.N., and Aivazyan, Yu. M. (1980). "Optical Rectification Effect and Its Applications (Review)," *Sov. J. Quantum Electron.* **10**, 1–16.

Nagel, L.W. (1975). "SPICE 2: A Computer Program to Simulate Semiconductor Circuits," Electron. Res. Lab., Univ. California, Berkeley, California, Ref. ERL-M520.

Namba, S. (1961). "Electro-optical Effect of Zincblende," *J. Opt. Soc. Am.* **51**, 76–79.

O'Connor, P., Flahive, P.G., Glemetson, W., Panock, R.L., Wemple, S.H., Shunk, S.C., and Takahashi, D.P. (1984). "A Monolithic Multigigabit/Second DCFL GaAs Decision Circuit," *IEEE Electron Dev. Lett.* **EDL-5**, 226–227.

Parker, D.G., Say, P.G., Hansom, A.M., and Sibbet, W. (1987). "110 GHz High-Efficiency Photodiodes Fabricated from Indium Tin Oxide/GaAs," *Electron. Lett.* **23**, 527–528.

Pei, S.S., Shah, N.J., Hedel, R.H., Tu, C.W., and Dingle, R. (1984). "Ultra High Speed Integrated Circuits with Selectively Doped Heterostructure Transistors," *Dig. Tech. Papers, GaAs IC Symposium, Boston*, 129–132.

Pockels, F. (1906). *Lehrbuch der Kristalloptik*, Part IV. Teubner, Leipzig, 492–510.

Ramo, S., Whinnery, J.R., and Van Duzer, T. (1965). *Fields and Waves in Communication Electronics*. Wiley, New York.

Rhoderick, E.H. (1978). *Metal-Semiconductor Contacts*. Clarendon, Oxford, England.

Riaziat, M., Zubeck, I., Bandy, S., and Zdasiuk, G. (1986). "Coplanar Waveguides Used in 2–18 GHz Distributed Amplifier," *Tech. Dig. IEEE MTT-S Int. Microwave Symp.*, 337–338.

Robins, W.P. (1982). *Phase Noise in Signal Sources*. Peregrinus, London.

Rodwell, M.J.W., Weingarten, K.J., Bloom, D.M., Baer, T., and Kolner, B.H. (1986a). "Reduction of Timing Fluctuations in a Mode-Locked Nd:YAG Laser by Electronic Feedback," *Opt. Lett.* **11**, 638–640.

Rodwell, M.J.W., Weingarten, K.J., Freeman, J.L., and Bloom, D.M. (1986b). "Gate Propagation Delay and Logic Timing of GaAs Integrated Circuits Measured by Electro-optic Sampling," *Electron. Lett.* **22**, 499–501.

Rodwell, M.J.W., Riaziat, M., Weingarten, K.J., Auld, B.A., and Bloom, D.M. (1986c). "Internal Microwave Propagation and Distortion Characteristics of Traveling-Wave Amplifiers studied by Electrooptic Sampling," *IEEE Trans. Microwave Theory Tech.* **MTT-34**, 1356–1362.

Rodwell, M.J.W., Bloom, D.M., and Auld, B.A. (1987). "Nonlinear Transmission Line for Picosecond Pulse Compression and Broadband Phase Modulation," *Electron. Lett.* **23**, 109–110.

Rodwell, M.J.W., Madden, C.J., Khuri-Yakum, B.T., Bloom, D.M., Pao, Y.C., Gabriel, N.S., and Swierkowski, S.P. (1988). "Generation of 7.8 ps Electrical Transients on a Monolithic Nonlinear Transmission Line," *Electron. Lett.* **24**, 100–101.

Rodwell, M.J.W., Bloom, D.M., and Weingarten, K.J. (1989). "Subpicosecond Laser Timing Stabilization," *IEEE J. Quantum Electron.* **QE-25**, 817–827.

Schmitt-Rink, S., Chemla, D.S., and Miller, D.A.B. (1985). "Theory of Transient Excitonic Optical Nonlinearities in Semiconductor Quantum-Well Structures," *Phys. Rev.* **B 32**, 6601–6609.

Shah, N.J., Pei, S.S., Tu, C.W., and Tiberio, R.C. (1986). "Gate Length Dependence of the Speed of SSI Circuits Using Submicrometer Selectively Doped Heterostructure Transistor Technology," *IEEE Trans, Electron Dev.* **ED-33**, 543–547.

Shelby, R.M., Levenson, M.D., and Bayer, P.W. (1985). "Guided Acoustic Wave Brillouin Scattering, *Phys. Rev.* **B31**, 5244–5252.

Shibata, T., Nagatsumo, T., and Sano, E. (1989). "Effective Optical Transit Time in Direct Electro-optic Sampling of GaAs Coplanar Integrated Circuits," *Electron. Lett.* **25**, 771–773.

Smith, P.R., Auston, D.H., and Augustyniak, W.M. (1981). "Measurement of GaAs Field-Effect Transistor Electronic Impulse Response by Picosecond Optical Electronics," *Appl. Phys. Lett.* **39**, 739–741.

Spitzer, W.G., and Whelan, J.M. (1959). "Infrared Absorption and Electron Effective Mass in n-Type GaAs," *Phys. Rev.* **114**, 59–63.

Stenersen, K., and Jain, R.K. (1984). "Direct Measurement of Picosecond Propagation Delays in Individual Logic Gates by a Differential Optoelectronic Technique," *IEEE Electron Dev. Lett.* **EDL-5**, 422–424.

Strid, E.W., Gleason, K.R., and Reeder, T.M. (1985). "On-Wafer Measurement of Gigahertz Integrated Circuits," in *VLSI Electronics: Microstructure Science* **11**, 265–287.

Sugeta, T., and Mizushima, Y. (1980). "High Speed Photoresponse Mechanism of a GaAs MESFET," *Jpn. J. Appl. Phys. Lett.* **19**, L27–L29.

Suzuki, N., and Tada, K. (1984). "Electrooptic Properties and Raman Scattering in InP," *Jpn. J. Appl. Phys.* **23**, 291–295.

Swierkowski, S., Mayeda, K., Cooper, G., and McConagy, C. (1985). "A Sub-200 ps GaAs Sample-and-Hold Circuit for a Multigigasample/Second Integrated Circuit," *Proc. IEEE Electron. Dev. Meeting, Washington, D.C.*, 272–275.

Sze, S.M. (1981). *Physics of Semiconductor Devices*, 2nd ed. Wiley, New York.

Takada, A., Sugie, T., and Saruwatari, M. (1986). "Transform-Limited 5.6 ps Optical Pulse Generation at 12 GHz Repetition Rate from Gain-Switched Distributed Feedback Laser Diode by Employing Pulse Compression Technique," *Electron. Lett.* **22**, 1346–1347.

Taylor, A.J., Wiesenfeld, J.M., Eisenstein, G., Tucker, R.S., Talman, J.R., and Koren, U. (1986a). "Electro-optic Sampling of Fast Electrical Signals Using an InGaAsP Injection Laser," *Electron. Lett.* **22**, 61–62.

Taylor, A.J., Wiesenfeld, J.M., Tucker, R.S., Eisenstein, G., Talman, J.R., and Koren, U. (1986b). "Measurement of a Very-High-Speed InGaAs Photodiode Using Electro-optic Sampling," *Electron. Lett.* **22**, 325–327.

Taylor, A.J., Tucker, R.S., Wiesenfeld, J.M., Burrus, C.A., Eisenstein, G., Talman, J.R., and Pei, S.S. (1986c). "Direct Electro-optic Sampling of a GaAs Integrated Circuit Using a Gain-Switched InGaAsP Injection Laser," *Electron. Lett.* **22**, 1068–1069.

Taylor, A.J., Wiesenfeld, J.M., Eisenstein, G., and Tucker, R.S. (1986d). "Timing Jitter in Mode-Locked and Gain-Switched InGaAsP Injection Lasers," *Appl. Phys. Lett.* **49**, 681–683.

Taylor, A.J., Tucker, R.S., Wiesenfeld, J.M., Eisenstein, G., and Burrus, C.A. (1986e). "High Repetition Rate Electro-optic Sampling with an Injection Laser," in *Ultrafast Phenomena V*, G.R. Fleming and A.E. Siegman, eds., Springer-Verlag, Berlin, 114–116.

Treacy, E.B. (1969). "Optical Pulse Compression with Diffraction Gratings," *IEEE J. Quant. Electron.*, **QE-5**, 454–458.

Tserng, H.Q., and Kim, B. (1985). "110 GHz GaAs FET Oscillator," *Electron. Lett.* **21**, 178–179.

Tucker, R.S., Taylor, A.J., Burrus, C.A., Eisenstein, G., and Wiesenfeld, J.M. (1986). "Coaxially Mounted 67 GHz Bandwidth InGaAs PIN Photodiode," *Electron. Lett.* **22**, 917–918.

Valdmanis, J.A., (1987). "1 THz-Bandwidth Prober for High-Speed Devices and Integrated Circuits," *Electron. Lett.*, **23**, 1308–1310.

Valdmanis, J.A., and Mourou, G.A. (1986). "Subpicosecond Electro-optic Sampling: Principles and Applications," *IEEE J. Quant. Electron.* **QE-22**, 69–78.

Valdmanis, J.A., Mourou, G., and Gabel, C.W. (1982). "Picosecond Electro-optic Sampling System," *Appl. Phys. Lett.* **41**, 211–212.

Valdmanis, J.A., Mourou, G., and Gabel, C.W. (1983a). "Subpicosecond Electro-optic Sampling," *IEEE J. Quantum Electron* **QE-19**, 664–667.

Valdmanis, J.A., Mourou, G., and Gabel, C.W. (1983b). "Subpicosecond Electrical Sampling," in *Picosecond Optoelectronics*, G. Mourou, ed. *SPIE Proc.* **439**, 142–148.

Van Stryland, E.W., Vanherzeele, H., Woodall, M.A., Soileau, M.J., Smirl, A.L., Guha, S., and Boggess, T.F. (1985). "Two Photon Absorption, Nonlinear Refraction, and Optical Limiting in Semiconductors," *Opt. Eng.* **24**, 613–623.

Wang, S.Y., and Bloom, D.M. (1983). "100 GHz Bandwidth Planar GaAs Schottky Photodiode," *Electron. Lett.* **19**, 554–555.

Wang, S.Y., Carey, K.W., and Kolner, B.H. (1987). "A Front-Side Illuminated InP/GaInAs/InP p-i-n Photodiode with a -3 dB Bandwidth in Excess of 18 GHz," *IEEE Trans. Electron Dev.* **ED-34**, 938–940.

Weingarten, K.J., Rodwell, M.J.W., Heinrich, H.K., Kolner, B.H., and Bloom, D.M. (1985). "Direct Electro-optic Sampling of GaAs Integrated Circuits," *Electron. Lett.* **21**, 765–766.

Weingarten, K.J., Rodwell, M.J.W., Freeman, J.L. Diamond, S.K., and Bloom, D.M. (1986). "Electro-optic Sampling of Gallium Arsenide Integrated Circuits," in *Ultrafast Phenomena V*, G.R. Fleming and A.E. Siegman, eds. Springer-Verlag, Berlin, 98–102.

Weingarten, K.J., Majidi-Ahy, R., and Bloom, D.M. (1987). "GaAs Integrated Circuit

Measurements Using Electro-optic Sampling," *Dig. Tech. Papers, 1987 GaAs IC Symp.*, 11–14.

Weingarten, K.J., Rodwell, M.J.W., and Bloom, D.M. (1988). "Picosecond Optical Sampling of GaAs Integrated Circuits," *IEEE J. Quantum Electron.* **QE-24**, 198–220.

White, I.H., Gallagher, D.F.G., Osinski, M., and Bowley, D. (1985). "Direct Streak-Camera Observation of Picosecond Gain-Switched Optical Pulses from a 1.5 μm Semiconductor Laser," *Electron. Lett.* **21**, 197–199.

Wiesenfeld, J.M., and Heutmaker, M.S. (1987). Unpublished results.

Wiesenfeld, J.M., and Heutmaker, M.S. (1988). "Frequency Response of an Internal Amplifier in a High-Speed Integrated Circuit Measured by Electro-optic Sampling," *Electron. Lett.* **24**, 106–107.

Wiesenfeld, J.M., Taylor, A.J., Tucker, R.S., Eisenstein, G., and Burrus, C.A. (1987a). "Electro-optic Sampling Using Injection Lasers," in *Characterization of Very High Speed Semiconductor Devices and ICs*, R.K. Jain, ed., *SPIE Proc.* **795**, 339–344.

Wiesenfeld, J.M., Tucker, R.S., Antreasyan, A., Burrus, C.A., Taylor, A.J., Mattera, V.D., and Garbinski, P.A. (1987b). "Electro-optic Sampling Measurements of High-Speed, InP Integrated Circuits," *Appl. Phys. Lett.* **50**, 1310–1312.

Wiesenfeld, J.M., Heutmaker, M.S., Bar-Joseph, I., Chemla, D.S., Kuo, J.M., Chang, T.Y., Burrus, C. A., and Perino, J.S. (1989). "Measurement of Multigigahertz Waveforms and Propagation Delays in Modulation-doped Field-Effect Transistors Using Phase-Space Absorption Quenching," *Appl. Phys. Lett.* **55**, 1109–1111.

Yariv, A., and Yeh, P. (1984). *Optical Waves in Crystals*. Wiley, New York.

Zhang, X.C., and Jain, R.K. (1986). "Measurement of On-Chip Waveforms and Pulse Propagation in Digital GaAs Integrated Circuits by Picosecond Electro-optic Sampling," *Electron. Lett.* **22**, 264–265.

Zhang, X.C., and Jain, R.K. (1987). "Analysis of High-Speed GaAs ICs with Electro-optic Probes," in *Characterization of Very High Speed Semiconductor Devices and ICs* R.K. Jain, ed., *SPIE Proc.* **795**, 317–338.

Zhang, X.C., Jain, R.K., and Hickling, R. (1987). "Electro-optic Sampling Analysis of Timing Patterns at Critical Internal Nodes in Gigabit GaAs Multiplexers/Demultiplexers," *Picosecond Electronics and Optoelectronics* **II**, F.J. Leonberger, C.H. Lee., R. Capasso, and H. Morkoc, eds. Springer-Verlag, Berlin, 29–32.

Zhu, Z.H., Weber, J.-P., Wang, S.Y., and Wang, S. (1986). "New Measurement Technique: CW Electro-optic Probing of Electric Fields," *Appl. Phys. Lett.* **49**, 432–434.

Zhu, Z.H., Pan, C.L., Lo, Y.H., Wu, M.C., Wang, S., Kolner, B.H., and Wang, S.Y. (1987). "Electro-optic Measurements of Standing Waves in a GaAs Coplanar Waveguide," *Appl. Phys. Lett.* **50**, 1228–1230.

Zhu, Z.H., Wu, M.C., Lo, Y.H., Pan, C.L., Wang, S.Y., and Wang, S. (1988). "Measurements on Standing Waves in GaAs Coplanar Waveguide at Frequencies up to 20.1 GHz by Electro-optic Probing," *J. Appl. Phys.* **64**, 419–421.

CHAPTER 6

Electron-Beam Probing

Graham Plows

CAMBRIDGE INSTRUMENTS
CAMBRIDGE, ENGLAND

335

I. Introduction: Pros and Cons of the Focused Electron Beam for Probing High-Speed Electrical Signals

Electron-beam testing or probing (usually known as E-beam testing) is possible because electrons hold a charge. Therefore, if an electron starts with a known kinetic energy and passes through spatially defined electric and magnetic fields to a point where its energy can be measured, its starting potential energy can be deduced. The electron in "electron beam" is the *primary* focused beam of electrons and it is the resulting *secondary* electrons from which the potential energy, the *voltage*, of the region of impingement is deduced.

The secondary-emission energy distribution cannot be known precisely in any practical situation. For example, the work function is unknown on a practical surface. The secondary-emission angular distribution also cannot be known because the surface is microscopically rough. The precise energy of each secondary electron cannot be measured. For all these reasons, the starting potential energy cannot be found by precise calculation.

Instead, the emitted secondary electrons are subjected to energy filtering. After such filtering, the different total numbers of electrons collected from a reference point and a test point respectively (or from the same point at different phases) give the voltage difference.

Fortunately the E-beam can be focused to a small enough diameter for E-beam testing using a primary electron energy that is low compared with the values normally used in either transmission or scanning electron microscopy. A primary beam in the energy range 500 electron volts to 1.5 keV causes negligible damage and negligible alteration of characteristics for most semiconductor devices in common use, even after several hours. Some care may be necessary in regions of the DUT (device under test) that depend for their operation on storing charge, but good results can usually be produced.

E-beam testing typically requires probe diameters of about 0.1 micrometer size. An electron current of the order of 1nA is the best that can be achieved using thermionic cathodes for this probe diameter. At 50-MHz bandwidth, the RMS primary-beam shot-noise will then be several volts, and voltage differences smaller than this cannot be discerned. To overcome this signal-to-noise problem, most E-beam test techniques use the sampling principle, as in the sampling oscilloscope, to cut down the bandwidth to, at most, a few Hertz.

A more easily understood analogy for stroboscopic images is the light stroboscope, although the same signal-to-noise considerations apply for images as for waveforms. A detailed analysis follows in this chapter.

The original ideas for E-beam testing were developed by research projects from the mid-1950s to 1969 in the University of Cambridge Engineering Laboratories. Later developments occurred at a rather leisurely pace in various laboratories. The pace was leisurely because there was little practical need for E-beam testing until device geometries dropped below 2 microns. However, when commercial application and commercial exploitation became practical and necessary, the equipment was developed and the applications refined very quickly. This was possible because of the highly developed nature of the commercial scanning electron microscope (SEM).

It was already possible, before serious applications of E-beam testing were justifiable, to produce fine-focused beams that could be positioned with great accuracy and used to produce high-quality images of the sample surface. These techniques were brought rapidly into service in E-beam testing, initially by using commercial SEMs with add-on equipment that adapted them for E-beam testing, and later by integrating a dedicated SEM into the E-beam tester. In this chapter, the term *E-beam tester* or EBT will be used only to describe the stand-alone dedicated instrument. Some of the assertions made about the E-beam tester apply also to the SEM with add-on.

A major advantage and a separate major drawback of the EBT result from the fact that the primary electron beam is an infinite-impedance current source. The advantage is that there is no capacitive loading of the transitions of the voltage waveform under study. The drawback is that an earth reference from the electron probe itself (and therefore a direct measurement of the absolute voltage at a single test point and at a single sampling phase) is not possible.

The EBT will not usefully measure magnetic fields. However, magnetic fields can be detected in some circumstances and can be a nuisance in others. It is not practical, therefore, to measure the current between nodes.

The EBT can be used to make measurements in the presence of dielectrics and can produce good-quality results on passivated devices, but extra care is always necessary in such cases. The best results always come from unpassivated devices (i.e., devices that have never received passivation), followed by depassivated devices (i.e., devices from which the passivation layer has been carefully removed), followed by passivated devices.

The use of E-beam testing is difficult to justify for large- (greater than 5-micron–) geometry devices on which a micromanipulated physical probe can readily be used. It comes into its own for submicron devices when the time taken to place a physical probe becomes long, when the damage produced by that probe is unacceptable, and when the capacitance loading produced by a physical probe is gross. Compensating for the major inconvenience of device insertion into and signal connection in vacuum is accurate time-difference measurement (in principle to 10 picoseconds or better); excellent risetime capability (down to 100 picoseconds in current commercial systems); good voltage-measurement resolution (10 millivolts or better); and great ease, accuracy, and speed of probe positioning. E-beam testing also offers various invaluable stroboscopic image modes with easily interpreted frozen pictures of the voltage distributions on the surface at any chosen phase within the high-frequency operating cycle, and therefore of logical events.

II. Background Science, Technology, and History

A. THE SCANNING ELECTRON MICROSCOPE

This instrument is too well known to require description. Its history and the basic principles of operation and design are described with great clarity in Oatley (1972).

Modern scanning electron microscopes can achieve spatial resolutions in the study of surface topography down to 2 nanometers or better. This is more or less irrelevant to E-beam testing because few features of modern IC devices that need to be probed have dimensions smaller than 0.1 micron. Indeed, the electron optical design approach needed to produce excellent spatial resolution at beam energies of 20 keV and above is rather different from that needed to produce high currents at low beam energies with stability and good alignment.

B. THE STROBOSCOPIC SEM

The stroboscopic SEM is the forerunner of the E-beam tester. It was described by Plows (1969) and by Plows and Nixon (1968). The additions required to the SEM were a controllable means of triggered high-rate pulsing of the primary electron beam, effective means of secondary electron extraction, a secondary electron energy filter, and some additional electronics for signal processing and display.

The essence of sampling to produce waveforms is shown in Figs. 1 and 3 and to produce stroboscopic images in Fig. 2. This basic single sampling principle (Plows, 1969) has remained unchanged since then, although minor details of application may vary. It is described shortly in more detail

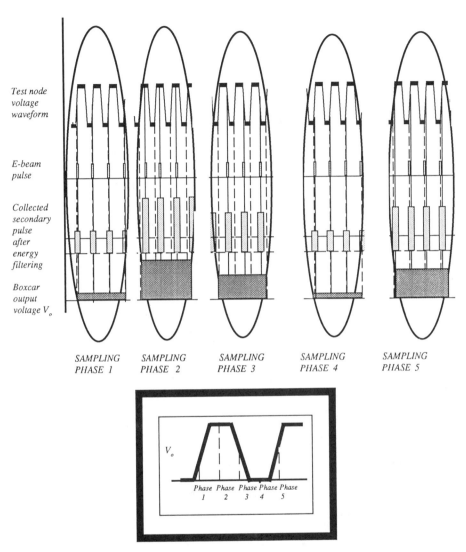

Test node voltage waveform

E-beam pulse

Collected secondary pulse after energy filtering

Boxcar output voltage V_o

SAMPLING PHASE 1   SAMPLING PHASE 2   SAMPLING PHASE 3   SAMPLING PHASE 4   SAMPLING PHASE 5

V_o

Phase 1   Phase 2   Phase 3   Phase 4   Phase 5

FIG. 1. The single sampling principle in E-beam testing. Many samples are taken at each phase before moving on to next phase. In the graph, wave form display shows lower sampled voltage amplitude as ordinate with phase at which the sample was taken as abcissa. See also Fig. 3.

C. ENERGY FILTERING

The energy filter originally shown by Plows (1969)—and developed by Gopinath and Sanger (1971) and Feuerbaum (1979), among others—is still used widely. It is a retarding field filter acting on the component of secondary-emission velocity normal to the surface. Its basic principle is

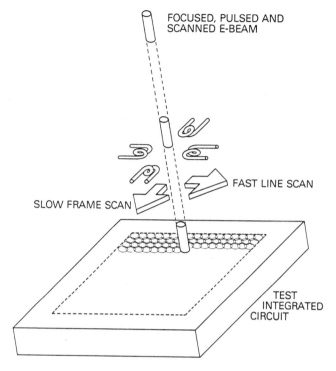

FOCUSED, PULSED AND
SCANNED E-BEAM

SLOW FRAME SCAN

FAST LINE SCAN

TEST
INTEGRATED
CIRCUIT

FIG. 2. The principle of stroboscopic imaging.

shown in Fig. 4. A planar mesh is interposed between sample and electron "collector" that is negative with respect to the sample. Only those electrons emitted with kinetic energies due to the normal component of emission velocity, which are greater than the potential energy difference produced by the voltage difference between sample and mesh, will penetrate the mesh and be collected. From the number collected and a knowledge of the secondary-emission distribution, the voltage difference and, therefore, the sample voltage, could be calculated.

Unfortunately there are several sources of error that make this technique only approximate. The energy–angular distribution is not constant. There exist also local electrical fields and sometimes magnetic fields that cause variations in the normal component of velocity. The curve of output signal against sample voltage is, in any case, inconveniently nonlinear.

A technique originally developed by Flemming and Ward (1970) for an electron-mirror instrument was further developed to make measurements much easier and somewhat more accurate. A negative feedback loop is used as shown in Fig. 5. This linearized the "S-curve" (the S-shaped inte-

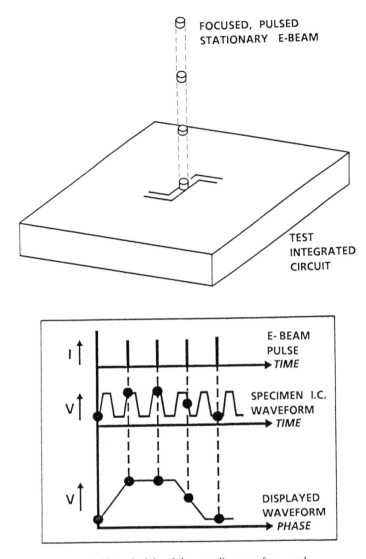

FOCUSED, PULSED
STATIONARY E-BEAM

TEST
INTEGRATED
CIRCUIT

E-BEAM
PULSE

SPECIMEN I.C.
WAVEFORM

DISPLAYED
WAVEFORM

Fig. 3. The principle of the sampling waveform mode.

gral of the secondary-emission energy distribution shown in Figs. 4, 6, and 7) and allowed the correct voltage waveform shape to be displayed directly. By this means the measurement error caused by nonlinearity with sample voltage change can be reduced to negligible values. This is true only if the S-curve shape remains constant with changing DUT voltage.

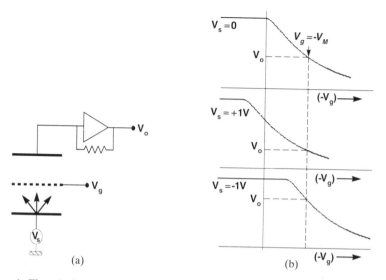

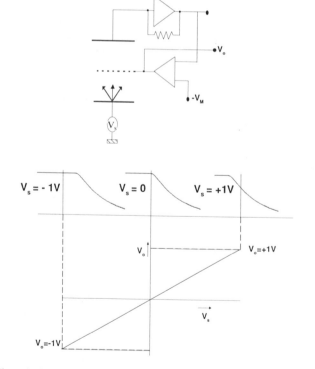

FIG. 4. The principle of electrostatic energy filtering in open-loop using a simple retarding field.

FIG. 5. The principle of closed-loop electrostatic energy filtering using a retarding field. For good linearization and measurement accuracy, the S-curve shape must remain constant.

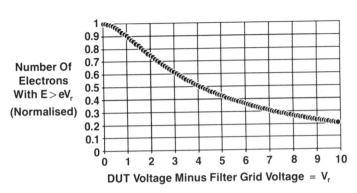

FIG. 6. Integral with respect to emission energy of the energy distribution.

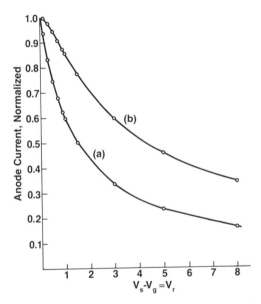

FIG. 7. Theoretical curves for anode current as a function of the voltage difference between grid and DUT surface for (a) filtering with respect to the component of velocity normal to the DUT surface and (b) filtering with respect to the total velocity in the direction of emission.

Hardware and software signal-processing algorithms can also be used to minimize other sources of error (to be described shortly).

D. Image Processing

EBTs always use image processors. The stroboscopic image is made up, as was shown in Fig. 2, from pulses of primary electrons deposited on the sample at the same phase in each trigger cycle. Because the detailed transition time of an event can be discerned only to a time resolution dictated by the electron-beam–pulse duration, this duration must be short with respect to the trigger period. Therefore, the beam duty cycle is small (sometimes as small as one part in one million). As a result, the average primary current, the average secondary current, and the average collected current used to form the stroboscopic image are low. A typical unpulsed current of one nanoamp will be reduced in most useful applications to less than one picoamp.

It is sometimes essential to work with TV scanned stroboscopic images. For example, charging effects are often reduced at TV scan rates, or a video recording may be needed. However, many applications benefit from using slow scan because a boxcar integrator can then easily be used. Using either TV or slow scanning, an image processor is essential for noise reduction. When combined with flexible, externally triggerable scanning electronics, the breadth of application is greatly increased.

E. Automation

Early E-beam test equipment required manual adjustments to both SEM and E-beam test parameters. These adjustments were tedious and led to errors of measurement. In modern equipment almost all of these adjustments are made automatically including automatic gain adjustments to bring up a waveform or an image immediately. Software control of every parameter of the EBT is desirable, and software control of a majority of parameters is mandatory. Without it, the results take much longer to obtain and are less accurate once obtained.

F. Signal Processing

Modern EBTs use various algorithms and much signal processing for noise and error reduction.

A typical early example of error reduction is described by Feuerbaum (1979) (though the method described is rather slow and inflexible, and more sophisticated schemes are now in use). Because the secondary-emission yield can drift during the recording of a waveform by reason, for example, of "contamination" (polymerization of an adsorbed layer of monomer by the E-beam) or charging, it is necessary to "renormalize" the signal amplitude at intervals. Feuerbaum did this by returning to zero phase before each increment of sampling phase. This zero-phase amplitude was used to correct the output voltage for each new sampling phase.

Noise reduction is routinely carried out during waveform recording by a variety of means such as recursive filtering and boxcar averaging. The degree of reduction is usually selectable. Fast algorithms can be used for quick results and slow ones for more accurate results.

For example, one satisfactory method of recursive filtering allows selection of the total number of successive phase sweeps, with additional selection by the user of the dwell time at each sampling phase. In this way, the degree of noise reduction can be traded off against the recording time over a wide range from a fraction of a second to several minutes.

III. Sampling by E-Beam Pulsing

A. THE SAMPLING PRINCIPLE AND ITS APPLICATION IN THE EBT

The Sampling Theorem (Shannon and Weaver (1949)) requires that the signal be sampled at a rate slightly higher than twice the highest signal frequency that is to be reproduced. This assumes delta function sampling pulses. The Sampling Theorem in the context of E-beam testing requires that there be $2f/F$ separate, regularly spaced sampling phases during the recording of a waveform with period $1/F$, where F is the fundamental frequency of the waveform being recorded and f is the highest frequency component of that waveform. The samples can be taken in either of two ways, by "single sampling" or by "multiple sampling" (Fig. 8).

Because multiple sampling has not been demonstrated as of the date of this writing, further discussion in this section applies to single sampling.

The E-beam pulse is triggered by an external pulse train of repetition rate F. (*Repetition rate* is used hereinafter for waveforms other than sine waves; *frequency* is reserved for sine waves.) A common requirement is that all signals whose repetition rates (nF) are multiples of F (and therefore contain only harmonics of the fundamental F) shall be displayed in the stroboscopic image or in the waveform. This is often necessary or convenient (for example, for timing measurements on logic or memory devices). In this case, the external trigger must be the lowest repetition rate signal of interest within the DUT.

Signals with $n > 1$ will be recorded with less than optimal signal-to-noise ratios. Because a sampling pulse is produced for each trigger (with the exception mentioned shortly) then for the nth harmonic, there will be only one pulse for every n cycles. The average beam current will be $1/n$ of the optimum and the signal-to-noise ratio will be a factor $(1/n)^{0.5}$ worse than it would be if the signal nF itself were used as the trigger. Care is needed to ensure that these nF signals do not lack information at the

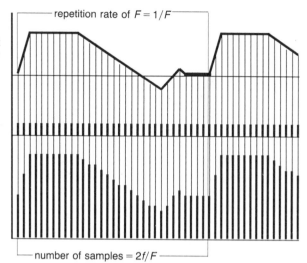

Voltage waveform on test node. Highest frequency component is f.

Primary E-beam pulse

Output voltage level derived from each secondary electron pulse using energy filter in negative feedback loop

FIG. 8. Single and multiple sampling for waveforms. In single sampling, one sample per cycle of F is taken and the sampling phase is advanced by increments of (F/f) π or less until samples have been taken at $2f/F$ different phases. In multiple sampling, $2f/F$ or more samples are taken in each cycle of F and are stored in $2f/F$ different storage locations. In each succeeding cycle, a new set of samples is "added in" and improves the signal-to-noise ratio.

higher-frequency components for large n. This can happen if there are insufficient sampling-phase steps or if the selected E-beam pulse duration is too long.

To ensure that the waveform display of nF does not lack information, the number of separate sampling phases must be $2f_n/nF$, where f_n is the highest-frequency component in nF. This can easily be brought about by correctly setting the time base (i.e., phase excursion).

Also, the E-beam pulse duration can be set (and, if necessary, reset) to smaller and smaller values until it is clear to the user that all the available information is being displayed.

These factors are usually dealt with automatically in a modern E-beam tester.

The input trigger also advances the sampling phase. The number of E-beam pulses for each sampling phase is (usually automatically) set as (NT/t) for the lowest-repetition-rate full-waveform display (period $T = 1/F$). This is done by using the ratio of selected sampling-phase sweep time t and the number N of separate sampling phases selected in a single phase sweep. (N should be greater than $2f/F$ as stated previously.)

This relationship does not apply at trigger frequencies F that are greater than the *maximum E-beam pulsing rate*, R_e. Above this rate the EBT counts down the trigger to give one E-beam pulse for every p trigger pulses. Then the number of E-beam pulses for each sampling phase is (NTp/t).

The *maximum trigger rate*, R_t (which is usually higher than maximum E-beam pulsing rate, R_e), is set by the capability of input devices that are usually emitter-coupled logic (ECL) and give maximum rates of 250 to 300 MHz. Again, it is possible to count down, or frequency-divide, a signal of rate greater than R_t and to produce a satisfactory waveform display. The signal bandwidth should be within the risetime capability of the EBT. The risetime is selected by choosing the E-beam pulse duration. In commercial equipment, it can be as low as 100 picoseconds and, in experimental equipment, as low as 10 picoseconds.

To restate, for clarity, the conclusion of the previous two paragraphs: a signal with repetition rate greater than the maximum trigger rate R_t must be counted down or frequency-divided. If the counted-down signal still has a repetition rate greater than the maximum E-beam pulsing rate R_e, the EBT input trigger circuitry will count down again to give one E-beam pulse for several trigger pulses. However, the original waveform will be faithfully reproduced in the display provided the risetime (which is selected by selecting the E-beam pulse duration) can be made small enough. Also, a small enough time per division (and therefore phase increment) must be chosen on the time base to satisfy the Sampling Theorem, and the jitter performance must not be limiting.

The time-discrimination or phase-discrimination capability of the EBT is usually much better than the risetime. For example, 10 picoseconds can be achieved. Thus, the timing difference between two transitions can be found to 10 picoseconds, even though the risetime may be 200 picoseconds. The time discrimination is set by jitter between the signal under test and the E-beam pulse. This can arise outside or inside the E-beam tester and depends upon both the EBT design and the jitter injected by the user between test signal and EBT trigger. There is averaging of this jitter (which improves the time discrimination in the limit to the standard deviation) during each sampling-phase dwell period.

It is quite often necessary to provide a predelay. For example, for a microprocessor the trigger may be provided well before the transition that is to be studied, because of minimum possible instruction loops. The trigger may have a period of milliseconds, whereas a node transition occurs in nanoseconds and therefore requires a time base of 5 or 10 nanoseconds per division. Most EBTs provide a predelay facility that uses the trigger to clock the delay for up to 10,000 trigger periods before starting the phase sweep. This is preferred to an accurate crystal-controlled digital delay, because the trigger relates directly to the DUT event that will have its own intrinsic jitter.

B. The E-Beam Pulse as Sampling Gate

In practice, the E-beam pulses are of finite duration, which causes significant loss of information only if the pulse duration is not negligible compared to the period of the highest-frequency Fourier component. The output voltage level for a given sampling phase is the average of the test-point voltage level during the sampling pulse.

A further error in sampling can arise from secondary electron transit-time effects. This is only significant for transit through the region of space within which the test-point voltage change produces a detectable change in the electrostatic or magnetic potential. The transit time out of this region is usually kept negligible by applying at least a small electric extraction field (even when using magnetostatic "extraction"). The risetime limits set by this and other effects are analyzed later in Section VC.

C. E-Beam Pulsing Methods

Although various other methods of E-beam pulsing exist, deflection of the E-beam by electrostatic plates has been used universally in E-beam testing. If the E-beam is given only a small deflection above the first condenser lens (see Fig. 9), this deflection is magnified by the second

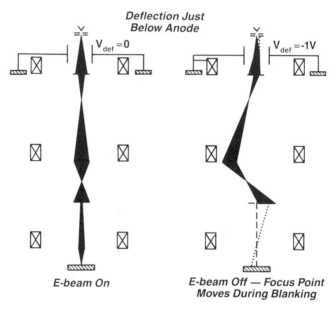

FIG. 9. E-beam pulsing by deflection blanking between anode and first lens. There is significant movement of the focused E-beam during beam pulsing.

condenser lens, and the E-beam can easily be deflected out of the exit pupil to cut off the beam.

The simplest method uses a voltage pulse that drives one of the deflection plates (the other being grounded) from a negative offset to 0 volts for the duration required of the E-beam pulse. Then it is driven back to the negative offset until the next pulse is required. The drawback of this method is that the virtual electron source is moved during deflection blanking, therefore the focused beam moves, leading to image blur or to the beam falling off the edge of the test node during waveform recording. For this reason, such systems minimize the virtual source movement by positioning the deflection plates directly beneath the anode.

Other variants of this approach use push-pull drive to both plates, which is said to mitigate chromatic aberration problems (Lischke et al., 1983). Two orthogonal sets of plates in succession can also be used. Their combined effect is to drive the beam along a square path. The E-beam pulse is produced as the beam passes across the exit pupil on one side of the square only. The pulse duration is decided by the slew rate of the rising edge of this side and can, in principle, be very short.

The most satisfactory method for good spatial resolution uses conjugate blanking. The deflection plates are positioned with their deflection center at the focus of the first (preferably) or the second condenser lens. Figure 10 shows that the virtual source and, therefore, the focused E-beam on the sample remain stationary during blanking. In practice, effectively zero blanking shift can be achieved using this method.

Time of flight past the plates must be carefully watched. A one-keV beam moves at 18.8×10^6 m/s and travels only 3.8 mm in 200 picoseconds. So the deflection plate length must be kept very short, which gives low sensitivity and calls for deflection voltage amplitudes of 1 to 10 volts, depending on the plate separation. Also a drive pulse of tens or hundreds of picoseconds duration must be adequately terminated to use fully the deflection power up to the highest frequencies. Careful plate design and testing using time-domain reflectometry is important for pulse durations below one nanosecond.

Thong et al. (1987) discuss methods for production of 10-picosecond pulses at repetition rates up to 50 GHz.

D. SINGLE AND MULTIPLE SAMPLING

In single sampling, only one electron pulse per cycle of the lowest repetition rate signal F has been assumed. If m electron pulses per cycle can be produced, stored separately, and used to reconstruct the waveform in the display, then the signal-to-noise ratio is improved by $(m)^{0.5}$ for the

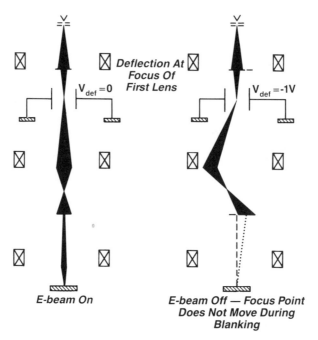

Fig. 10. E-beam pulsing by conjugate blanking between first and second lens. There is no movement of the focused E-beam during beam pulsing.

same recording time. Alternatively, the recording time can be reduced by $(m)^{0.5}$ for the same signal-to-noise ratio.

From the Sampling Theorem it is clear that samples should be taken at regular phase intervals to provide the maximum information. Therefore a "multiple sampling system" is engineered to generate a burst of E-beam pulses at rate f_b on receipt of each trigger. In an idealized case,

$$f_b = 2f/F \tag{1}$$

and fulfills the requirements of the Sampling Theorem with no need for a phase sweep. Each sample is required to be energy-filtered, amplified, and processed in the usual way. In addition, means have to be found of separating it from its neighboring samples and storing it in a separate location.

There is a practical upper limit to frequency f_b, and a mixed system that includes both phase sweep and "burst sampling" is required if *both* the full risetime capabilities and the operational flexibility of present-day EBTs are to be preserved.

Multiple sampling can also be applied to images, when it is known as *burst mode imaging*. Several to many samples are taken at n fixed phases within each DUT cycle. The E-beam is rastered in the usual way for an image and all the samples for each phase are diverted into a single image store and stored each in its correct pixel location. In this way, n stroboscopic images are recorded simultaneously. The elapsed time to record n images with the same signal-to-noise ratio is therefore the same as that normally required for one conventional stroboscopic image.

IV. Electron Energy Filtering in E-Beam Testing

A. RETARDING FIELD ENERGY FILTERING

Secondary electrons are emitted with (1) an energy distribution that varies in shape and (2) total yield that varies in amplitude, from point to point within a typical DUT. Causes of variation would include atomic number and work function for a perfectly clean flat surface. For a practical surface, variations also occur with work function of adsorbed layers and of polymerized contaminants, with surface texture and with topography.

Figure 11 shows a typical energy distribution in the low-energy *secondary* range, to be distinguished here from back-scattered or inelastically scattered electrons.

The angular distribution of emission of a perfectly clean, perfectly flat surface would probably be a cosine or $\cos^2$ distribution (Jonker, 1951). Any practical surface is microscopically rough and the distribution may have peaks in any direction within a solid angle of 2π. These variations are always present and any method for measuring DUT voltage from the secondaries must take them into account.

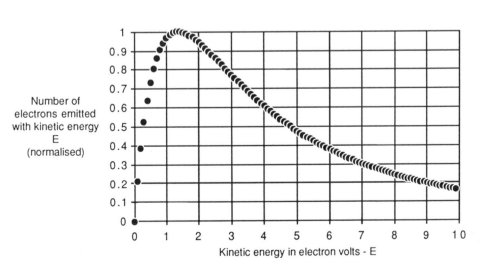

FIG. 11. Typical low-energy secondary electron emission energy distribution.

Most workers start by assuming that a hemispherical retarding filter with the measurement point at the center is best because electrons emitted at any angle can be subjected to the same retarding field. Unfortunately, no test device is a mathematical particle; devices rather have finite extent and therefore give rise to nonradial field lines, so perfect filtering is not possible without some means of focusing the secondaries to the center of the hemisphere. If the radius of the sphere is made larger to make the DUT relatively small, then so is the working distance (if the filter is below the objective lens) and therefore the EBT's performance becomes poor in spatial, voltage, and time resolutions (Gorlich et al., 1987).

It is usually more practical in "below-the-lens" operation to filter the normal velocity components of the emitted secondary. *Normal* means parallel to the primary beam axis and normal to a planar DUT. In EBTs, the test device is always perpendicular to the primary electron beam axis and is not tilted as in the SEM. This allows the use of a planar retarding mesh held parallel to the DUT surface. Such an electrode is convenient to make and can be quite close to the DUT, thereby helping to keep the working distance reasonably small.

Figure 6 shows the integral of the energy distribution. Figure 4(a) shows a simplified planar retarding field structure with DUT at voltage V_s, planar retarding mesh at V_g, and a schematic planar anode as electron collector. (The anode would be held at a small positive voltage.)

In this structure, when $V_g < V_s$, electrons with a normal component of emission velocity v_n such that

$$v_n < \sqrt{2e/m \ (V_s - V_g)} \qquad (2)$$

will be turned back by the mesh and will not reach the anode to contribute to the collected current. Curve (b) in Fig. 7 shows the general shape of the curve of anode collected current against $(V_s - V_g)$. This is commonly known as the S-curve. If the total emission velocity (rather than just its normal component) had been filtered, then curve (a) would have been obtained. This is the integral $n(E)dE$ shown also in Fig 6. With a planar retarding grid filter, the point of inflexion is taken out.

Figure 4 shows the basis of *open-loop* DUT voltage measurement. In 4(b) is shown V_o, the output voltage of the anode transresistance amplifier, against $(-V_g)$ for $V_s = 0$, which is identical with Fig. 6. If V_s is changed to $+1V$, the whole curve is shifted by one volt to the left. For $V_s = -1V$, the whole curve is shifted to the right.

Also shown in Fig. 4 is the value of V_o for $V_g = -V_M$, taken at a point of high slope on the curve. V_o (V_M) takes a smaller value for $V_s = +1V$ and a larger value for $V_s = -1V$. V_o follows the voltage V_s, although the waveform $V_o(t)$ will be distorted because the curve V_o (V_s) is nonlinear.

This distortion can be removed by operating in *closed loop* (Fig. 5). From the "anode" a transresistance amplifier is followed by a high-gain voltage amplifier. The output is fed back to the retarding mesh, superimposed on DC level $V_g = -V_M$. This is a negative feedback loop that forces the retarding mesh voltage to follow the DUT voltage with a linearity error that can be made small by making the voltage gain large. This is conventional negative feedback but for one thing. The unusual aspect is that secondary electrons moving between the retarding mesh and the anode form a part of the loop. Therefore the primary-beam current amplitude and the secondary emission yield affect the loop gain and the DUT voltage measurement error.

This method is successful. Practically, the linearity error can be reduced to well below one percent provided the S-curve shape remains constant with changing V_s, as shown in Fig. 5. In modern commercial EBTs, digital linearization and error-reduction schemes are more usual.

B. Extraction Fields and Their Effects

Electrostatic Extraction

Figure 12 shows that a planar surface with a voltage difference of 10 volts, held parallel with a planar "anode" at 10 volts, generates a field on the positive side of the "junction" that is retarding for emitted secondaries.

This retarding field velocity filters secondaries emitted in the relatively positive region. It is the main cause of voltage contrast in the SEM equipped with the conventional Everhart-Thornley electron detector. Lower-energy electrons are returned to the specimen surface; higher energies are deflected by this retarding field.

Because this "self-filtering" cannot be made quantitative, it is a source of error in the E-beam tester and must be minimized. By increasing the anode voltage (applying a high extraction field), the region within which it applies and the deflections produced by it can be reduced as first shown by Plows (1969). However, even for fields of 1000 volts per millimeter, "local field effects" are felt over several microns.

In the simplified electrostatic filter of Fig. 13, electrostatic extraction is added by inserting a positive mesh between the DUT and the filter mesh. The primary purposes of extraction are to bring emitted secondaries to the filter mesh for velocity filtering and to remove them quickly from the DUT surface so that transit-time limits are not imposed on the risetime of the EBT. For these purposes only a few tens or hundreds of volts are needed. Figure 14, shows electron paths in this type of structure.

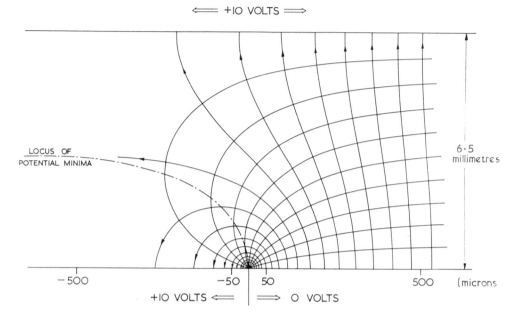

FIG. 12. Electrostatic equipotential and flux lines on the DUT surface, at the boundary between 0-and 10-volt regions showing that a retarding field above the positive side leads to "self-velocity filtering," the main cause of voltage contrast in the SEM (Plows, 1969).

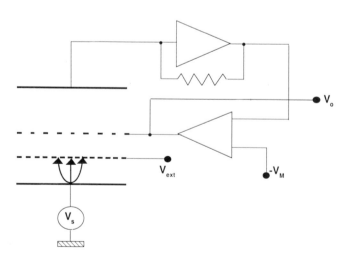

FIG. 13. The principle of closed-loop electrostatic energy filtering with the addition of a positive extraction grid for reduction of the local field effect.

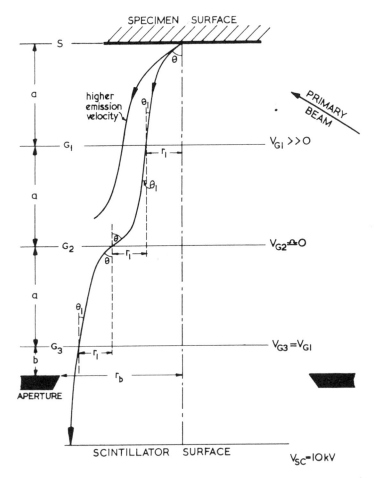

FIG. 14. Electron paths in the first voltage measuring electron collection system (spectrometer) to use the principles illustrated in Figs. 12 and 13. (Plows, 1969)

The secondary, but crucially important purpose of the extraction field is to reduce the self-velocity–filtering effects of local fields. As shown later, a magnetic field can also be used.

The quantitative effects of self-velocity–filtering are discussed shortly. In qualitative terms, secondaries can be totally lost before reaching the filter mesh or they can be accelerated, retarded, or deflected by the local field, i.e., the field local to the impingement point of the primary beam on the DUT, so that the normal component of velocity at the filter mesh varies with the local field. This variation of normal component is added to that which occurs because of changes of potential difference between the point of emission and the filter mesh and, therefore, causes an error in the voltage measurement at the test point due to variations in the voltage of closely neighboring regions. Both loss of electrons and addition or subtraction of velocity by local fields are called local field effect (LFE) or, confusingly, crosstalk.

For adequate reduction of LFE, electrostatic extraction fields as high as 1 to 3 volts per micron are sometimes used with modern DUTs. Such high fields, when applied by a mesh held close to the DUT, can be hazardous because of field sharpening and breakdown to the DUT. When applied using a tube or aperture method, such breakdown does not occur. However, the use of a high-voltage extractor can be a problem when looking at passivated DUTs and can possibly produce errors in measurement on depassivated devices because of local field effects caused by charge retention during imaging of thick field oxide which neighbors the measurement node. A dielectric surface charges toward the potential of the nearest electrode for E-beam voltages around one keV.

Menzel and Kubalek (1983), Nakamae et al. (1981), and Gorlich et al. (1987) analyze the theoretical and experimental results in various configurations.

2. Magnetostatic "Extraction" and Collimation

The primary purpose of extraction, to get the emitted electrons to the filter mesh, can be fulfilled by a suitable magnetic field in combination usually, with a low electrostatic field. The secondary purpose, to reduce errors caused by self-velocity–filtering by local electric fields, can also be fulfilled.

Ideally a method of energy filtering is needed that (1) will filter with respect to the full velocity of secondary emission rather than its normal component and (2) is invariant with emission angle. In a sense, the self-filtering or crosstalk effect of the local field, caused by conductors neighboring the test point, merely changes the "effective angle of emission" (i.e., the angle to the normal after leaving the local field region). Those electrons with an effective angle of emission between $\pi/2$ and $3\pi/2$ (those electrons that are returned by the local field to the DUT surface) cannot be rescued. The effect of this loss can be minimized by working some way down the S-curve with a filter voltage of, say, -7 volts. If a method can be found that brings all electrons emitted above the DUT plane (that is, with "effective angles of emission" between $\pi/2$ and $-\pi/2$) to the planar filter mesh with conserved momentum and with the transverse velocity reduced to small value, then crosstalk will be much reduced. Also, the energy resolution and, therefore, the voltage resolution will be improved.

Hsu and Hirshfield (1976). Beamson et al. (1980), and Kruit et al. (1981)

have presented the principle of an extractor or "parallelizer" that uses an inhomogenous magnetic field to achieve this. Garth et al. (1985) has presented some results of an extractor–energy filter using a snorkel lens that uses this principle.

The DUT is immersed in the field. The Lorentz force $e(\mathbf{v} \times \mathbf{B})$ causes each emitted electron to spiral around a magnetic field line. The magnetic field (i.e., flux density) is made high at the DUT and low near the filter mesh. The rate of change of the field is designed to cause the electron's spiral path to be unwound and "parallelizes" the electron trajectories so they approach the planar filter mesh almost normally.

Figure 15 shows the principle and the type of magnetic field needed. It is necessary that the angular momentum of the electron be conserved if the DUT voltage measurement is to be accurate. This condition is fulfilled if

$$r_f/r_i = (B_i/B_f) \tag{3}$$

where r_i is the high-field radius, r_f is the low-field radius of the electron's spiral trajectory, and B_i and B_f are the corresponding magnetic fields. This also means that the electron trajectory is bounded by the same field lines throughout.

This principle has been used in two ways. Garth et al. (1985) used a snorkel lens below the conventional objective lens of the SEM. Rao (1987) used an asymmetric lens of unconventional design in which the maximum of the axial magnetic field falls beyond the final pole piece, thereby allowing the DUT to be immersed. To increase the scanned area Rao also designed for "variable-axis" operation. Richardson and Muray (1987) followed

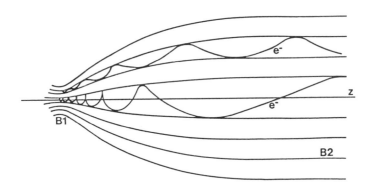

FIG. 15. The principle of magnetic collimation used for efficient secondary electron collection together with a reduction in local field effect. B1, a strong magnetic field, changes to B2, a weaker magnetic field, in such a way that the electron's angular momentum is conserved as the helical trajectory unwinds.

Rao, although using a smaller scanned area and apparently producing a
curved (rather than planar) focal surface resulting in defocusing within the
scanned field at low magnification.

C. BELOW-THE-LENS ENERGY FILTERS

Two main types of below-the-lens energy filters have been used, one
without and the other with actual physical meshes to define the energy-
filtering surface. Plows (1969) (Fig. 16) used a highly tilted DUT in the
original work. This allowed the primary beam to enter between the extraction mesh and the DUT surface. It also allowed the use of fine-pitch,
high-transmissivity electro-formed planar meshes that did not require a
hole in the center. Therefore, good planarity was achieved both for extraction and for energy filtering. DUT-to-mesh distance was 6.5 mm, and a
high tilt angle was required to achieve a reasonable scanned area. Extraction fields up to 300 volts/mm could be used before serious scan distortion occurred. The S-curve, probably the first ever produced, is shown in
Fig. 17.

Feuerbaum (1979) also used meshes to define the extraction and filtering
surfaces. In this case, the meshes were electroformed with a hole in the
center through which the primary beam could pass. The DUT surface and
the extraction and filter meshes were normal to the E-beam.

This allowed extraction fields up to 300 volts per mm. Small areas of
several hundred microns could be scanned through the central mesh holes.

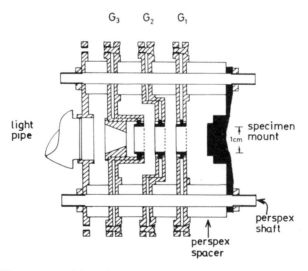

FIG. 16. The structure of the voltage measuring electron collection system of Fig. 14.

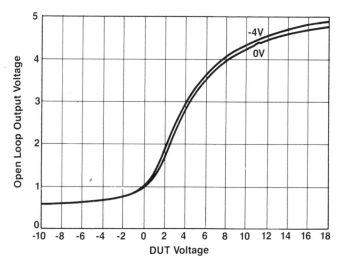

FIG. 17. A historical S-curve showing the variation of collected electron current with DUT voltage and the effect on this S-curve of the voltage of the neighboring DUT interconnect.

Larger scans showed the mesh bars in the image. Secondaries were deflected sideways after filtering and on to a scintillator. Because the equipotentials would be distorted in the region of the central hole, and because at high extraction voltages the diameter of the secondary electron beam will be only a few hundred microns, it is unclear that the filtering equipotential surface is planar.

Feuerbaum's working distance (from lens final surface) is about 13 mm. In one commercial version of this detector, astigmatism caused by the secondary collection fields limited image resolution severely, although in other respects the image quality was judged reasonable.

Between 1980 and 1984, Lintech Instruments supplied commercially an energy-filtering arrangement that included an annular scintillator and that did not use meshes. Extraction fields of 1000 to 2000 volts/millimeter could be used, and although the working distance was 25–30 millimeters, successful results were recorded on device geometries down to 0.5 microns.

All below-the-lens detectors are limited to collection efficencies of 60 to 70 percent at most, either because electrons are lost to the walls in travelling through the detector or because the meshes intercept secondaries. Menzel and Kubalek (1983) present a thorough review of below-the-lens electron spectrometers that use electrostatic extraction.

Garth et al. (1987) have described an energy filter that is both below the lens and in the lens. Intended for use with an established design of SEM column, it is placed below the objective lens of the SEM but also includes another magnetic lens as well as arrangements both for electrostatic extraction and for magnetic collimation. First, electrons are extracted by an electrostatic field (the magnetic field being maintained low at the DUT surface) and approximately focused to a point in the back bore of the "spectrometer lens." Next, they are subjected to an adiabatically reducing magnetic field produced by a set of microgaps in the back bore which field is said to collimate them before they encounter a planar mesh retarding electrostatic velocity filter.

This arrangement is reported to improve upon the earlier Garth snorkel-lens design by removing the strong magnetic field from the vicinity of the DUT. This magnetic field had been found to lead to various practical problems (for example, with lead frame devices in which there are ferrous components).

Evans (1986) had previously pointed out that some probe cards used for die-on-wafer testing include ferrous elements and this could lead to severe problems when using magnetic "extraction." This may be a weakness of all below-the-lens and in-the-lens magnetic extraction arrangements as described by Garth et al. (1985), Rao (1987), Richardson and Muray (1987), and Frosien and Plies (1987).

D. THROUGH-THE-LENS AND IN-THE-LENS ENERGY FILTERS

The concept of extracting secondary electrons using electrostatic or magnetic fields generated below the objective lens, and energy filtering these extracted secondaries using filter arrangements above the lens gap, in the back bore of the lens, was first discussed some years ago (Plows and Paden, private communication, 1982). Several realizations of this through-the-lens or in-the-lens concept now exist. Most are capable of both electrostatic and magnetic extraction with differing degrees of efficiency. Electrostatic energy filtering using planar or hemispherical meshes in the back bore is common to all of them.

The design criteria will be clear from the previous discussion, with the additional, if rather obvious, requirement that the imaging quality of the extraction–lens–energy filter–scanning arrangements should be adequate for E-beam testing. Given that EBTs, whether commercial or experimental, may be required in many cases to test DUTs with geometries down to one quarter micron, the spatial resolution available in real time using beam currents of about one nanoamp should be about 0.1 micron (100 nm) and should be desirably better than 50 nm for lower currents.

Detailed design information about the early through-the-lens and in-the-lens arrangements (as used for example by the commercial instruments sold by ABT and by Lintech Instruments Ltd.) has not been published as of early 1989. Some information has been published about such systems, for example by Frosien and Plies (1987) and Richardson and Muray (1987).

V. The Existing Limits of Performance

A. SIGNAL-TO-NOISE RELATIONSHIPS IN WAVEFORM DISPLAY

In a conventional SEM, the dominant source of noise is always shot-noise in the primary electron beam. Other sources of noise are negligible if the arrangements for secondary electron collection are efficient and when using a scintillator followed by a photomultiplier to produce low-noise–high-gain first-stage amplification.

In an EBT, the same comment usually applies. Given that the instrument has adequate stability over a period of a few minutes, recording conditions can be selected that make shot-noise the limiting factor even in low-duty cycle image or waveform recording. Sackett and Spicer (1986) and Gopinath (1977) have analyzed the shot-noise limits in waveform recording.

The voltage difference between filter electrode and DUT is

$$(V_s - V_g) = V_m + R \tag{4}$$

when operating in closed loop. R is the linearity error, which is negligible for present purposes; therefore,

$$(V_s - V_g) = V_m. \tag{5}$$

V_m is chosen to satisfy two criteria: (1) the slope should be adequate to maintain loop gain and (2) the local field effects can be minimized.

As discussed, the shape of the S-curve will vary with the choice of extraction–filter arrangement. For this calculation, extraction is assumed to deliver all secondaries with effective angles of emission between $-\pi/2$ and $+\pi/2$ to a planar filtering equipotential, normally and with conserved momentum. Therefore, the S-curve is the integral of the secondary-emission energy distribution.

Chung and Everhart (1974) give a theoretical equation for $N(E)$ that agrees well with experimental data:

$$N(E) = \frac{kE}{(E + \Phi)^4} \tag{6}$$

where Φ is the work function. This is plotted in Fig. 18 in units of E/Φ for convenience. The maximum lies at $E = \Phi/3$ and commonly lies between

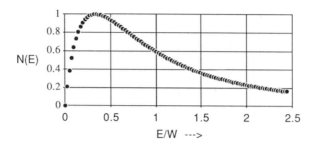

F<small>IG</small>. 18. Theoretical secondary electron energy distribution.

one and three volts. For aluminium, $\Phi = 4.1$ volts; however, the surface is usually covered by adsorbed layers and contaminants that will change Φ.

The criterion of maximum loop gain, therefore, would require

$$V_m = \Phi/3. \tag{7}$$

However, even using perfectly adiabatic magnetic extraction, some low-energy electrons would be trapped by the local fields and returned to the DUT. Therefore, V_m is usually chosen to be higher, so that operation is in a region of the S-curve less affected by this trapping. Most DUTs operate with $Vcc = 5V$ and electrons emitted with more than 5V energy will escape with an effective angle of emission in the range $-\pi/2$ to $+\pi/2$ (i.e., above the DUT). Therefore, under our assumptions on filtering and extraction methods, the S-curve would be unaffected by local field trapping for $V_m > = 5V$.

For this analysis, a choice is made of

$$V_m = \Phi, \tag{8}$$

which makes the calculation more straightforward and should give results with acceptable errors.

The S-curve shape is

$$N(E)dE = \frac{-k(3E + \Phi)}{6(E + \Phi)^3}, \tag{9}$$

which is shown in Fig. 19 in units of E/Φ.

Sackett and Spicer (1986) use the RMS fractional error $\dfrac{1}{\sqrt{n}}$ in the detection of n electrons together with the slope of the S-curve to give a formula

$$\Delta V_R = \frac{4Q}{3\sqrt{n}} \tag{10}$$

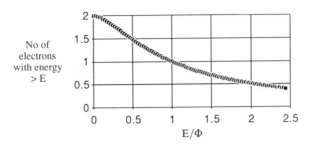

FIG. 19. The integral with respect to energy of the distribution of Fig. 18: the theoretical S-curve.

where Q is a constant and ΔV_R, the voltage resolution, is equated with the RMS voltage-measurement error.

Using a primary beam current density J_0 and the beam diameter D, and inserting values for the secondary emission yield Y_S and the collection efficiency E_T, the number of electrons arriving per second in the primary beam is calculated as

$$n_p = 1.96 D^2 J_0 10^{10} \tag{11}$$

and the required number of E-beam pulses per sampling phase as

$$P = 1 + \frac{\text{ABS}(2.94\tau_p \times 10^{-9})}{V_{\text{res}}^2 D^2 J_0} \tag{12}$$

where τ_p is the pulse duration.

Figure 20 shows the waveform measurement time (assuming 200 separate sampling phases) for different E-beam pulse rates F_E. From this figure it can be deduced, for example, that for

$$V_{\text{res}} = 10 \text{ mV}$$

$$F_E = 1 \text{ MHz and}$$

$$t_p = 200 \text{ ps,}$$

the time to record a waveform will be 20 seconds. Some caution is needed in using these figures because any analysis of this type involves the use of "fudge factors," which can only be deduced experimentally and which are perforce carried across from one apparatus to another.

The current density should be higher than that assumed when a modern EBT with LaB_6 (instead of Tungsten) source is used at one kV. It would probably be lower for an SEM with below-the-lens detector at 700 volts.

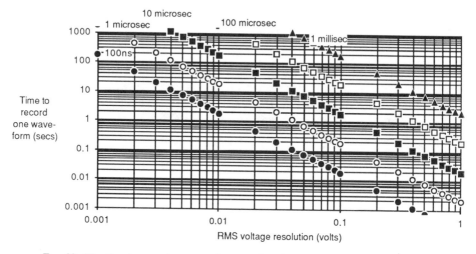

FIG. 20. The time to record one waveform as a function of the required voltage resolution for various repetition rates at a typical Tungsten filament brightness.

Evidence suggests that Fig. 20 provides reasonable guidance for experimental equipment. Some commercial equipment (e.g., from Cambridge Instruments, ICT, and Hitachi) provides significantly better performance.

EBTs can also be used in real time, either by recording a single-shot waveform or by real-time averaging by recording many waveforms to improve the signal-to-noise ratio. Multiple sampling is another mode mentioned earlier.

Real time modes are curiosities that are not in common use. A single-shot waveform cannot be recorded with more than a few kilohertz bandwidth for reasonable voltage and spatial resolutions. Real-time averaging and multiple sampling can both give waveforms in a much shorter recording time (assuming the same beam current) than the sampling method. Real-time averaging bandwidth is dictated by the video bandwidth, which for current EBTs is 1–5 MHz but might be increased to 50 MHz. Multiple sampling retains the bandwidth of the sampling method and places strenuous requirements on both beam pulsing and the early stages of video processing and amplification.

Both real-time averaging and multiple sampling cause rapid DUT contamination if large beam currents are used (as in the sampling method). Their importance is not mainly in improving signal-to-noise ratio but in lessening the waveform acquisition time.

B. Relationship between Spatial Resolution, Voltage Resolution, and Duty Cycle in Images

For a primary beam that is not pulsed, the spatial resolution depends primarily upon the beam diameter. This is dictated by Gaussian demagnification, spherical aberration, chromatic aberration, and diffractions, each to a greater or lesser degree. Which factors are limiting and what the performance will be under different circumstances have been analyzed in detail for the SEM and will not be discussed here. A primary beam current density of $J_0 = 1.3$ A/cm^2 is used in the following analysis.

When the beam is pulsed, with duration τ_p and rate F_E, the average primary current I_p is reduced to

$$I_p = I_0 \tau_p F_E \tag{13}$$

Therefore, if the signal-to-noise ratio is to be preserved, the time to record an image must be increased by $(\tau_p F_E)^{-1}$, the inverse of the duty cycle. Because, in most applications, $\tau_p F_E$ lies between 10^{-2} and 10^{-6}, this is impractical.

Fortunately most DUTs do not, as of the late 1980s, have features that require a beam diameter less than 0.1 micron for any other mode than real-time high-magnification imaging, so much larger values of I_0 are possible than those used for high-resolution scanning electron microscopy. EBTs commonly work with beam currents I_0 between 0.1 and 10 nA, though the availability of currents as high as 100 nA is useful when fast events with very long repeat times must be studied. An E-beam duty cycle of a few parts per million is not uncommon, especially in microprocessor work. (When changing modes from real-time imaging to stroboscopic imaging, the EBT should increase the probe current automatically, and vice versa. Otherwise in the real-time mode, heavy contamination, charging, or even DUT damage may occur.)

For voltage waveforms a simple definition was possible for voltage resolution, based upon the RMS shot-noise only. In essence, the smallest voltage change that can be detected in a waveform is defined equal to the RMS voltage noise (due to primary-beam shot-noise) in that waveform. For images, a definition is more difficult.

A simple and fundamental criterion is that the change in brightness caused by the smallest detectable change in voltage (the voltage resolution) should equal the change in brightness caused by noise. Now the change in brightness due to noise is set by, or affected by, at least, the following factors:

1. The voltage signal applied to the display tube and the noise level in that signal.

2. The line and frame scan rates. If TV rate is used, for example, the persistence time of the phosphor integrates the signal and increases the signal-to-noise ratio. If slow scan is used, then the noise bandwidth for the display in line and frame directions is set by the line and frame scan rates. For example, for a line scan rate of f_1, the highest frequency component that appears in the display (and therefore that sets the upper limit of the display noise bandwidth) is approximately $n_1 f_1/2$ where n_1 is the number of distinguishable pixels in the line direction, a characteristic of the display system.

3. Display tube parameters; the black level setting (i.e., the video level that can be independently selected by the user at which the screen has zero brightness); the gamma relationship for conversion of video signal to displayed brightness; and the choice of eye or photograph (and, if photograph, Polaroid or otherwise). These choices set the saturation white level.

4. Whether or not an image processor is used to improve the signal-to-noise ratio and the type of averaging, filtering, or summing used in that image processor.

5. The degree and type of contrast in the image caused by other factors such as atomic number and topography.

Conditions can be optimized for the display of small voltage differences. Almost always these optimized conditions run counter to common sense in that no topographic or atomic-number contrast can then be seen, so a waveform display would be more informative. A good rule of thumb for images is that a voltage difference of 0.2 volts can be resolved with reasonable care in an image with low contrast due to atomic-number variation and topographic features. Visibility of a 0.1-volt difference requires a high magnification image of a featureless surface that bears no voltage differences greater than 0.5 volts. This treatment assumes henceforth a voltage resolution limit of 0.2 volts in the real-time image. Figure 21 shows these limits in a graphical form.

C. LIMITS ON RISETIME

The risetime of the EBT is set by the E-beam pulse duration. Deflection pulsing of the E-beam produces a trapezium-shaped pulse for durations greater than a few picoseconds. The beam is deflected (using any of the methods) into and out of an aperture that is bigger than the beam diameter at the point of blanking. This must be the case since the physical blanking aperture is the exit pupil of the electron optical system. The secondary electron emission current is an identical function of time.

If an electric extraction field is applied to make sure the transit-time

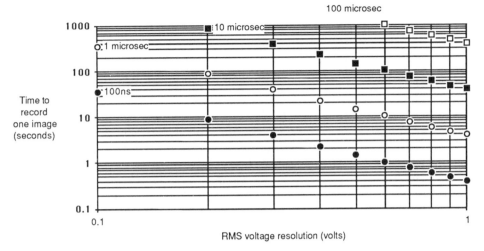

FIG. 21. The time to record one stroboscopic image as a function of the required voltage resolution for various repetition periods as a typical Tungsten filament brightness.

spread in the local field region is negligible, then the electron current that reaches the scintillator surface has an amplitude time function

$$I_c(t) = I_{bs} + k_F V_s(t) I_0(t) \tag{14}$$

where I_{bs} is the base current mainly due to backscattered electrons, which is therefore unaffected by energy filtering; $I_0(t)$ is the pulsed primary current; and k_F is a constant which includes the secondary emission yield, the efficiency of electron collection, the slope of the S-curve, and the linearity error in the loop. (This last is assumed to have a negligible variation with V_s and t.)

In this case, provided the duration τ_p of the primary current pulse is short compared with $1/f$. The period of the highest frequency component of $V_s(t)$, then $V_s(t)$ will be faithfully reproduced at the output even though the risetime of the scintillator and/or all later elements of the amplification and signal-processing chain are much greater than $1/f$.

Limits on risetime then can result from:

1. Inability of the deflection drive arrangements to generate a short enough voltage pulse (for deflection into an aperture) or a fast enough slew rate (for deflection across an aperture);
2. Inability of the deflection plates to use the deflection drive fully (for example, because of inadequate matching or termination or because they are too long and the transit time of the beam past the plates is limiting);

3. Lengthening of the E-beam pulse after deflection because of primary-beam energy spread and electron optical path differences (i.e., aberrations);

4. Distortions arising through local field energy filtering while the secondary electron pulse is travelling through the local field region.

Limits 1 and 2 are mainly a matter of engineering design and can be eliminated. Though available drive devices may eventually force a choice between high repetition rates and fast risetimes, it is feasible to generate less than 50-picosecond pulses at repetition rates up to 50 MHz using straightforward devices and deflection across an aperture.

Lengthening of the primary E-beam pulse (limit 3) occurs during transit through the column after the deflection plates. The distance between deflection plates and DUT, in an EBT electron optical column, may be as much as 300 mm. An axial electron emitted from the cathode with zero emission velocity (i.e., its kinetic energy is derived only from the accelerating voltage) will, for a one-kV beam, take 16 ns to make this journey. If a one-volt spread is assumed for emission energies, then the transit-time spread from deflection plates to DUT will be 8 picoseconds. This neglects the effects of retardation and/or acceleration in any electrostatic filtering or extraction arrangements.

To this must be added the transit-time spread caused by electron optical path differences. Chromatic and spherical aberrations (C_C and C_S, respectively) in the final lens only (not in the condenser lenses) are significant if the column is well aligned. The electron optical path difference W for semiangle α will be approximately

$$W = \sqrt{C_S^4 + C_C(V/V_E)}, \qquad (15)$$

for $C_C = 1.4$ cm, $C = 2.7$ cm, and $\alpha = 0.02$ rads (which are reasonable figures for a conventional SEM lens at one-nA beam current and working distance 5 mm). Then, from Eq. (15) $W = 1.76 \times 10^{-5}$ m and $\tau = 0.94$ ps.

There may also be some broadening of the pulses caused by electron–electron interactions. However, because beam deflection occurs well after the gun, electron–electron interactions need only be considered at the foci of lenses 2 and 3 (or lens 2 only in a two-lens column). At the source brightness typical of thermionic emitters, this broadening is insignificant compared with the transit-time spread caused by the spread of thermal emission energies. For field emitters and photoemitters, electron–electron interactions may be significant.

Lischke et al. (1983) analyzes the situation in systems using nonconjugate blanking, where the process of blanking can increase the energy spread and make the chromatic limit less acceptable.

Limit 4 is concerned with distortions arising through local field-energy filtering. Assuming perfect magnetic extraction (i.e., electrons that escape with effective angles of emission, after the local field, between $\pi/2$ and $-\pi/2$ reach the velocity filter plane normally), only electrons that are trapped by the local fields affect the voltage measurement accuracy.

For electrostatic extraction, the transit-time effect is calculated by Monte Carlo simulation (May et al., 1987) to be a few picoseconds for extraction fields between 100 V/mm and one kv/mm on submicron geometries.

A good rule-of-thumb figure for the limiting risetime of a system using thermionic emission is 10 ps. However, most commercial systems are not designed to approach this limit.

D. LIMITS ON TIME DISCRIMINATION

Time discrimination, the ability to distinguish between two identical rising or falling edges, is not determined by the duration of the E-beam pulse. It is set rather by:

1. The sampling phase increment that is selected;
2. The linearity of the sampling phase sweep;
3. The variation in phase position of the center of the E-beam pulse, averaged over the number of samples taken at each phase; and
4. The variation in the width of the E-beam pulse averaged over the number of samples taken at each phase.

Items 1 and 2 can always be made negligible by good engineering and sensible use. If a very small sampling-phase increment is required, then a small total phase excursion is set up. This, in turn, makes the linearity of the sampling phase better.

Items 3 and 4 set the attainable limits. Item 3 is the mean deviation of the center of the E-beam pulse with respect to the DUT pulse for the total number of pulses used at that sample. Item 4 is the mean deviation of the width of the E-beam pulse with respect to the DUT pulse for the total number of pulses used at that sample.

There will be a shot-noise contribution both to pulse-position variation and to the pulse-width variation. However, this will usually be negligible after integration of all the samples, compared with jitter caused by triggering arrangements from trigger source to EBT and internal to the DUT. These will vary from case to case and cannot be analyzed here. A reasonable practical figure is 10 ps for commercial EBTs.

E. LIMITS FOR THERMIONIC EMISSION AND FIELD EMISSION

Both cold-field emitters and built-up thermal-field emitters (e.g., zirconiated tungsten) have been proposed for use in EBTs. Room-temperature field emitters are possibly not suitable since EBTs generally use electron-probe diameters of 0.05 microns or more. For these probe diameters LaB_6 or Schottky emitters can provide as much or more beam current and the expense and difficulty of using room-temperature field emitters is much greater.

The low-frequency–noise spectrum of room-temperature field emitters is also a matter of great concern. Because the real-time bandwidth of the signal channel from DUT to display (set by the sampling phase sweep time) is typically DC to 500 Hz, noise in this range will come through to the waveform display.

Setting this aside, the higher brightness achieved with room-temperature field emission gives both advantages and disadvantages. A real advantage over both LaB_6 and tungsten emitters is that higher-resolution real-time micrographs can be recorded, because the higher brightness can be used to advantage at beam diameters of 10 to, say, 50 nm (depending upon the beam voltage). However, Schottky (built-up) emitters can provide the same advantages in this range and have an acceptable low-frequency–noise spectrum.

The small energy spread of room-temperature field emitters is an advantage, since the chromatic aberration and transit-time spread due to electron optical-path differences is less. However, Schottky emitters also provide this advantage to some extent and the effects of electron–electron interactions become comparable with chromatic effects at brightnesses of the order of 10^8 A/cm^2 steradian (May et al., 1987). No detailed analysis has been found. However, the preceding figures suggest that a cold field-emission system may have a limiting risetime of 2 to 4 ps.

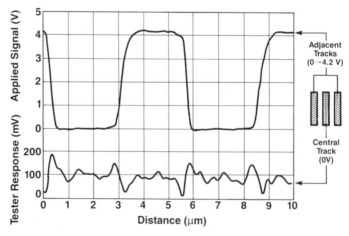

FIG. 22. Typical local field effect (crosstalk) performance.

Given the greater complexity and expense of experimental room-temperature field-emission equipment and the substantially higher price of commercial room-temperature field-emission equipment, it appears mainly applicable to DUTs with risetimes of a few picoseconds. This is true only if the problems caused by the low-frequency–noise spectrum can be circumvented.

F. Voltage-Measurement Limits Imposed by Local Fields

A great deal of work (to give only a few examples widely spaced in time, Plows (1969), Nakamae et al. (1981), and de Jong and Reimer (1986)) has been devoted to the analysis of local-field effects, also known as crosstalk. This is partly because it is satisfyingly academic. The effects can be analyzed endlessly for a variety of rather unreal situations. It becomes truly embarassing when a voltage cannot be measured except at a point where small geometry interconnects run parallel and closely spaced in the same plane or in different vertically separated planes and when it is impossible to find the same waveform elsewhere on the DUT. A real concern is caused on two or three metal (layer) devices when it may not be known (for example, by a failure analyst who has no access to the CAD date) that signal tracks are in close vertical proximity. It is difficult to be precise about the voltage-measurement limits imposed because a variety of non-comparable and artificial devices have been used in the published papers. A performance that (as of the late 1980s) is typical of good commercial equipment is shown in Fig. 22.

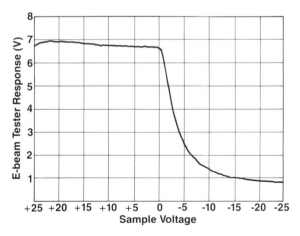

FIG. 23. A typical S-curve in modern equipment. Compare with theoretical curve shown in Fig. 7a and historical S-curve in Fig. 17.

Figure 23 shows the S-curve of the through-the-lens energy filter–detector used to take these results. Compare this with curve (a) in Fig. 7, for agreement with theory, and with Fig. 17 to note the improvements since 1968.

VI. Useful and Less Useful "Modes" (Including Brief Application Descriptions)

A. REAL-TIME IMAGES, WITH AND WITHOUT VOLTAGE CONTRAST

Good-quality real-time images are essential to find the test point accurately. It is possible to use a link to the CAD data base to "navigate" into the right area of the DUT. The accuracy that can be achieved is 1–2 microns if a stage movement is required (this is not good enough without further manual beam positioning) or about 0.2 microns if location is purely by beam movement.

As of early 1989, no commercial equipment uses laser interferometer stages to improve the accuracy of stage positioning (though it will shortly be available). Nor is there any commercial equipment that uses pattern recognition to deduce the position error, from which a fine adjustment to beam position can be made. Experimental equipment with such capabilities does exist (Tamama and Kuji, 1985) but it uses a great deal of expensive mainframe CPU time to achieve variable success. The problem is probably solvable. The analogy with pattern recognition using images derived by optical microscopes and TV cameras (for example, in IC bonders) suggests that the task would be eased greatly by a good signal-to-noise ratio.

A spatial resolution of 0.1 microns or better is mandatory for modern VLSI devices. Commercial E-beam testers with 0.3-microns resolution exist, but it is difficult to understand why.

Most E-beam testers subdue topographic contrast compared with the conventional SEM because the DUT is held normal to the beam and because it is necessary to achieve isotropic secondary collection if possible. The images look a little flat.

B. VOLTAGE CODING

Invented by Lukianoff and Touw (1975), this technique starts with a real-time image and adds to it a particular type of synchronization. The scanned line is triggered by the DUT signal and then free runs until after the line flyback, when the DUT signal is used to trigger the next line. Therefore, for the duration of each line, the transitions from logic high to low and back again will occur at a distance along the line that is decided by the line scan speed and the frequency and phase of the signal on that line. As a result, superimposed on the image of each interconnect appear stationary bright–dark stripes whose width and spacing depend upon the repetition rate and the mark–space ratio of the signal on that line.

Voltage coding is mainly used for an initial check upon the connectivity of a device. It is especially useful with "static" devices that do not depend upon charge storage and that can be run at rates giving conveniently visible (i.e., separated by at least a few pixels) dark–bright bars.

C. STROBOSCOPIC IMAGES (VOLTAGE MAPS)

In Fig. 2, the E-beam is scanned as for a real-time image. In addition, the E-beam is pulsed using an external trigger derived from the DUT signal that is to be studied or from a subharmonic of that signal. Often the clock is unsuitable because it is the highest frequency present on the DUT and a frequency divider or hold-off circuit is used, with due attention paid to the jitter consequences.

Usually the phase of the E-beam pulse relative to the trigger is maintained constant or swept slowly. With constant phase, those parts of the DUT that are relatively positive at that phase appear dark and those that are relatively negative appear bright. Therefore, the logic states of the whole of the scanned area are seen at a glance. By changing the phase manually or by sweeping the phase slowly, the changes in logic states can be followed through the cycle and logical sequences followed through the image area. This mode is used mainly for analysis of the operations of digital devices, which give clear dark–bright transitions.

In some contexts, a picture is worth a thousand waveforms.

D. LOGIC-STATE MAPS

Starting with a stroboscopic image, logic-state maps add a further facility. The sampling phase is swept in synchronism with the frame scan or line scan of the display. In the original definition of logic-state mapping, the corresponding frame or line scan of the E-beam itself was stopped. The effect of this was to display an image in which contrast in this direction was produced only by logic-state transitions. This is a powerful method of representing pictorially the logic transitions on an interconnect. It is most powerfully and most commonly used in bus regions where many lines run parallel and the relative timing is clearly demonstrated.

Sometimes the E-beam frame scan is left running, thereby retaining the image, allowing the test area to be precisely visualized, and avoiding heavy line contamination. This evolved mode is also commonly called logic-state mapping.

E. Voltage Waveforms by Single Sampling

This mode is exactly analogous to the original form of the sampling oscilloscope. The physical probe of the oscilloscope is replaced by the primary E-beam. The sampling gate of the sampling oscilloscope is replaced by the pulse of primary electrons.

At first sight, this is the most valuable of the available modes because of its familiarity (a time-domain voltage waveform is produced) and of its exceptional performance. Although the limits of the resolution performance in all factors cannot be achieved simultaneously, it is well worth the expense and inconvenience of vacuum to be able to position a probe of diameter less than 100 nanometers to an accuracy of 20 nanometers or better. It is also possible to use this probe to measure, with timing accuracy of better than 10 picoseconds, voltage waveforms with risetimes in the region of 200 picoseconds and repetition rates that may vary from a few Hertz to a Gigahertz (at the time of writing). All this is possible with a voltage resolution that, in some cases, is better than 10 millivolts.

F. Voltage Waveforms by Multiple Sampling

The E-beam tester is a single probe instrument. Its measurements are serial. Also, each waveform takes at least several seconds to record. Anything that can be done to speed up the recording of waveforms is welcome.

The most obvious measure is to take several samples per trigger cycle, each at a fixed phase in the trigger cycle. Samples for each separate phase are stored in a separate location and the displayed waveform is constructed from the averaged or integrated values in each location. In the simplest form, no phase sweep would be necessary because there would be a separate phase for each point on the waveform display. This is practicable at trigger rates up to 100 kHz. At higher rates, some phase sweep is necessary because of practical limits on the rate of E-beam pulsing. Above 20 to 50 Mhz, single sampling must be employed.

Multiple sampling requires some engineering compromises. Because the samples must remain separable until they can be routed to their separate storage locations, the scintillator used must be very fast, as must be the photomultiplier and the first stages of electronic amplification and integration. Plastic scintillators with decay times of about 2 nanoseconds are used that are less robust than the high-brightness long–decay-time YAG or P47 which are often used. The requirement for a low transit-time spread in the photomultiplier tube also constrains it choice. Care must be taken with video amplifier and integrator design. However, it is practical to achieve multiple sampling rates up to at least 20 MHz.

G. TIMING DIAGRAMS

Timing diagrams show an idealized "logic waveform" in which the transition between exact logic levels is shown occurring instantaneously.

As of the late 1980s, the main benefit of timing diagrams is that this simplified display of the logic transitions allows a clear display of the timing relationships between different nodes.

H. REAL-TIME WAVEFORMS AND LOGIC STATES

It is argued by some that real-time waveforms, with no use of sampling, have widespread value. There are three limited cases where value may be discerned:

1. For low repetition rates, up to several hundred Hertz, the noise bandwidth is usefully low when a single-pass waveform is recorded. This may be used, for example, in the study of low-temperature charge coupled devices (CCDs) in both research and quality control of liquid crystal display (LCD) screens, and in testing of low-temperature infra-red (IR) diodes.

2. For all repetition rates, if many passes are used to integrate to a bearable signal-to-noise level. The upper frequency limit here is placed by the bandwidth of the scintillator–photomultiplier tube–head amplifier, but within this limit waveforms can be derived much more quickly by this means than by single sampling. As compared with multiple sampling, there is a small advantage in waveform collection time and a considerable disadvantage in risetime capability.

3. For digital devices using a single pass, if a very wide tolerance is allowed in threshold detection to allow for a poor signal-to-noise ratio.

It is rather difficult to make a strong case for real-time methods for any but these most specialized applications, given the all-round high performance of single-and multiple-sampling methods. It is quite illusory to imagine that one off-random event or pseudorandom events can be detected. If that were so, then there would be a clear benefit over stroboscopic and sampling methods in some cases. Unfortunately, the signal-to-noise ratio is so appalling for single-pass real-time waveforms in the normal digital range of repetition rates that such events could never be discerned.

VII. Software Control

A. WHY?

Commercial E-beam testers are heavily software controlled and will become more so. There are many parameter adjustments to make in normal use. The great majority can be made either by restoring the parameter value that was used the previous time a given mode was used or by allowing a software or hardware algorithm to pick the best value. No E-beam tester or SEM add-on system is worthwhile that does not possess an integrated 16- or 32-bit controller and a well-developed and-evolved software control and processing system.

B. WHAT?

Operations fall into three main categories: housekeeping or "autonomic" operations, mode and parameter control, and CAD and tester interfacing. Different E-beam testers provide different mixes, but all aim to make control and display as immediate as possible.

C. CAD/CAE LINKS

Finding the test node can be a significant "navigational" problem. Fortunately, proprietary software exists that enables the EBT, when linked to a CAD/CAE workstation, to move the DUT or the E-beam to bring the E-beam to within a micron or so of the node of interest.

The most straightforward way of doing this uses a mask layout display on the workstation screen. A mouse is used to position a cursor on the mask layout. This cursor defines on the mask layout the point of interest. The EBT stage movement is linked to this cursor so that the corresponding point on the DUT and an area around this point are imaged on the EBT image screen.

However, most IC designers prefer to work from the schematic (circuit diagram) if possible, and, if this is not possible, they will settle for the netlists. Because the CAD vendors and users have not yet defined adequate standards, display of schematics from all of the commonly used CAD packages is not easily achievable. The EDIF formulation is developing in the right direction, but is still some way from universal application.

The IC designer who has access to the design data base can now work in the following way. A netlist window allows selection of a net number. The corresponding physical net is highlighted on the mask layout display. A cursor is positioned, as previously described, on a point within that highlighted net and the EBT stage moves to bring up an image of the corresponding area of the DUT on the image screen, thereby completing the chain of control from netlist to DUT–E-beam positioning. By this means, waveforms can be quickly derived and displayed even from a device with several hundred thousand nodes.

Once these waveforms are obtained, a comparison with simulated waveforms is straightforward. The simulation data is loaded and a waveform is constructed for graphical comparison on the workstation screen with the real waveform from that test point.

This CAD linkage, which is a powerful aid to the IC designer, is of little use to the failure analyst who will not usually have access to the design data. Care must therefore be taken by EBT manufacturers not to design the user interface in a way that ignores or makes relatively inconvenient other techniques of fault detection (Savart and Courtois, 1987). Dynamic fault imaging (DFI), described shortly, does not require such access. DFI itself, as described by May et al. (1984), has very severe drawbacks in any practical situation. However, the thinking behind DFI was seminal.

D. Burst Mode Imaging, Dynamic Fault Imaging, and Image-Comparison Techniques

A waveform records the voltage variation at a single test point. An image stored in a 512×768 image store can record the voltage levels at 393,216 points (provided the scan rate is set slow enough to accommodate the limited bandwidth of the measurement-mode electronics). A set of n stroboscopic images, taken at n regular phase increments through the cycle of the lowest frequency of interest, can record faithfully (on a synchronous device) the waveforms on all the nets within the scanned area, provided that n is big enough and that the sampling pulse is short enough to fully sample all the frequencies present. In that case, the set of n stroboscopic images also can be conceptually treated as a complete set of all the waveforms with that set of inputs. If the truth table were gone through completely, then all possible logic states could be exercised. The data collected could be compared with a complete circuit simulation.

This would be more than laborious, and the amounts of storage required are too large to be practical at the time of writing. Dynamic fault imaging, as it was originally described by May et al. (1984), propounded a semimystical approach based on this sort of thinking. By replaying the images in "movie mode," faults could be seen to propagate outward from their source. This mitigated the storage requirements by reducing the spatial-resolution requirements. It was not thought necessary to image the nets at full resolution in order to trace the faults.

The drawbacks of DFI as originally described are several. One is the sheer cost and complexity. Another example is the difficulty of guaranteeing that the same scan distortions will be maintained over substantial periods (necessary to achieve accurate image overlay). For these and other reasons, DFI has not fulfilled its early apparent promise and, although manufacturers of commercial equipment claim to provide DFI capability, this usually indicates an ability to process, store, and compare smaller sets of images than DFI originally required. In itself, this more limited and practicable capability is useful.

Comparison of two images by subtraction of one from the other or by repeated alternate display of two images to display a slight difference between the two as a flickering element (a technique imported from astronomy) is well known and requires the storage of a few images (say, four) for convenient use.

Burst mode imaging is the imaging analog of multiple sampling. (See Fig. 24.) Instead of one E-beam pulse per cycle, n pulses per cycle are used and the resulting signals are stored in n separate DRAM-based image stores. Full noise reduction is carried out and n images at the different phase points can be recorded in the same time as one image using conventional stroboscopic imaging. This is the crucial and fundamental technique for the future, more powerful uses of imaging in device debugging.

These imaging comparison and matching techniques carry within them the potential for much faster debugging than the use of waveforms. In the limit mentioned at the beginning of this section, the sets of images can equally be regarded as sets of waveforms or timing diagrams.

E. IC-TESTER AND MEMORY-TESTER LINKS

IC testers are used primarily for production testing and inspection. They cost typically $1–3 million, so the economic rationale for linking them to E-beam testers is not at first obvious. However, there is such a rationale in the development and quality control of highly complex commodity devices. It is that the test vectors can be applied to devices in the E-beam tester by the usual means and without writing special and additional test programs, and that the IC-tester diagnostics and the E-beam–tester diagnostics can easily work together to give a quick and thorough diagnosis.

Interfacing requires mechanical matching of test head to EBT, convenient electrical connection that maintains the timing performance usually achieved by the test heads when operating the devices in a more standard fashion, and if possible a degree of software interfacing.

F. FUTURE TRENDS

E-beam testing is a maturing technology. After its conception, it was built rapidly upon the established foundations of scanning electron microscope technology. Later, its growth slowed considerably and has been set

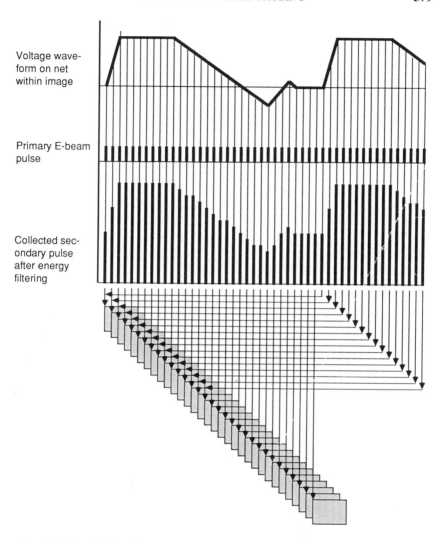

Voltage wave-
form on net
within image

Primary E-beam
pulse

Collected sec-
ondary pulse
after energy
filtering

Fig. 24. The principle of burst mode imaging. Resulting signal pulse goes into separate image store for each separate phase, so up to 32 noise-reduced images are produced in the time taken to produce one with conventional stroboscopic operation. Any one image chosen randomly can be displayed instantly, a difference image between two stored images can be displayed, or in "movie mode" the sequence of images or a sequence of difference images can be replayed at leisure.

not by the rate of arrival of inspiration but by the development funds available and justifiable for the development of complex and expensive instrumentation which sold only into a small market. Many of the instrumental advances made within the last five years have waited 10 to 20 years from first inception to realization.

For many years, the results expected from E-beam testers were modest, especially in the accuracy of voltage measurements. These expectations rose sharply during the late 1980s and are still rising. This change in expectations results from a change in the qualifications and the experience of the typical user. In earlier years, the user was typically a failure analyst with a very strong background in SEM technology. Now the user is more likely to be an IC designer who expects greater ease of use along with higher capability and immediate application to specific device diagnostic problems.

The capabilities of modern commercial E-beam testers have expanded to include convenient interfacing to CAD data bases. These interfaces enable easy and friendly navigation within the DUT and comparison of real waveforms and timing diagrams with the simulations. There has also been progress in integrating these two elements with the IC or memory tester. There is some dialogue about whether full automation of E-beam testing is possible or desirable. See Wolfgang (1987) and Cocito and Melgara (1987).

The human observer is still essential. First, the E-beam probe cannot be placed on the DUT to the necessary accuracy of 0.1 microns or better. Stage positioning accuracy is not adequate. Second, the effects of local fields still produce errors that, in some instances, fall outside the testing tolerances usually placed on IC and memory testers.

Possible solutions to probe-placement inaccuracy are to use laser interferometer stage positioning or to use pattern recognition to determine the relative positioning of the probe and DUT, and to correct this by deflecting the E-beam. Both of these methods are used in E-beam lithography equipment. The use of laser interferometer positioning in E-beam testing would push up the price significantly but is straightforward.

The use of pattern recognition is less straightforward. Alignment marks as in E-beam lithography cannot be used. Instead, an unpredictable feature or area of the device pattern may have to be learned (say, from the CAD data or from a comparison device or comparison mask). If that feature has changing voltage contrast on or near to it (as in stroboscopic images), then the feature may have to be relearned for each different state and so on. Pattern recognition in such circumstances is not always easy and may be of an unpredictable degree of difficulty. This suggests worst-case design and may result in the need for either very large and fast mainframes or expensively developed hardware or firmware algorithms. The hope must be that the ascending capability and descending cost of workstation–mini-computer hardware will allow an economical solution soon. Tamama and Kuji (1985) and Stentiford and Twell (1987) describe software-based systems.

E-beam testing of two and three metal devices (that is, devices in which some signal interconnects may be buried under one, two, or three layers of dielectric plus one or two layers of interconnect) could be difficult. New EBTs under development as of 1989 may be able to offer a focused ion beam column which will etch a "via" to a buried signal interconnect and will also be capable of depositing a conducting track from the interconnect to the top surface where voltage measurements with the E-beam will not be so strongly affected by local field effects or by charging. This could be valuable in itself. If the ion beam can also be controlled by the CAD interface so that location of the buried interconnect from the net list and mask layout is quick and easy, then the EBT's diagnostic capabilities can be enhanced and its applications widened to more complex devices for many years to come.

What are the likely trends in fundamental "scientific" capabilities? The theoretical limits on risetime are probably around 2 picoseconds due to the various causes outlined in Section V.C. The advantages of multiple sampling and burst mode imaging have been described rather than demonstrated. There is further work, as always, in magnetic-collimation and in energy-filtering systems intended to reduce the local field effect even further. But as I said at the beginning of this section, E-beam testing is becoming useful and also middle-aged.

References

Beamson, G., Porter, H.Q., and Turner, D.W. (1980). *J. Phys. E. Sci. Instrum.* **13**, 64–66.

Chung, M.S., and Everhart, T.E. (1974). *J. Appl. Phys.* **45**, 707–710.

Cocito, M., and Melgara, M. (1987). *Microelectronic Engineering* **7**, 235–241.

De Jong, J.L., and Reimer, J.D. (1986). *Scanning Electron Microscopy A.M.F. O'Hare*, 933–942.

Evans, R. (1986). Private communication.

Feuerbaum, H.P. (1979). *Scanning Electron Microscopy 1979, SEM Inc, AMF O'Hare*, 285–296.

Flemming, J.P., and Ward, E.W. (1970). *Scanning Electron Microscopy 1970*. IITRI, Chicago, 465–470.

Frosien, J., and Plies, E. (1987). *Microelectronic Engineering* **7**, 163–172.

Garth, S.C.J., Nixon, W.C., and Spicer, D.F. (1985). *J. Vac. Sci. Technol. B Jan/Feb 1986*, 217.220.

Garth, S.C.J., Sackett, J.N., and Spicer, D.F. (1987). *Microelectronic Engineering* **7**, 155–161.

Gopinath, A. (1977). *J. Phys. E: Sci Instr.* **10**, 911–913.

Gopinath, A., and Sanger, C.C. (1971). *J. Phys. E: Sci. Instr.* **4**, 334–336.

Gorlich, S., Kessler, P., and Plies, E. (1987). *Microelectronic Engineering* **7**, 147–154.

Hsu, T., and Hirshfield, J.L. (1976). *Rev. Sci. Instrum.* **47**, 236–238.

Jonker, J.H. (1951). *Phil. Res. Rept.*, **6**, 372.

Kruit, P., Kimman, J., and van der Wiel, M.J. (1981). *J. Phys. B: At. Mol. Phys.* **14**, 1, 597–602.

Lischke, B., Plies, E., and Schmitt, R. (1983). *Scanning Electron Microscopy/1983/III*, 1177–1185.

Lukianoff, G.V., and Touw, T.R. (1975). *Scanning Electron Microscopy, IITRI, 1975*, 465–571.

May, P., Halbout, J.-M., and Chiu, G. (1987). *SPIE Review on High Speed Testing.*

May, T.C., Scott, G.L., Meieran, E.S., Winer, P., and Rao, V.R. (1984). *IEEE/IRPS*, 95–108.

Menzel, E., and Kubalek, E. (1983). *Scanning* **5**, 151–171.

Nakamae, K., Fujoika, H., and Ura, K. (1981). *J. Phys. D: Appl. Phys.* **14**, 1939–60.

Oatley, C.W. (1972). *The Scanning Electron Microscope.* Cambridge University Press.

Plows, G.S. (1969). Ph. D. dissertation, Cambridge University, Cambridge, England.

Plows, G.S., and Nixon, W.C. (1968). *J. Phys. E: Sci. Instrum.* **11**, 595–600.

Rao, V.R. (1987). Private communication.

Richardson, N., and Muray, A. (1987). *Proc. Int. Symp. on Electron, Ion and Photon Beams.*

Sackett, J., and Spicer, D.F. (1986). *J. Vac. Sci. Technol.* **B4** (I), 213–216.

Savart, D., and Courtois, B. (1987). *Microelectronic Engineering* **7**, 259–266.

Shannon, C.E., and Weaver, W. (1949). *The Mathematical Theory of Communication.* University of Illinois Press, Urbana.

Stentiford, F.W.M., and Twell T.J. (1987). *Microelectronic Engineering* **7**.

Tamama, T., and Kuji, N. (1985). *Proc. IEEE Int'l Test Conf.*, 643–649.

Thong, J.T.L., Garth, C.J., Nixon, W.C., and Broers, A.N. (1987). *SPIE, Florida*, March 1987.

Wolfgang, E. (1987). *Microelectronic Engineering* **7**, 435–444.

CHAPTER 7

Photoemissive Probing

A.M. Weiner
R.B. Marcus

BELLCORE
RED BANK, NEW JERSEY

I. Introduction

The photoemissive probe uses a pulsed optical beam to probe a signal on a metal line on a semiconducting or nonsemiconducting substrate. The optical beam causes photoelectrons to be emitted from which the waveform of the signal is derived. The sampling principle that derives the waveform is shown in Fig. 1. The optical-beam pulses must be significantly shorter than the pulses of the signal being measured. Also, the phase difference between the beam and the signal must vary in a continuous and known way in order to effectively sample the signal.

The photoelectron energy distribution is fixed by the wavelength of the illuminating optical beam and the work function of the emitting metal line,

Copyright © 1990 by Bell Communications Research.
ISBN 0-12-752128-3

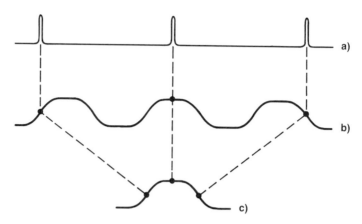

Fig. 1. The principle of signal sampling. Probe pulses (a) interrogate the signal (b), and a reconstructed waveform (c) is synthesized from the response.

and the distribution shifts with the potential of the emitter. The photoelectrons are analyzed by measuring the spectral shift of the distribution with a high-pass or narrow-pass energy filter. This method is similar to that used in electron-beam probing. (See Chapter 6.)

A photoemissive probe experiment contains five functional parts as illustrated in Fig. 2: (1) an optical system for generating and focusing light pulses onto the device under test (DUT) (2) a means of generating a drive signal and of synchronizing this signal with the probe pulses, (3) a vacuum

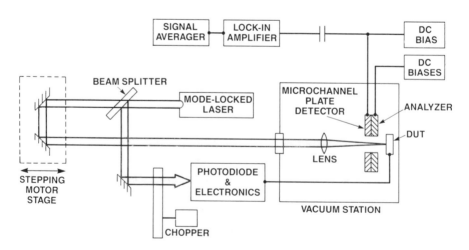

Fig. 2. Block diagram of experimental apparatus for photoemissive sampling.

chamber, (4) an electron energy analyzer and spectrometer, and (5) instrumentation for data processing. Some of the concerns relating to functional parts 1 and 3 are similar to those that relate to electro-optic and photoconductive probe methods discussed in Chapters 3 through 5. Similarly, the concerns relating to functional parts 2, 4, and 5 are similar to those pertaining to electron-beam probing discussed in Chapter 6 with one important exception: the significantly higher temporal resolution offered by photoemissive probing adds an additional complication caused by the relatively slow transit time of photoelectrons from the DUT to the analyzer. The high speeds of signals on the DUT as well as on adjacent metal lines create time-varying electric fields that can perturb the photoelectron energy during transit. This problem must be overcome.

II. Photoemission

A. GENERAL

The photoelectric effect was first described by Lenard (Lenard, 1900) and Thompson (Thompson, 1899), and Einstein was awarded the Nobel Prize in physics in 1922 for his explanation of this effect as a quantum event (Einstein, 1905). In this process, a material absorbs light with sufficient energy to eject an electron out of the surface of the material and into a medium on the other side, usually a vacuum. This process is illustrated by Fig. 3, which shows a photon of energy $h\nu$ ejecting an electron originally at level E_i into the vacuum at a final level E_f. The kinetic energy of the emitted electron is E_k. Alternatively, a fast succession of three photons each of energy $h\nu/3$ can produce the same electron transition as shown in Fig. 3.

All emitted electrons must overcome an energy barrier given by the work function $\Phi = E_{VAC} - E_F$, where E_F is the Fermi level and E_{VAC} is the vacuum level. Work functions for clean gold and aluminum surfaces are, respectively, 5.1 and 4.3 eV (*Handbook of Chemistry and Physics*, 1986), although contaminated surfaces can have significantly lower values. The kinetic energy of the photoemitted electron is given by

$$E_k = h\nu - (\Phi + E_b) \tag{1}$$

where $h\nu$ is the photon energy and E_b is the electron binding energy. E_k can have a range of values from 0 to $(E_{f,max})$, depending in part on the energy level of the initial state. For electrons initially at the Fermi level $(E_b = 0)$, the photoelectron kinetic energy is at a maximum at

$$E_{k,max} = h\nu - \Phi = E_{f,max} - E_{VAC}. \tag{2}$$

The photoelectron spectrum is given by the energy-distribution curve

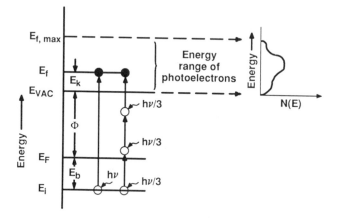

FIG. 3. Energy-level diagram for photoemission with photons of energy $h\nu$ and $h\nu/3$. E_F is the Fermi level, Φ the work function, and E_b the binding energy of an electron at energy level E_i. E_k is the kinetic energy of the photoelectron. Absorption of a single photon of energy $h\nu$ can elevate electrons from the initial to the final state as indicated; successive absorption of three photons each of energy $h\nu/3$ can also elevate an electron to the same energy level, through with significantly lower quantum efficiency. A hypothetical energy-distribution curve is shown at upper right.

(EDC), which reflects the distributions of both the initial and final states. A hypothetical EDC is shown at the right of Fig. 3.

The process of photoemission of an electron from a solid into the vacuum is easily viewed as a sequence of three events: optical excitation of the electron, transport of the electron to the surface, and emission into the vacuum (Berglund and Spicer, 1964; Cardona and Ley, 1978, Chapter 1).

Optical excitation is largely controlled by the joint density of states (filled and unfilled) as given by the band structure. The energy distribution of the joint density of states (EDJDS) can be described by the excitation probability summed through the Brillouin zone,

$$EDJDS(E) = \sum_{f,i} \int_{BZ} d^3k\, \delta(E_f - E_i - h\nu)\delta(E - E_f), \qquad (3)$$

where k is the momentum vector, E is the final electron energy, and $h\nu$ is the photon energy. Equation 3 is derived from an assumption of direct (momentum-conserving) transitions between intial (filled) and final (unfilled) states. It is reasonably successful in describing the behavior of metals such as gold (Cardona and Ley, 1978), which is a metal commonly employed on GaAs devices and, therefore, of concern to photoemissive sampling.

Transport to the surface is controlled by optical absorption (described by the optical-absorption coefficient α) and by inelastic electron scattering (described by the electron mean free path l_e):

$$T(E) = \frac{\alpha l}{1 + \alpha l} \qquad (4)$$

where $T(E)$ is a transport function. Photoemissive-sampling studies performed as of the late 1980s have used photons ranging in wavelength from 248 to 625 nm. Absorption coefficients for gold at 248 and 625 nm are, respectively, 0.071 and 0.064 nm^{-1}. The electron-electron scattering length of photoelectrons generated in gold by these two light sources (three-photon photoemission is assumed for irradiation by 625-nm light, as is discussed shortly) is about 5 nm (Krowlikowski and Spicer, 1970), and $T(E)$ (Eq. (4)) is about 0.25. Because only about half of the photogenerated electrons are moving in the correct direction to reach the surface, only 10–15% of the electrons generated reach the surface with no collisions. Those that suffer inelastic collisions on their way to the surface contribute a smooth secondary-electron background to the signal, peaking at the low-energy end of the EDC. This background only begins to appear for photon energies more than a few eV above the work function. The EDCs shown in Fig. 4 were obtained from photon bombardment of gold in the energy

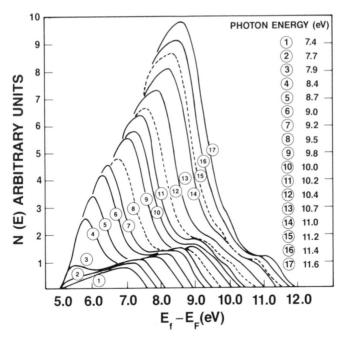

FIG. 4. Energy-distribution curves for gold for photons in the energy range 7.4–11.6 eV. (Krowlikowski and Spicer, 1970)

range 7.4–11.6 eV (Krowlikowski and Spicer, 1970). While the rapidly rising peak at the left edge for curves 4 and above has a growing contribution from secondary electrons, there is a negligible contribution to the EDC from inelastically scattered electrons for photon energies of less than a few eV above the work function.

Once photoelectrons arrive at the surface, they must surmount the energy barrier given by the work function. Assuming an isotropic distribution of electrons, only those with a kinetic energy component normal to the surface greater than Φ will escape. That fraction is given by

$$D(E) = \frac{1}{2}\left(1 - \sqrt{\Phi/E}\right). \tag{5}$$

Finally, the total photoelectron spectrum is given by the product of Eqs. (3) through (5),

$$P(E) = EDJDS' \times T(E) \times D(E) \tag{6}$$

where $EDJDS'$ is a slight modification of Eq. (3) achieved by weighting each transition by the appropriate matrix element. Because Eqs. (4) and (5) add no structure to the photoelectron spectrum, the spectrum is essentially a distorted replica of $EDJDS$ (Cardona and Ley, 1978, Chapter 1).

A good agreement between theory and experimental measurements of photoelectron spectra is demonstrated for several systems (Smith, 1971), in particular gold (Smith, 1972) and aluminum (Koyama and Smith, 1970). For gold, the first peak in the experimental spectrum corresponds to a d-band transition from a state initially at a level 2.7 eV below the Fermi level. This peak is the one that emerges at the left in the series of curves shown in Fig. 4 and appears only with photon energies more than 2.7 eV above the gold work function. For photoemission studies with light of wavelengths greater than 175 nm (single-photon photoemission assumed), experimental spectra from gold are expected to be relatively structureless.

A close agreement is also found between experimental EDCs of aluminum (Wooten et al., 1966) and calculations based on a direct transition model (Koyama and Smith, 1970). Here, the first peak corresponds to a transition initially at a state about one eV below the Fermi level, and structure is, therefore, not expected in spectra for incident light with a wavelength greater than 240 nm.

Because the photon energies used for photoemissive probe studies to date are within one or two eV of the work function, Φ, the photoemission spectrum is relatively structureless. Analyzers of the type described in Chapter 6 can be used, because the same principles of retarding field analysis apply.

The total photoelectron yield is obtained by integrating Eq. (6) over the

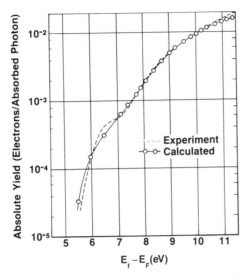

FIG. 5. Quantum yield for single-photon photoemission from gold. (Krowlikowski and Spicer, 1970)

energy range of the photoelectrons

$$Y = \int_{\Phi}^{h\nu} P(E)\, dE. \tag{7}$$

This yield can be roughly expressed as (Broudy, 1970)

$$Y = C(h\nu - \Phi)^n \tag{8}$$

where the variable parameter n has values from 1 to 2.5. The number of photoelectrons per incident photon, the quantum yield, is shown in Fig. 5 for gold over a photon energy range of 5.5 to 11.5 eV. An incident average laser power of at least 20 nW is needed to generate more than 0.1 pA photoelectron current at 5.5 eV.

B. MULTIPHOTON PHOTOEMISSION

Up to this point in the chapter, we have assumed that the photon energy is greater than the work function, and that a single photon has a finite probability of ejecting a photoelectron from the sample. Photons of energy less than the work function may also eject photoelectrons through a process of multiphoton photoemission. Multiphoton photocurrent density is described by the equation (Anisimov et al., 1977)

$$J = K_n I_0^n \tag{9}$$

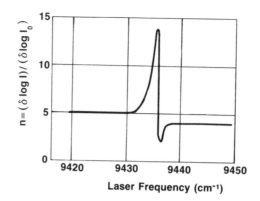

FIG. 6. Variation of the slope $n = \dfrac{\delta \log I}{\delta \log I_0}$ as a function of the laser frequency. (Lompre et al., 1978)

where J is the photoelectron current density, I_0 is the photon intensity, and n is the next integer larger than $\Phi/h\nu$. Multiphoton photoelectron emission has been demonstrated for gold with $n = 2$ and 3 (Logothetis and Hartman, 1967; Marcus et al., 1986) and with $n = 4$ and 5 (Lompre et al., 1978). Although n is generally interpreted as representing the number of photons involved in the emission of one photoelectron, in the vicinity of the transition from $n = m$ to $n = m + 1$, n changes in a complex way. Figure 6 shows a transition from $n = 5$ to $n = 4$ as the laser source is tuned through the transition wavelength of 1.06 μm (Lompre et al., 1978).

The constant K in Eq. (9) contains all factors entering into the physical process and has been measured for two- and three-photon photoemission for gold; data are shown in Table I. While the reason for the large discrepancy between nanosecond and subpicosecond excitation in the three-photon photoemission data is not understood, data from two-photon photoemission from silicon (Bensoussan et al., 1981) show that quantum yield increases with decreasing n for the same incident photon flux.

C. Photoemission and Surface Texture

Photoemission increases from a roughened surface. The effect is similar to surface-enhanced Raman scattering and is caused by enhancement of the optical power density due to plasmon resonance. This enhancement can reach a value as high 10^7 for 5-nm spheres of silver (Kerker, 1985). Measurements on 2-nm silver particles in an aerosol show a factor of 100 enhancement (increase in the parameter C, Eq. (8)) for single-photon photoemission (Schmidt-Ott et al., 1980). Single-photon photoemission

TABLE I: Typical Parameters for Ultrashort Pulse Photoemission

n (# photons)	Photon energy (eV)	Photoelectron current (pA)	Photon fluence (s^{-1})	Photon pulse duration	I_0 (MW/cm²)	P_0 (mW)	$K \left(\dfrac{A/cm^2}{[MW/cm^2]^n} \right)$	Y	Reference
1	5	2.0	1×10^{13}	1.5 ps		.008		10^{-6}	Clauberg et al., 1986
2	3.5			40 ns	0.3×10^{-3} 2.1×10^{-3}		2.35×10^{-3}	2×10^{-11} 2×10^{-9}	Logothetis and Hartman, 1967
3	1.8			40 ns	.24 .98		1.02×10^{-7}	1.1×10^{-14} 1.8×10^{-13}	Logothetis and Hartman, 1967
3	2.0	0.1	7.5×10^{15}	80 fs	500	2	6.1×10^{-15} 6.1×10^{-11}	8×10^{-11} 8×10^{-7}	Marcus et al., 1986[a] Weiner et al., 1986[b]
3	2.3	1.0	9×10^{16}	0.5 ps	300	30	1.8×10^{-14}	6×10^{-11}	Bokor et al., 1986.

[a] Smooth gold surface.
[b] Rough gold surface.

from gold-metal lines is also enhanced (by a factor of 10) when photoemission occurs at the edge of the line rather than on the flat surface (Halbout et al., 1987).

A significantly larger increase in photoemissive yield has been found from roughened gold using three-photon photoemission (Marcus et al., 1986), as shown in Table I. In these experiments, roughened gold was prepared in three ways:

1. 300–nm gold was deposited on a lithographically defined, two-dimensional feature array (period = 300 nm) on a silicon substrate (Liao et al., 1981).

2. 500–nm gold was deposited on unpolished GaAs with sharp features with radii smaller than 10nm.

3. 500–nm gold on polished GaAs was electrochemically roughened to produce features with dimensions of 50 to 100 nm.

All three preparations behaved similarly, giving a yield 10^4 times higher than that obtained from a nominally smooth gold surface. A tenfold enhancement is also found at edges of gold lines with three-photon photoemission.

D. Space-Charge Limitation on Photoemissive Current

The yield of photoelectrons can be calculated from Eqs. (8) and (9) for single and multiphoton photoemission, respectively. The total photoelectron current available to the analyzer, however, may be less than the calculated amount because of a space-charge limitation at the cathode. Electrons leaving the cathode in sufficient numbers form a negative cloud immediately above the cathode, repelling all electrons below a critical energy and preventing them from reaching the analyzer. A similar problem is unlikely to occur at the anode, which (in sampling experiments) is an extended structure compared with the cathode source. It is interesting to compute the magnitude of this space-charge limited current for photoelectrons emitted during one laser pulse, and to compare this value with the measured photocurrent for various photoemission experiments.

The space-charge limited current can be calculated from Poisson's equation by deriving the relationship between space charge and voltage drop across the charge cloud. Integrating Poisson's equation in one dimension

$$\frac{d^2V}{dz^2} = -\frac{\rho}{\varepsilon_0} \tag{10}$$

yields

$$V = A + Bz - \frac{1}{2}\frac{\rho}{\varepsilon_0}z^2 \qquad (11)$$

where V is the potential at a distance z from the cathode, ρ is the electron charge density (assumed constant), and ε_0 is the permittivity of free space. A one-eV electron travels a distance $s = 5.9 \times 10^{-6}$ cm in 0.1 ps (the duration of the photon pulse used in the photoemission studies of Marcus et al., 1986). Assuming a negligible applied field over this distance, the voltage drop due to space charge is

$$\Delta V = \frac{Qs}{2A\varepsilon_0} \qquad (12)$$

where Q is the emitted charge, s is the extent of the electron cloud along the z direction, A is cross-sectional area (area of emission), and $\rho = Q/sA$. The space-charge limited current (I_{max}) becomes

$$I_{max} = Qf = \frac{2\Delta V A \varepsilon_0 f}{s}, \qquad (13)$$

where f is the repetition rate (frequency) of the pulses. Assuming a 2-μm diameter spot, a voltage change ΔV of 1 V over the z distance of the electron cloud, and a repetition rate f of 10^8 s^{-1},

$$I_{max} = \frac{9 \times 10^{-21}}{\tau} \qquad (14)$$

where τ is the time duration of the laser pulse in seconds and I_{max} is in amperes.

Photoemissive measurements have been made of electrical signals on gold with laser pulses of 80 fs (Marcus et al., 1986), 0.5 ps (Bokor et al., 1986), and 1.5 ps (Clauberg et al., 1986); the corresponding space-charge limited current values are 110, 18, and 6 nA, respectively. Photoemissive currents have generally been well below these threshold limits.

III. Signal Detection and Waveform Measurement

A block diagram of typical experimental apparatus is shown in Fig. 2. The apparatus consists of five subsystems: (1) a source of ultrashort laser pulses and an optical system for focusing these pulses onto the DUT, (2) some means for providing electrical drive signals to the DUT and for synchronizing these drive signals with the laser pulses, (3) a vacuum system, (4) an electron-energy analyzer and detector, and (5) instrumentation for processing the measured waveform. In the following, each subsystem is discussed individually.

A. The Ultrashort Pulse Laser

The laser must provide pulses sufficiently short to ensure an adequate temporal resolution and sufficiently intense to yield a detectable photoelectron current. Either ultraviolet lasers (single-photon photoemission) or visible lasers (multiphoton photoemission) may be used for this purpose. Ultraviolet laser pulses generally provide higher quantum efficiency for photoelectron generation and better spatial resolution. Visible lasers offer shorter pulse durations and higher peak intensities. To date, three laser systems have been used for photoemissive sampling studies; typical experimental parameters are listed in Table I.

Single-photon sampling experiments have been performed using a specially designed laser system that provides 1–2-ps pulses with a 5-eV photon energy (Clauberg et al., 1986). The laser system consists of a 2.5-eV synchronously pumped dye laser, cavity dumped at a 4-MHz pulse-repetition rate, and a second harmonic generation cell (Kash et al., 1987). The system delivers approximately 10^{13} UV photons/s. Focused onto aluminum, these pulses produce photoelectron emission with a quantum efficiency of 10^{-6}, for an average photoelectron current of 2 pA.

Multiphoton photoemissive sampling measurements have been performed using a pulse-compressed, frequency-doubled Nd:YAG laser and a colliding-pulse–mode-locked (CPM) ring dye laser. The pulse-compressed Nd:YAG system (Johnson et al., 1984) produces pulses of less than 500-fs duration, at a photon energy of 2.3 eV, a repetition rate of 100 MHz, and an average power of 200 mW. In typical experiments (Bokor et al., 1986), 30-mW power was focused onto a gold surface, producing a one-pA photoelectron current. Similar pulse-compressed Nd:YAG laser systems are available commercially.

The CPM laser, invented in 1981 (Fork et al., 1981), has produced pulses as short as 27 fs, the shortest pulses to date generated directly from a laser (Valdmanis et al., 1985). In sampling experiments (Marcus et al., 1986; Weiner et al., 1987a), pulse durations of approximately 80 fs were used, with a 2.0-eV photon energy, a 117-MHz repetition rate, and a few mW average power. These parameters yielded photocurrents of approximately 0.1 pA. Higher photoelectron currents (10–100 pA) were obtained with the increased peak intensity available from a cavity-dumped CPM laser operating at a one-MHz repetition rate (Weiner et al., 1986). The

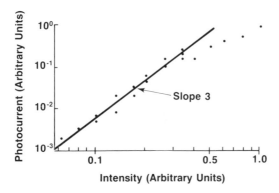

FIG. 7. Photoelectron current from a roughened gold surface as a function of optical intensity. The slope of the curve (slope = 3) is consistent with the three-photon photoemission expected for 2.0 eV photons. (Marcus et al., 1986)

third-order dependence of photocurrent on optical intensity, shown in Fig. 7, confirms the three-photon photoemission process.

In experiments, pulses from any of these lasers are admitted into the vacuum chamber through a window and focused by a microscope objective lens onto the sample. The same objective lens could be used as part of a system for viewing the surface of the DUT. In a more sophisticated setup, the probe beam could be rastered over the surface of the DUT using high-speed beam deflectors or by mechanical motion of the sample stage.

B. THE ELECTRICAL DRIVE SIGNAL

All sampling measurements require a repetitive test signal synchronized to the gating waveform. In photoemissive sampling, the drive signals for the DUT must be synchronized with the ultrashort laser probe pulses. Synchronism may be achieved by deriving the drive signals from a high-speed photodiode illuminated by the ultrashort pulses or by using a common clock to drive the DUT and laser.

The block diagram shown in Fig. 2 assumes synchronism is obtained using a high-speed photodiode mounted outside the vacuum chamber. In its simplest embodiment, electrical pulses from the photodiode are fed into the chamber and connected via cable directly to the DUT. If a more complex drive sequence is required, the photodiode signal triggers user-supplied electronic circuitry, the outputs of which are then fed into the vacuum chamber. High-speed photodetectors with impulse responses below 5 ps have been demonstrated (Wang and Bloom, 1983). However,

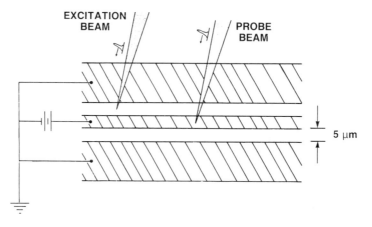

FIG. 8. Excitation and probe geometry for a 50-Ω, gold coplanar stripline on polished GaAs. (Weiner et al., 1987a)

such pulses would broaden to several tens of picoseconds in the cable, which feeds from the photodiode to the vacuum chamber.

The electrical drive signal may be generated photoconductively, using a photoconductive switch on or close to the DUT. Photoconductive generation of ultrafast electrical pulses is reviewed in Chapter 3. Figure 8 illustrates the geometry used for photoemissive sampling measurements of electrical transients generated photoconductively on a 5-μm coplanar transmission line on GaAs (Weiner et al., 1987a,b). Specifically, a 100-mV electrical step is generated using sliding contact geometry (Ketchen et al., 1986), in which the photoconductive excitation beam is focused between the biased center line and one of the ground planes. The photoemissive probing beam was focused onto the center line at a spot 40 μm distant from the source of the photoconductive excitation.

This approach is advantageous, since subpicosecond electrical risetimes can be obtained and because the close proximity of excitation and probing locations minimizes dispersive broadening of the test signal. A drawback to this approach is that special structures must be designed into the DUT to permit photoconductive excitation and that specially treated (e.g., ion-bombarded) materials are required to obtain picosecond electrical pulses for impulse-response measurements. (For step-response measurements, no special material treatment is necessary.) Alternatively, signals can be generated using a photoconductive switch on a separate chip adjacent to the DUT (Valdmanis and Mourou, 1986).

Synchronism may also be obtained by mode-locking the laser and driving the DUT with signals derived from a common clock. This approach has been used in electro-optic sampling measurements performed with compressed Nd:YAG laser pulses and is described in detail in Chapters 4 and 5. In this approach, timing jitter between the clock and actual laser pulses can degrade temporal resolution; therefore, timing jitter must be minimized.

C. The Vacuum System

The vacuum system must enclose the DUT, the electron-energy analyzer and detector, and (optionally) optics used to focus the laser beam. Optical and electrical access into the chamber must be provided. Only moderate vacuum levels are required for photoemissive sampling measurements; contamination caused by electron-beam–induced hydrocarbon polymerization (a problem with electron-beam sampling methods) is not a problem with photoemissive sampling. To date, experiments have used vacuums in the range 10^{-6} to 10^{-7} Torr.

D. The Electron-Energy Analyzer

In photoemissive sampling, voltage measurements are based on shifts in the photoelectron-energy distribution as a function of voltage applied to DUT. Hence, the electron-energy analyzer is a crucial component of the measurement system. Because photoelectron spectra used for probe studies are structureless (see Section II.A) and are otherwise similar to secondary-electron distributions used in electron-beam probing, electron-energy analyzers used for the two types of probe studies are also similar. Several types of analyzers were investigated in connection with scanning electron microscopy and electron-beam probing; these analyzers are discussed in Chapter 6. Although similar functions are required for analyzers used in electron-beam and photoemissive probing, there are differences. Photon optics requires that an objective lens be placed in close proximity to the DUT (usually within one cm) to obtain μm or sub-μm spatial resolution. This requirement, which does not apply to electron-beam probing, places constraints on the analyzer geometry. Next we discuss two analyzer designs that have been used with photoemissive sampling.

1. Retarding-Field Analyzer

A compact, parallel-grid, retarding-field analyzer is pictured schematically in Fig. 9 (Bokor et al., 1986). An extraction grid is placed approximately 750 μm from the sample and biased positive as high as 4 kV to provide an extraction field up to 53 kV/cm. A retarding grid placed behind the extraction grid performs the retarding-field energy analysis. A collector

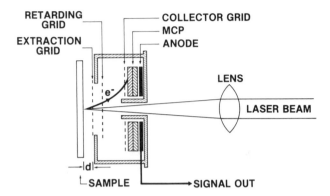

FIG. 9. Schematic diagram of retarding-field electron-energy analyzer. (Bokor et al., 1986)

grid directs the remaining electrons to a microchannel plate (MCP) detector for electron multiplication and detection. The laser beam is focused through a center hole in the MCP and through the grid spaces in the retarding and extraction grids.

The characteristic analyzer "S-curves" are shown in Fig. 10 (Bokor et al., 1986). The retarding-field analyzer acts as a high-pass energy filter. For low sample voltages, electrons are passed by the retarding grid; but for high sample voltages, electrons are rejected. The retarding-grid volt-

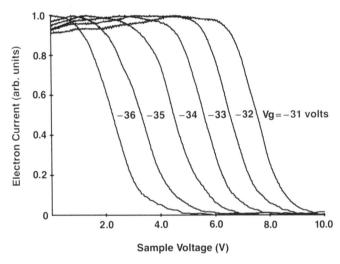

FIG. 10. S-curve characteristics of the retarding-field electron-energy analyzer. Photoelectron current is plotted versus sample potential, for various retarding grid voltages V_g. (Bokor et al., 1986)

age needed to extinguish the photoelectron current should ideally be equal to the sample voltage plus the width of the electron-energy distribution (~2 ev). Experimentally, there was a significant offset caused by a slight penetration of the extraction field through the retarding grid. However, the relationship between the sample voltage and the extinguishing retarding-grid voltage remained linear.

2. Electrostatic-Deflection Analyzer

A compact analyzer device based on an electrostatic deflection filter (Wells and Bremer, 1969) is depicted in Fig. 11 (Weiner et al., 1987a). A 500-mesh (500 holes/inch) wire grid, biased positive up to 3 kV and held approximately 320 μm above the DUT, establishes an extraction field. A 100-μm square hole is opened in the mesh to permit passage of the optical beams. Two deflection electrodes are used to deflect electrons away from the objective lens and sideways toward the microchannel plate detector. One deflection electrode is mounted opposite the MCP and biased negative at several hundred volts; the second, which includes a transparent conduction tin–indium oxide film on glass substrate, is mounted directly below the objective and biased positive at several hundred volts. A positively biased attractor grid directs the deflected photoelectrons to the MCP.

The sample potential is derived by using the influence of the sample potential on the required deflection voltages. Figure 12 shows the dependence of the collected photocurrent on the sample potential for various

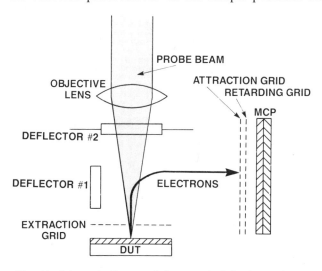

FIG. 11. Schematic diagram of electrostatic deflection analyzer.

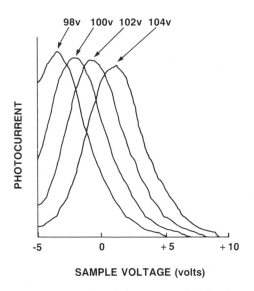

FIG. 12. Bandpass filter characteristics of the electrostatic deflection analyzer. Photoelectron current is plotted versus sample potential for various transparent deflector voltages. (Weiner et al., 1987a)

transparent deflection electrode voltages. The filter is bandpass with a fullwidth of 5V, and the photocurrent versus sample voltage curve shifts linearly with transparent deflector voltage. Improvements in the design and construction of the analyzer should yield a filter width limited only by the width of the photoelectron distribution. With a retarding grid mounted in front of the MCP, performing a retarding-field energy analysis was also possible. Time-resolved data (described shortly) obtained by retarding-field analysis were similar to those obtained by electrostatic deflection analysis.

E. Voltage Measurement and Waveform Acquisition

Derivation of the sample potential may be accomplished in two ways. In the first approach, the analyzer is adjusted so the photoelectron current detected at the MCP varies linearly with sample potential. The optical excitation beam used to generate the test waveform is chopped (see Fig. 2), and the resultant modulation of the probe photoelectron current is detected after the MCP using a sampling resistor and a lock-in amplifier. The voltage is determined from the measured current modulation and from the slope of the photocurrent versus sample potential curve. A linear response is obtained when the voltage swing is less than the width of this curve.

To obtain a linear voltage measurement over a wider dynamic range, a constant current feedback circuit similar to that used in e-beam probing may be used (Bokor et al., 1986; Clauberg et al., 1986; Gopinath and Sanger, 1971). As implemented in Bokor et al. (1986), the circuit varies the retarding field to maintain operation at the inflection point of the analyzer S-curve. The feedback signal to the retarding grid is used as output. To reduce sensitivity to laser-intensity fluctuations, a 30-kHz voltage modulation was added to the retarding grid voltage; a lock-in amplifier demodulates the MCP current modulation at twice this frequency. The feedback circuit attempts to maintain a null output from the lock-in.

Waveform acquisition requires repetition of the sampling operation with the laser probe arriving at different relative phases with respect to the electrical waveform being measured. The timing of the probe is typically adjusted by using a stepper motor-driven translation stage. Time-resolved measurements are acquired by repetitively scanning the translation stage and storing the lock-in output in a signal averager synchronized to the stage. Alternatively, the time delay of the probe pulse may be scanned by running the DUT at a frequency equal to a harmonic of the laser repetition frequency, plus a small offset. In this case, the relative phase of probe and waveform will scan automatically at a rate determined by the frequency offset.

IV. Performance of Photoemissive Sampling

A. VOLTAGE-SENSITIVITY

The ultimate voltage-sensitivity that may be achieved with photoemissive sampling is limited by shot-noise. The minimum detectable voltage V_{min} is, therefore, measured as

$$V_{min} = \Delta V \sqrt{\frac{2e\,\Delta f}{\eta I_0}}, \tag{15}$$

where I_0 is the average photoelectron current, Δf is the measurement bandwidth, and ΔV is the width of the measured electron-energy distribution. η accounts for the quantum efficiency of the detector and for any electrons lost in the analyzer. The voltage-sensitivity, usually described in units of $mV/\sqrt{Hz}$, may be optimized by increasing I_0 and by minimizing ΔV. I_0 typically ranges between 0.1 and 10 pA. ΔV is determined by the photoelectron energy distribution and the analyzer resolution; in the best case, ΔV is equal to the width of the energy distribution, typically 1–2 volts. An experimental sensitivity of one $mV/\sqrt{Hz}$ was reported in Clauberg et al. (1986), with $I_0 \approx 1.6$ pA and $\Delta V \approx 1$V. In comparison, the shot-noise equation gives $V_{min} \approx 450\,\mu V/\sqrt{Hz}$. The shot-noise limit, Eq. (15), determines the best voltage-sensitivity possible. Probe-laser-intensity fluctuations can increase the photoelectron current noise above the shot-noise limit and degrade the sensitivity. The effect of laser-intensity fluctuations is more severe for multiphoton photoemissive sampling than it is for the case of single-photon emission. The noise spectrum of mode-locked lasers often approaches the shot-noise limit at frequencies of several MHz and above; therefore, by using high-frequency modulation techniques (Heritage and Allara, 1980), it should be possible to achieve shot-noise–limited voltage-sensitivity. High-frequency modulation techniques have also been used to improve signal-to-noise ratio in electro-optic sampling.

B. Temporal Resolution

The temporal resolution of photoemissive sampling is determined by the laser pulsewidth, timing jitter between the probe and the waveform under test, the generation time for photoelectrons, and the photoelectron transit time.

As noted in Table I, the pulse durations of lasers used for photoemissive sampling have been approximately one ps and below. Hence, the laser pulse duration does not seriously limit temporal resolution.

Timing jitter can be made negligibly small by deriving the electrical drive signal from a photodiode or photoconductive switch triggered by a portion of the probe-beam energy. Timing jitter can be a problem when synchronism is established by mode-locking the laser and driving the DUT with signals derived from a common clock. One solution is to build a phase-locked loop to reduce the laser jitter. In this way, the timing jitter of compressed Nd:YAG laser pulses has been reduced to less than one ps (Rodwell et al. 1986).

The generation time for photoelectrons depends on the time for incident photons to be absorbed and the time required for an electron generated within an escape depth of the surface to move to the surface. In Menzel and Kubalek (1983), the escape time is estimated to be on the order of 10^{-14} to 10^{-13} seconds from a depth of 150 Å.

Temporal resolution is limited primarily by the transit time of photoelectrons out of the region of time-varying local electric fields. If the sample potential changes during transit, the electrons will see a time-varying electric field; the energy of photoelectrons arriving at the analyzer will be related to some average sample potential. The transit-time effect was analyzed theoretically in connection with electron-beam testing (Fujioka et al. 1985) and observed experimentally in an electron-beam measurement of a one-GHz sine wave (Nakamae et al., 1986). With proper parameters, the transit time can be reduced to the picosecond range, as observed in the first photoemissive sampling measurements (Marcus et al., 1986; Bokor et al., 1986).

1. Transit-Time Effect: Simple Model and Experimental Results

We first discuss the transit-time effect in terms of an "effective" transit time, defined as the time required for electrons to traverse the region of space in which the field is sensitive to the sample potential. The effective transit time T is given by

$$T = \frac{\sqrt{2m}}{e\mathscr{E}} (\sqrt{e\mathscr{E}S + E_0} - \sqrt{E_0}), \tag{16}$$

where $\mathscr{E}$ is the extraction field, m and e are the electron mass and charge, S is the characteristic distance over which the field due to the sample extends, and E_0 is the initial electron kinetic energy in the direction of the extraction field. The transit time may be reduced by increasing the extraction field, by using photoelectrons with higher initial energies, and by minimizing the distance S. In a real circuit, the spatial extent of the local electric fields depends on the widths and spacings of the metal lines being probed; the characteristic distance S may be much smaller than the distance to the extraction electrode (Weiner et al., 1986, 1987a,b; Fujioka et al., 1985; Clauberg et al., 1987). Thus, the transit time is reduced and the temporal resolution improved when examining structures with signal lines spaced close to ground planes or return lines. Figure 13 shows a plot of the transit time for one-eV electrons, as a function of $\mathscr{E}$ and S.

Several experiments verified the improvement in temporal resolution as a function of increasing extraction field (Marcus et al., 1986; Bokor et al., 1986). The maximum permissible value for the extraction field is set by tolerance to capacitive loading, device-sensitivity to an imposed field, electrical breakdown between the test line and the extraction grid, and is generally less than 100 kV/cm. The effect of increasing extraction fields up to 53.3 kV/cm on the waveform measurement is shown in the data of Fig. 14, which shows photoemissive sampling measurements of 60-ps voltage pulses on a gold microstrip transmission line (Bokor et al., 1986). The sample to extraction grid spacing is 750 μm. For comparison, Fig. 14d shows a measurement of the waveform performed by photoconductive sampling. The deconvolved transit times for Fig. 14a–c are 121, 54, and 39 ps, respectively, for fields of 6.7, 26.7, and 53.3 kV/cm. These results are in agreement with the transit times predicted by Eq. (16).

The characteristic transit distance S may be decreased either by moving

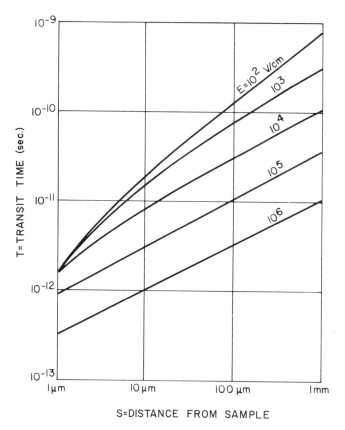

S=DISTANCE FROM SAMPLE

FIG. 13. Transit time as a function of distance from sample calculated from Eq. (16) with $E_0 = 1$ eV. (Weiner et al., 1987b)

the extraction grid close to the DUT or by using circuit geometries with signal lines close to ground planes. The first approach was used in Blacha-et al. (1987). With a sample-grid separation of 70 μm and an extraction field on the order of 100 kV/cm, the temporal resolution was estimated to be below 10 ps. This was sufficient to observe dispersion of photoconductively excited voltage pulses on microstrip transmission lines. However, because the extraction grid to sample spacing was smaller than the transmission-line width, the extraction grid may load the transmission line and contribute to the anomalously high dispersion observed. This possibility remains a sub-ject for further investigation.

In a circuit with closely spaced signal and ground lines, S can be much smaller than the distance to the extraction grid (Weiner et al., 1986, 1987a,b; Fujioka et al., 1985; Clauberg et al., 1987); this circumvents the

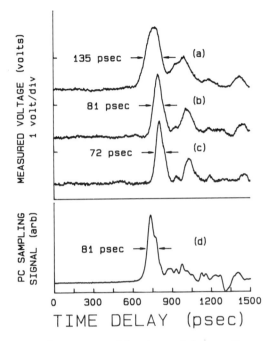

FIG. 14. Voltage waveforms measured by photoemissive sampling on a gold microstrip line, with extraction fields of (a) 6.7 kV/cm, (b) 26.7 kV/cm, and (c) 53.3 kV/cm. The waveform in (d) was measured by photoconductive sampling. (Bokor et al., 1986)

need for a closely spaced extraction grid. The effect of a particular sample geometry on S and on the transit time T is illustrated by Fig. 15 and 16 (Weiner et al., 1987c). We consider a hypothetical circuit, which consists of two metal lines, each one μm wide and separated by one μm; and we assume an extraction field of 100 kV/cm. The equipotential lines for this coplanar microstrip geometry have been calculated and are plotted in Fig. 15. The solid lines are the equipotential lines corresponding to electrodes at 0V and 1V; the dashed lines correspond to both electrodes at 0V. The effect of the changing electrode voltage diminishes as the distance from the circuit increases. Figure 16 shows the apparent voltage change sensed by an electron Z micrometers away from and directly above the ($Y = 1 \mu$m) electrode that experiences the 1V potential change. At one μm away from the circuit, the influence of the changing sample potential is reduced by more than 40%; at $Z = 2 \mu$m, the effect is reduced to 20%; and at 5μm, to less than 10%. Figure 16 also shows the time required for a 0.01-eV electron, emitted from the $Y = 1 \mu$m metal line in Fig. 15, to travel a fixed distance away from the surface of the circuit. A transit time of

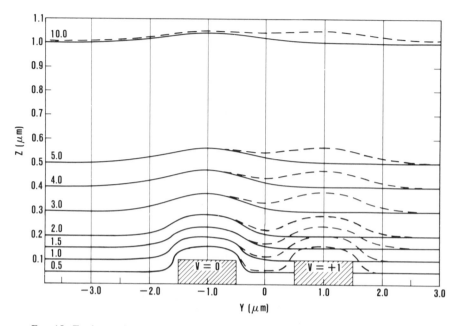

FIG. 15. Equipotential lines for one-μm–wide coplanar lines separated by one μm. The extraction field is 100 kV/cm. Solid curves correspond to electrodes at 0V and 1V, respectively; dashed curves correspond to both electrodes at 0V. (Weiner et al., 1987c)

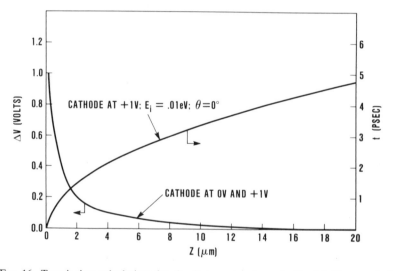

FIG. 16. Transit-time calculations for circuit geometry shown in Fig. 15. The transit time for an electron to arrive Z μm above the surface is shown, together with the apparent voltage change sensed by an electron at that point. (Weiner et al., 1987c)

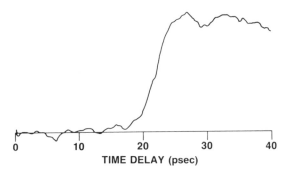

FIG. 17. Time-resolved measurement for 100-mV electrical step, with extraction field of 85 kV/cm. The risetime is 5.3 psec. (Weiner et al., 1987a)

1.5 ps corresponds to $Z = 2$ μm and a voltage uncertainty of 20%; a transit time of 2.5 ps corresponds to $Z = 5$ μm and a measurement accuracy better than 90%.

Photoemissive sampling measurements of voltage transients on coplanar transmission lines confirm that temporal resolution is improved compared to measurements on microstrip lines (Weiner et al., 1986, 1987a,b). The best results were obtained with electrical-transients generated photoconductivity on a 5-μm, gold coplanar stripline on gallium arsenide (Weiner et al., 1987a,b) (See Fig. 8). A time-resolved measurment is shown in Fig. 17 for an extraction field of 85 kV/cm and a grid spacing of 320 μm. The 10–90% risetime is 5.3 ps. With weaker extraction fields, the risetimes are slower; for example, at 30 kV/cm, the risetime is 7.5 ps. The risetimes are in reasonable agreement with transit times calculated according to Eq. (16), using S in the range 10–15 μm. These results confirm that temporal resolution is improved when the dimensions of the structure under examination are reduced. Subsequent measurements with the photoelectron–electron-beam sampling technique (May et al., 1987) obtained similar risetimes, also using small dimension coplanar lines.

A final way to decrease the transit time would be to increase E_0, the initial photoelectron kinetic energy. Figure 18 shows a plot of transit time as a function of extraction field and initial kinetic energy, with S fixed at 5 μm. It is assumed that the initial electron velocity is entirely in the direction of the extraction field. As E_0 is increased, the extraction field required to obtain a given transit time is reduced drastically. For example, with $E_0 = 3$ eV, a 5-ps transit time can be achieved with essentially zero field. Initial energies in the range 2–4 eV could be obtained by using compressed and frequency-tripled (or quadrupled) Nd:YAG laser pulses to stimulate two-photon emission from gold or aluminum lines.

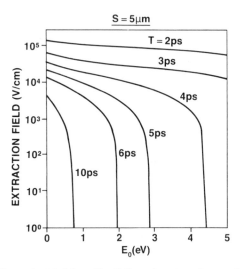

Fɪɢ. 18. Transit time calculated from Eq. (16)—a function of extraction field—and initial kinetic energy, with S = 5 μm.

2. Transit-Time Effect: Numerical Simulations

Numerical stimulations of the transit-time effect in electron-beam sampling and in photoemissive sampling were performed respectively by Fujioka (Fujioka et al., 1985) and by Clauberg (Clauberg et al., 1987; Clauberg, 1987). In both analyses, photoelectron (or secondary-electron) trajectories were computed in the presence of the quasi-static electric field caused by to the DUT and the extraction voltage. In Fujioka et al. (1985), a distribution of initial secondary-electron energies and angles was used; in Clauberg et al. (1987) and Clauberg (1987), electrons with fixed initial energies, ejected in a direction perpendicular to the DUT, were assumed. The effects of extraction field, sample geometry, initial electron energy, signal risetime, and analyzer type were investigated in these simulations. The effects in electron beam and in photoemissive sampling are similar, except that initial electron energies are slightly higher in the electron-beam approach. The numeric results are generally consistent with the simple model presented in the previous section.

The calculation of the transit-time effect and the influence of sample geometry depend on potential distribution between the DUT and the extraction grid. According to Clauberg et al. (1987), the potential $\phi(\bar{r}, t)$ can be written as a linear superposition of individual potential distributions $\phi_i(\bar{r}, t)$ due to the extraction grid and to the various distinct conductors on the DUT. That is,

$$\phi(\bar{r}, t) = \sum_i \phi_i(t) R_i(\bar{r}) \tag{17}$$

where $\phi_i(t)$ is the time-dependent potential on conductor i and $R_i(\bar{r})$ is the potential distribution that occurs when the potential is unity at conductor i and zero at all other conductors.

The effect of sample geometry on the response functions R_i is illustrated by Figure 19a (Clauberg et al., 1987). Response functions are shown for a macroscopic geometry with an infinitely large plate sample, for a one-μm–wide microstrip line over a ground plane, and for a coplanar stripline consisting of a signal line placed between two neighboring ground lines, with 1-μm widths and spacings. The extraction grid is 200 μm above the

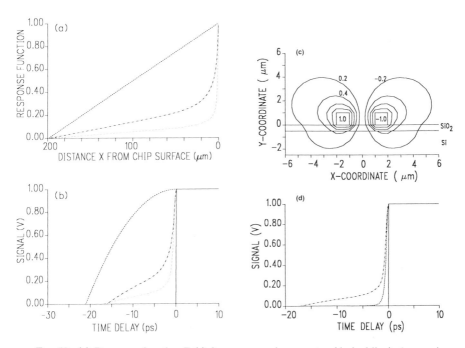

FIG. 19. (a) Response function $R_i(r)$ for macroscopic geometry (dashed-line), 1-μm microstrip line (dash-dot), and coplanar three-line structure with 1-μm lines and spacings (double-dotted). (b) Simulated sampling measurement of an abrupt electrical step on the three structures listed in (a). The input signal here and in (d) is shown as a solid line. (c) Equipotential lines for one-μm–wide coplanar lines separated by 2 μm. (d) Simulated sampling measurements of an electrical step on coplanar microstrip line (two-line structure: dashed) and coplanar three-line structure (dash-dot). (Clauberg et al., 1987; Clauberg, 1987)

DUT. Simulated sampling measurements of an electrical step are shown in Fig. 19b for an extraction field of 50 kV/cm (Clauberg et al., 1987). As seen, the temporal resolution improves for Ri which decay rapidly away from the DUT. Similar calculations were performed by Fujioka (Fujioka et al., 1985), who found that increased temporal resolution could be achieved as the dimensions of coplanar striplines were reduced. Further improvements are obtained by considering a coplanar microstrip line, consisting of two one-μm electrodes spaced by one μm. Figure 19c shows the two-dimensional response function for this structure, and Fig. 19d shows the simulated sampling measurement. The calculated risetime for the coplanar microstrip is below one ps, and compared to the three-line coplanar stripline, the foot on the leading edge is greatly reduced. Similar calculations for coplanar microstrip lines were shown earlier in Figs. 15 and 16.

The simulations confirm that an increasing extraction field increases temporal resolution, as discussed in Section IV.B.1. They also reveal that the transit-time effect can cause an apparent temporal shift of a measured signal, even when no degradation of the risetime is observed. Figure 20a

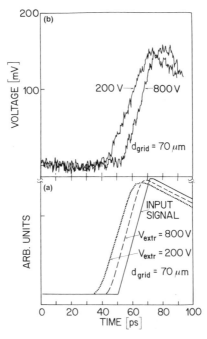

FIG. 20. (a) Simulated sampling measurement of input pulse with assumed 17-ps risetime (on microstrip transmission line). (b) Experimental data performed under conditions similar to those in (a) (Blacha et al., 1987).

shows simulated measurements of an input pulse with an assumed 17-ps rise and 150-ps fall on a microstrip transmission line, for extraction voltages of 200 and 800V. Although the risetimes are approximately the same as that of the input signal, the measured wave form is shifted substantially. Voltage waveforms measured under similar conditions are shown in Fig. 20b; these data demonstrate the predicted temporal shift (Blacha et al., 1987).

C. Spatial Resolution

The spatial resolution of the photoemissive sampling technique is governed by the spot size of the focused laser beam and by the effect of local electric fields near the surface of the DUT. These topics are addressed separately next.

1. Spot Size

The laser spot size must be sufficiently small that it can address a single desired metal line, without impinging on neighboring lines. Furthermore, light from the periphery of a spot larger than the metal linewidth could cause photogeneration of carriers and charge injection into active regions of nearby devices. This also must be avoided.

The diameter of the focused laser beam is limited by the wavelength of light. In the case of uniform illumination of a circular aperture lens, the diameter D of the first Airy disk is given by Hecht and Zajac (1976):

$$D = \frac{1.22\lambda}{\text{NA}},\qquad(18)$$

where λ is the optical wavelength and NA is the numerical aperture of the lens. The maximum numerical aperture for a lens in air (or vacuum) is unity, and $D \simeq 0.75$ μm for 2-eV photons and 0.3 μm for 5-eV photons.

Clearly, submicron spot sizes are easier to achieve with single-photon photoemission (e.g., 5-eV photons) than with multiphoton emission (e.g., 2-eV photons). In the case of multiphoton emission, however, the addressing capability is better than the spot size (Weiner et al., 1987c). The nonlinear dependence on optical intensity ensures that only the most intense portion of the focused spot will contribute effectively to photoemission. For three-photon emission and a Gaussian beam profile, the effective spot radius $r/\sqrt{3}$ is smaller than the actual radius r.

The possibility of charge injection into the semiconductor substrate depends on the laser spot size and the metal linewidth. The restrictions imposed by the requirement to avoid charge injection will be discussed in Section V.B.1.

To achieve high spatial resolution, the electron-energy analyzer must be compatible with the short working distance (<1 cm) of a high-power optical objective lens. As of the late 1980s, the smallest reported structure probed by photoemissive sampling was a coplanar stripline with 5-μm linewidths (see Fig. 8) (Weiner et al., 1987a,b). In that experiment a second, "excitation" laser beam was focused onto the gap between the signal line and the adjacent ground plane; the photoelectron current due to the probe beam exceeded the background photoelectron current generated unintentionally by the excitation beam by more than three orders of magnitude.

2. Local-Field Effects

An additional factor that affects spatial resolution is the presence of local fields above the surface of the DUT. Microfields, arising from different potentials at neighboring lines on the integrated circuit, cause crosstalk and measurement inaccuracy. Local-field effects may be categorized as static or dynamic.

Effects caused by static local electric fields are well documented in the electron-beam–probing literature (Menzel and Kubalek, 1983; Nakamae et al., 1981). There are two main problems:

1. A potential barrier, which can reflect low energy electrons, can form above positively biased lines. This modifies the electron energy distribution.

2. Local fields can affect the angular distribution of emitted electrons. If the energy analyzer is sensitive to the angular distribution of incident electrons, this will influence the measurement.

Both effects can lead to measurement errors of up to several volts (Menzel and Kubalek, 1983; Nakamae et al., 1981).

The first problem can be solved by the use of a high extraction field to eliminate the possibility of a potential barrier above positively biased lines (Menzel and Kubalek, 1983). The extraction field already being used in photoemissive sampling to reduce the transit time is sufficient for this purpose.

The second problem can be addressed by using an energy analyzer that is insensitive to the electron angular distribution. A specially designed analyzer that combines a uniform, planar extraction field and a hemispherical analyzing field accomplishes this goal. With this new analyzer, the accuracy of e-beam voltage measurements is improved by a factor of 10 compared to measurements performed with a planar retarding-field analyzer (Nakamae et al., 1985). A second approach uses a magnetic extraction field to collimate the emitted electrons in the direction normal to the DUT. A planar retarding field may then be used for accurate measurements of the energy distribution. Using this technique, waveforms were measured to 2% accuracy on lines with submicron spacings (Garth et al., 1986).

Dynamic crosstalk effects, which arise because of changing local electric fields during the transit of the photoelectron away from the DUT, have been predicted and analyzed by Clauberg (Clauberg et al., 1987; Clauberg, 1987). These effects occur when the time variation of the local field is linked to changes in the potentials of lines neighboring the emitting line; the potential at the emitting line is assumed fixed. Dynamic crosstalk will be observed only when potentials on neighboring lines change on the scale of the electron transit time or faster.

Clauberg calculated the dynamic crosstalk for the three-line structure discussed in Section IV.B.2. Photoelectrons are emitted from one of the outside ground lines, which remains at zero potential. The potential on the center line rises linearly from 0 to 1V, with the risetime as a parameter. Simulated time-resolved measurements are shown in Fig. 21 (Clauberg et al., 1987). As seen, a 0.25V crosstalk signal is obtained when the signal voltage rises instantaneously; when the signal risetime is slower, the amplitude of the crosstalk is reduced.

Dynamic crosstalk may be understood as follows. The energy gained by a photoelectron emitted at conductor A as it travels to extraction grid B is determined by the line integral of the electric field along the electron's trajectory. If all electric fields remain fixed, the energy gain is equal to the potential difference. Now, suppose that the potential of a neighboring line changes abruptly when the electron is at a point C (between A and B) in its trajectory. Although the path integral from A to B remains fixed, the

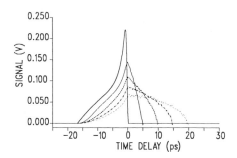

FIG. 21. Simulated crosstalk signals for input signals that rise linearly from 0V to 1V in 0 ps (solid line), 5 ps (dotted), 10 ps (dashed), 15 ps (dash-dot), and 20 ps (double-dot). The DUT consists of three one-μm lines, with a center signal line separated by one μm from ground lines on either side. (Clauberg et al., 1987)

potential difference between points C and B changes. As a result, the energy gained by the electron in its transit from A to B is modified, and a crosstalk signal is observed.

V. Discussion

A. SINGLE-PHOTON VERSUS MULTIPHOTON APPROACH

Differences between the application of multi- and single-photon photoemission to high-speed sampling are caused (1) by differences between the single and multiphoton processes and (2) by effects of variations in photon energy on the sampling process. Photon energies for single-photon photoemission must exceed the work function of the metal (4.3 and 5.1 eV for clean aluminum and gold, respectively), while the multiphoton approach can involve energies down to 2 eV for three-photon emission (Marcus et al., 1986) and following text.

Quantum yield (the number of photoelectrons per incident photon) is significantly lower for the multiphoton case. (See Table I.) The lower quantum yield for multiphoton emission means that higher photon fluences are needed to produce useful photoelectron current levels, and these high photon currents may create unwanted charge injection when narrow lines are examined. (See Section V.B.1.)

Multiphoton photoemission is an advantage when exciting electrical pulses or steps photoconductively with part of the laser probe beam. The nonlinear dependence of the photocurrent on the light intensity (Eq. (9)) means that electrical excitation can be performed at sufficiently low intensities to avoid simultaneous generation of photoelectrons. The same intensity of higher-energy photons (above the work function) would be more likely to generate photoemission at that site.

High spatial resolution is needed for locating and probing narrow metal lines on a device surface and also to avoid charge injection when the linewidth is comparable to the laser spot size. The laser spot size is determined by the wavelength and the numerical aperture of the lens. (See Eq. (18)). The minimum spot size is 0.75 μm for 2-eV photons and 0.3 μm for 5-eV photons.

A smaller spot clearly can be generated by a single-photon photoemissive source. The smaller spot and the higher quantum yield generated with the shorter-wavelength laser light, needed for single-photon photoemission, combine to permit easier interrogation of narrow linewidths near active device regions. However, as discussed in Section IV.C.1, the advantage of using a higher photon energy is partially offset because for multiphoton photoemission, the effective spot size is less than the nominal spot size. For instance, the minimum spot radius used for single-photon photoemission from gold at a photon energy of 5 eV can be 0.4 times the minimum radius for three-photon photoemission using 2-eV photons based on the wavelength difference alone. But for three-photon emission, the useful intensity falls within a spot radius $r/\sqrt{3}$ that is smaller than the nominal radius r. Thus, the effective spot size for 5-eV photons is really 0.7 times the radius for 2-eV photons.

Laser light used in multiphoton sampling is in the visible region of the spectrum, an asset in aligning the optics and manipulating the sample under the incident beam. In the absence of a second light source for illumination, (visible) laser light reflected from the sample can be imaged and used for optical alignment.

For measurements approaching the temporal resolution limits imposed by the transit-time problem, a higher initial photoelectron energy is an advantage, as discussed in Section IV.B.1. The maximum photoelectron energy for multiphoton photoemission is (after Eq. (2))

$$E_{k,\max} = nh\nu - \phi. \tag{19}$$

A wavelength can be chosen such that $nh\nu$ is a few eV larger than the work function Φ. Coupled with a high extraction field, this may permit sub-ps measurements. For this purpose, multiphoton photoemission has a significant advantage over single-photon photoemission, which becomes increasingly difficult as one moves into the deep UV region of the spectrum.

B. SPECIAL CONCERNS OF PHOTOEMISSIVE SAMPLING

1. Charge Injection

Photons can inject carriers in an active region of a device through photogeneration of electron–hole pairs. Room-temperature bandgaps of the common semiconductors are in the energy range 1.0–1.5 eV, which is less than the energy used for three-photon photoemission, so carrier injection is ensured. Because photon fluences of 10^{13} s^{-1} to 10^{17} s^{-1} are used in sampling experiments and typical device currents are in the 1–10-μA range, carrier injection clearly should be avoided.

Absorption coefficients for both gold and aluminum are above 0.06 nm^{-1} in the photon-energy range 2–5 eV. In this energy range, 95% of all light is absorbed within the first 21 nm of an aluminum film and the first 42 nm of a gold film. A typical metal-film thickness of 500 nm would clearly block all light from reaching the semiconductor. Light from the periphery of a spot larger than the metal feature size could, however, give an injected current. The relationship between spot size, metal-feature size, and injected current is now calculated.

For a Gaussian beam, the spatial-intensity distribution is

$$I(r) = \frac{P_0}{\pi w^2} e^{-r^2/w^2} \tag{20}$$

where P_0 is the total power and w is the radius at which the intensity falls to $1/e$ of its peak. The power (P_{in}) that can be focused onto a metal line of width D is given by

$$P_{in} = P_0 erf(D/2w). \tag{21}$$

When the linewidth equals the beam diameter ($2w$), 84% of the optical power can be focused onto the line. The fraction of power (P_{out}) that cannot be focused onto the line is:

$$P_{out} = P_0 erfc(D/2w), \tag{22}$$

where $erfc$ is the complementary error function. This equation can be used to compute the relation between spot size and metal linewidth in order to ensure injection of less than a critical carrier density. For example, using a photon energy and fluence of 2 eV and 7×10^{15} s^{-1}, respectively, for three-photon photoemission from gold, the linewidth must be greater than 4.02 μm to maintain the injected current at a level below 0.1 μA for a diffraction-limited spot radius of 0.75 μm (NA = 0.5). On the other hand, assuming a photon energy and fluence of 5 eV and 1×10^{13} s^{-1}, respectively, for the single-photon photoemissive sampling experiment described in Clauberg et al. (1986) and a diffraction-limited radius of 0.3 μm, features as narrow as 0.8 μm may be probed without the injected photocurrent rising above 0.1 μA. The larger quantum yield and smaller spot radius for single-photon photoemission is clearly an advantage in permitting the sampling of small features on active regions of devices.

2. Device Loading

One solution to the transit-time problem is to locate the extraction electrode close to the device so that the region above the electrode is shielded from the time-varying field of the device signal. This electrode, however, may introduce a loading on the device. The capacitance of such a system is given by

$$C = \frac{\varepsilon_0 A}{d} \tag{23}$$

where A is the electrode area, ε_0 the permittivity, and d the spacing between the parallel electrode and device surface. If the electrode is a circular disk of diameter D, and d and D are both given in μm, then

$$D_{max} = \sqrt{1.44 \times 10^{17} C_{max} d} \tag{24}$$

where C_{max} is a maximum permissible capacitance (F) and D_{max} is the corresponding maximum electrode diameter. A safe upper limit for probe capacitance for most devices is 10 fF. To allow the extraction electrode to permit temporal resolutions of better than 10 ps at a high extraction field, d should be in the region of 10 μm. An electrode spacing of 10 μm requires that the electrode diameter be less than 120 μm.

3. Limitations on Extraction Field

Two related limitations on the maximum permissible extraction field are the onset of field emission and interference with device performance. Field emission of electrons from asperities on an electrode surface begins at applied fields greater than 10^5 V/cm (Orloff, 1987). The field is enhanced at regions of high curvature. Because it is difficult to avoid asperities on any machined metal surface, 10^5 V/cm may be considered an upper limit when such surfaces are used. Smoother surfaces may permit the use of extraction fields up to an order of magnitude higher.

Field-effect devices operate with maximum fields (between source and drain) also in the 10^5 V/cm range. In sampling experiments where an extraction field is applied directly over a source-drain region without a passivating surface dielectric layer, the extraction field (reduced by the dielectric constant of the semiconductor) competes with the source-drain field. The result is that drain voltages may have to be increased to achieve current saturation. In this case, the extraction field may have to be lowered. If the extraction field is applied only to a metal line over a one-μm-thick dielectric away from an active region, then an extraction field of 10^5 V/cm can safely be used. In a real sampling experiment, however, it is likely that the electrode covers a significant amount of device area so that, no matter where sampling is performed, sensitive regions are exposed to the field. The importance of this problem depends on the device characteristics and on the presence of a passivating layer that would offer a dielectric shield against the extraction field.

VI. Summary and Future Trends

The photoemissive technique is a method for probing high-speed signals on metal lines on any material by stimulating photoelectron emission with short pulses of light and analyzing the energies of the emitted electrons. Temporal resolutions of 5 ps have been achieved with an ultimate resolution of one ps or below, depending on sample geometry, strength of extraction field, spacing between extraction electrode and DUT, and initial photoelectron velocity. A voltage-sensitivity of one mV/$\sqrt{\text{Hz}}$ has been obtained, with an ultimate shot-noise–limited sensitivity of 0.5 mV/$\sqrt{\text{Hz}}$.

Spatial resolution is determined by the spot size, which is wavelength-limited to better than one micron.

As with other probe techniques described in this book, the photoemissive method is ideal for certain applications and less suited for others. Because the probe is applied to a metal line on the surface of a chip or wafer, it may be more suitable for studies of complex devices and circuits than other methods that require clear optical paths to and from an active region. In some devices, irradiation of a large area of a device with laser light for viewing purposes is less intrusive than a rastering electron beam; photoemissive probing may therefore be more suitable for analysis of highly radiation-sensitive MOS circuits and devices than electron-beam probing. For other devices, stray light at the periphery of the probe spot must be prevented from entering the semiconductor and creating excess electron–hole pairs. Photoemission probing is not yet suitable for problems requiring sub-ps temporal resolution.

Future development and refinement of this method will probably move along three directions: improvements in the optical system, improvements in electron analyzers, and the addition of rastering capability.

Both ultraviolet lasers (single-photon photoemission) and visible lasers (multiphoton photoemission) have been used for photoemission probing. These lasers are costly and quite bulky, and replacement with a solid-state diode laser would be a significant improvement in both senses. The problem of generating short diode laser pulses of sufficient intensity at the right wavelength is formidable. Mode-locked diode lasers, suitable for either single- or multiphoton photoemissive sampling, do not yet exist.

Improvements in electron analyzers will undoubtedly be spurred on by improvements in the ability to generate shorter electron-beam pulses in electron-beam–testing programs. Electron-beam sampling technology is well advanced, and the problems in analyzer design are similar. A main concern is the limitation in temporal resolution imposed by the transit-time problem. This problem can be partly solved by design of an extraction electrode of low cross-sectional area near the DUT. In addition, for photoemissive probing, the analyzer must be optically transparent.

Crosstalk and other problems introduced by local microfields can be reduced by use of a magnetic lens arranged to maximize the field at the sample surface (Garth et al., 1986). Incorporation of such a magnetic field source in an analyzer assembly would be a major improvement.

Finally, we may briefly address the problem of generating information on the two-dimensional distribution of potentials on a device or circuit surface photoemissively. "Logic-state mapping" is a useful method for interrogating the logic states of an array area. Potential differences of 0.2V or larger can be discriminated on a video screen using electron-beam–

probe methods. (See Chapter 6.) A similar sensitivity to voltage differences is expected with photoemissive probing. A raster action can be achieved with piezoelectrically or galvanometrically driven mirrors operating on the optical beam, or by mechanical motion of the sample stage.

References

Anisimov, S.I., Benderskii, V.A., and Farkas, G. (1977). *Usp. Fiz. Nauk* **122**, 185.

Bensoussan, M., Moison, J.M., Stoesz, B., and Sebenne, C. (1981). *Phys. Rev. B* **23** (3), 992.

Berglund, C.N., and Spicer, W.E. (1964). *Phys. Rev.* **136**, A1030.

Blacha, A., Clauberg, R., and Seitz, H.K. (1987). *J. Appl. Phys.* **62**, 713.

Bokor, J., Johnson, A.M., Storz, R.H., and Simpson W.M. (1986). *Appl. Phys. Lett.* **49**, 226.

Bokor, J., Johnson, A.M., Storz, R.H., and Simpson, W.M., (1986), in *Ultrafast Phenomena V*, G.R. Fleming and A.E. Siegman, eds., Springer-Verlag, Berlin, pp. 123–126.

Broudy, R.M. (1970). *Phys. Rev. B* **1** (8), 3430.

Cardona, M., and Ley, L., eds. (1978). "Photoemission on Solids I. General Principles," in *Topics in Applied Physics*, **26**. Springer, Berlin, New York.

Clauberg, R. (1987). *J. Appl. Phys.* **62** (5), 1553.

Clauberg, R., Seitz, H.K., Blacha, A., Kash, J.A., and Beha, H. (1986). In *High Speed Electronics*, B. Kallback and H. Beneking, eds. Springer-Verlag, Berlin, pp. 200–203.

Clauberg, R., Blacha, A., and Beha, H. (1987). In *Characterization of Very High Speed Semiconductor Devices and Integrated Circuits*, Ravi Jain, ed., *SPIE* **795**, 207–213.

Einstein, A. (1905). *Ann. Phys.* **17**, 132.

Fork, R.L., Greene, B.I., and Shank, C.V. (1981). *Appl. Phys. Lett.* **38**, 671.

Fujioka, H., Nakamae, K., and Ura, K. (1985). *J. Phys. D: Appl. Phys.* **18**, 1019.

Garth, S.C.J., Nixon, W.C., and Spicer, D.F. (1986). *J. Vac. Sci. Technol.* **B4**, 217.

Gopinath, A., and Sanger, C.C. (1971). *J. Phys. E* **4**, 334.

Halbout, J.-M., Chiu, G., and May, P. (1987). IBM Research Labs, Yorktown Height, New York. Unpublished. II *Handbook of Chemistry and Physics*, 67th ed. (1986). CRC Press, Boca Raton, Florida, E-89.

Hecht, E., and Zajac, A. (1976). *Optics* Addison-Wesley, Reading, Massachusetts, 352.

Heritage, J.P., and Allara, D.L. (1980). *Chem. Phys. Lett.* **74**, 507.

Johnson, A.M., Stolen, R.H., and Simpson, W.M. (1984). *Appl. Phys. Lett.* **44**, 729.

Kash, J.A., Blacha, A., Salemink, H., Watt, R., and Mitchel, G. (1987). IBM, Zurich, unpublished.

Kerker, M. (1985). *J. Colloid and Interface Science* **105** (2), 297.

Ketchen, M.B., Grischkowsky, D., Chen, TC., Chi, C.C., Duling, I.N. III, Halas, N.J. Halbout, J.M., and Li, G.P. (1986). *Appl. Phys. Lett.* **48**, 751.

Koyama, and R.Y., Smith, N.V. (1970). *Phys. Rev. B* **2** (8), 3049.

Krolikowski, W.F., and Spicer, (1970). *Phys. Rev. B* **1** (2), 478.

Lenard, P. (1900). *Ann. Physik* **2**, 359.

Liao, P.F., Bergman, J.G., Chemla, D.S., Wokaun, A., Melngailis, J. Hawryluk, A.M., and Economou, N.P. (1981). *Chem. Phys. Lett.* **82**, 355.

Logothetis, E.M., and Hartman, P.L., (1967). *Phys. Rev. Lett.* **18** (15), 581.

Lompre, L.A., Mainfray, G., Manus, C., Thebault, J., Farkas, Gy., and Horvath, Z. (1978). *Appl. Phys. Lett.* **33** (2), 124.

Marcus, R.B., Weiner, A.M., Abeles, J.H., and Lin, P.S.D. (1986). *Appl. Phys. Lett.* **49** (6), 357.

May, P., Halbout, J.-M., and Chiu, G. (1987). *Appl. Phys. Lett.* **51**, 145.

Menzel, E., and Kubalek, E. (1983). *Scanning* **5**, 103.

Nakamae, K., Fujioka, H., and Ura, K. (1981). *J. Phys. D.* **14**, 1939.

Nakamae, K., Fujioka, H., and Ura, K. (1985). *J. Phys. E: Sci. Instrum.* **18**, 437.

Nakamae, K., Fujioka, H., and Ura, K. (1986). "Measurement of Transit Time Effect on Voltage Contrast in the Stroboscopic Scanning Electron Microscope," *Proc. XIth Int. Cong. on Electron Microscopy*, Kyoto.

Orloff, J. (1987). Oregon Graduate Center, unpublished.

Rodwell, M.J., Weingarten, K.J., Bloom, D.M., Baer T., and Kolner, B.H. (1986). *Opt. Lett.* **11**, 638.

Schmidt-Ott, A., Schurtenberger, P., and Siegmann, H.C. (1980). *Phys. Rev. Lett.* **45** (15), 1284.

Smith, N.J. (March 1971). *CRC Critical Reviews in Sol. State Science*, 45.

Smith, N.V. (1972). *Phys. Rev. B* **5** (4), 1192.

Thompson, J.J. (1899). *Phil. Mag.* **48**, 547.

Valdmanis, J.A., and Mourou, G. (1986). *IEEE J. Quantum Electron.*, **QE-22**, 69.

Valdmanis, J.A., Fork, R.L., and Gordon, J.P. (1985). *Opt. Lett.* **10**, 131.

Wang, S.Y., and Bloom, D.M. (1983). *Electron. Lett.* **19**, 554.

Weiner, A.M., Marcus, R.B., Lin, P.S.D., and Abeles, J.H. (1986). In *Ultrafast Phenomena V*, G.R. Fleming and A.E. Siegman, eds. Springer-Verlag, Berlin, 127–130.

Weiner, A.M., Lin, P.S.D., and Marcus, R.B. (1987a). *Appl. Phys. Lett.* **51**, 358.

Weiner, A.M. Lin, P.S.D., and Marcus, R.B. (1987b). *Picosecond Electronics and Optoelectronics II*, F.J. Leonberger, C.H. Lee, F. Capasso, and H. Morkoc, eds., Springer-Verlag, Berlin, 56–60.

Weiner, A.M., Lin, P.S.D., and Marcus, R.B. (1987c). In *Characterization of Very High Speed Semiconductor Devices and Integrated Circuits*, Ravi Jain, ed., SPIE **795**, 292–298.

Wells, O.C., and Bremer, C.G. (1969). *J. of Physics E* **2**, 1120.

Wooten, F., Huen, T., and Stuart, R. (1966). In *Proc. Int. Colloq. on Optical Properties and Electronic Structures of Metals and Alloys, Paris, 1965*, F. Abeles, ed. North-Holland, Amsterdam, 333.

Glossary of Symbols

α	optical absorption coefficient
B	optical impermeability
c	velocity of light in vacuum
D	diffusion constant; electric displacement
E	energy
E_f	Fermi energy
e	charge of electron
ϵ	dielectric constant
$\mathscr{E}$	electric field
f	frequency (electrical)
B_f	bandwidth
G	conductivity
Γ	coefficient of reflection; phase retardation
$h\nu, \hbar\omega$	photon energy
I	current (DC current); optical intensity
i	current (noise, AC current)
J	current density
L	diffusion length
Λ	electron mean free path
λ	wavelength
m_0	free electron mass
m^*	effective electron mass
μ_n, μ_p	electron, hole mobility
$N_t(N_d, N_a)$	trap (donor, acceptor) concentration
n	index of refraction
n_e	electron density
n_h	hole density
n_i	intrinsic carrier density
ω	frequency
P	electric polarization
p	dipole moment
Φ	work function
ϕ	built-in potential
Q, q	electrical charge
R	reflectivity
ρ	density; charge density
$S_{1,2}$ etc.	S-parameter(s)
σ	secondary electron yield
σ_c	capture cross section
$\sigma^2(t)$	signal variance

t	time (variable)
τ	time (constant, or characteristic)
τ_c	capture time
V	potential
V_{min}	voltage resolution
$v(t), s(t)$	time-varying signal
v, or $\mathbf{v}$	velocity of particle
$\langle v_{th} \rangle$	mean thermal carrier velocity
w	radius of optical beam
Y	yield, quantum yield

Index

423

Contents of Previous Volumes

431